# GOAL
# PROGRAMMING
# FOR
# DECISION
# ANALYSIS

AUERBACH Management and Communication Series

George T. Vardaman, General Editor

SANG M. LEE    *Goal Programming for Decision Analysis*

# GOAL
# PROGRAMMING
# FOR
# DECISION
# ANALYSIS

## SANG M. LEE

COLLEGE OF BUSINESS
VIRGINIA POLYTECHNIC INSTITUTE
AND STATE UNIVERSITY

First Edition

AUERBACH®
publishers

philadelphia
new york
london

AUERBACH Publishers Inc.

Philadelphia, 1972

*658.4*
*L 48l*

Library of Congress Catalog Number: 72-81816
International Standard Book Number: 0-87769-144-4

First Printing

Printed in the United States of America

**Library of Congress Cataloging in Publication Data**

Lee, Sang M      1939–
  Goal programming for decision analysis.

  (Management and communications series)
  Includes bibliographical references.
  1.  Decision-making.  I.  Title.
HD69.D4L43        658.4'03        72-81816
ISBN 0-87769-144-4

**DEDICATED TO**

my parents,
my wife, and
my daughter

# Contents

## PART 4. FINAL REMARKS

# Preface

The traditional concept of "economic man" postulates that he is a "rational" individual. As a rational decision maker, "economic man" is assumed to know not only what he wants but exactly how to get it. As an economic optimizer, his primary goal is the maximization of profit. In this regard, it is assumed that he has a knowledge of the relevant aspects of decision environment, possesses a stable system of preferences, and also has the skill to analyze the alternative courses of action. However, recent developments in the theory of the firm have cast considerable doubt on whether the concept of economic man can be applied to the decision maker in today's complex organizations.

According to broad empirical investigation, there is no evidence that any one individual is capable of performing a completely rational analysis for complex decision problems. Also, there is considerable doubt that the individual value system is exactly identical to that of the firm in determining what is best for the organization. Furthermore, the decision maker in reality is often quite incapable of identifying the optimum choice, because of either his lack of analytical ability or the complexity of the organizational environment.

The concept of the individual profit maximizer in classical economic theory does not sufficiently provide either a descriptive or a normative model for the decision maker in an organization. To be sure, the concept of profit is difficult, if not at all impossible, to define. Even when it is defined, it is extremely difficult to measure profit. Another problem is that there exists considerable doubt, based on empirical analysis, as to whether profit maximization is indeed the sole objective of the firm. As a matter of fact, a wide range of political, social, economic, and ethical aspects have been treated as organizational goals by business firms.

In this context, a management theorist has suggested that in today's complex organizational environment the economic man is not trying to maximize; instead he tries to satisfice. The decision maker is then, in reality, one who attempts to attain a set of goals to the fullest possible extent in an environment of conflicting interests, incomplete information, limited resources, and limited ability to analyze the complex environment. The primary difficulty in modern decision analysis is the treatment of multiple conflicting objectives. The question becomes one of value trades in the social structure of conflicting interests. A formal decision analysis that is capable of handling multiple conflicting goals through the use of priorities may be a new frontier of management science.

Goal programming appears to be an appropriate, powerful, and flexible technique of decision analysis for decision problems with multiple conflicting objectives. Unlike traditional numerical analysis, goal programming allows ordinal solution to a system of complex multiple objectives. This book presents goal programming as a modern decision analysis technique for problems with multiple goals under complex environmental constraints.

This book is written to introduce the underlying concepts, solution methods, and applications of goal programming. It is the first book which devotes its entirety to goal programming. Many aspects of goal programming presented here—for example, graphical solution method, a FORTRAN program, and postoptimal sensitivity analysis—are original works of the author, who has tested the material thoroughly over the past three years. Some of the topics discussed in the book are still at the exploratory stage, and these are so described in the text.

Evidently, the author of an introductory text covering a relatively new technique must exercise his own judgment in determining the discussion topics, depth of analysis, and emphasis of the text. Since the purpose of this book is to introduce goal programming to students and practitioners of decision analysis, the author attempts to explain the content in the simplest possible manner. Consequently, complex mathematical notations and analyses are avoided whenever possible. The basic emphasis of the book is the application of goal programming to real-world problems. Therefore, seven chapters of the text, in addition to numerous practical examples and problems in Chapter 3, 4, 5, and 6, are devoted to application examples.

This book consists of four parts: Part I, concepts of goal programming; Part II, methods and processes of goal programming; Part III, applications of goal programming; and Part IV, implementation of goal programming. Part I (Chapters 1-2) discusses the philosophic background and mathematical foundation of goal programming. Part II (Chapters 3-7) is devoted primarily to solution methods of goal programming. Special emphasis is placed on the goal programming model formulation from verbal descriptions of decision problems; Chapter 3 is entirely devoted to this topic. Goal programming solution procedures based on the graphical method, modified simplex method, and computer-based analysis are thoroughly discussed for the first time. Chapter 7 presents some exploratory ideas in order to point out future research areas of goal programming. Part III (Chapters 8-14) presents application examples of goal programming. First, four chapters present applications to decision problems in business firms. The selected topics are production, finance, marketing, and overall corporate planning problems. Chapters 12, 13, and 14 present contemporary decision problems in institutions of higher education, government, and health-care organizations. Part IV (Chapter 15) discusses some philosophic viewpoints concerning the ordinal solution technique, implementation of goal programming, and epilogue.

This book is written primarily for college students who have career aspirations for staff, management, or administrative positions in business, government, or nonprofit organizations. The book can be used as a text or supplement for a management science, operations research, or decision theory course at junior, senior, or graduate level. Because the book places special emphasis on the application to real-world problems, it can also be used as a reference book for practicing management scientists, operations researchers, or professional managers in various organizations.

If the book is adopted for a junior-level course, it is suggested that the instructor provide general background knowledge of linear programming before discussing goal programming. Actually, it is possible to introduce both linear programming and goal programming simultaneously by analyzing their similarities and differences.

The author's experience is that an effective instructional program can be composed by covering linear programming, the first six chapters and Chapter 15 of this text, and an independent project that applies goal programming to practical problems. A senior-level course should include at least three application chapters in the text, in addition to the chapters and project suggested for a junior-level course. At the graduate level, the entire book should be studied, and in addition it is recommended that the course include outside readings and an independent research project that applies goal programming to a complex decision problem by using the computer program. The logical organization of the chapters of the book is presented graphically in the figure.

This book represents a large portion of the author's research work during the past three years. Some of the ideas presented are exploratory and may need further analysis. It is hoped that further research will achieve this. In writing the book, I have relied on the contributions of many scholars. I have benefited greatly from the works of Professors A. Charnes, W. W. Cooper, Yuji Ijiri, and Veikko Jaaskelainen. I am especially grateful to Veikko Jaaskelainen, an inspiring friend, who gave me many new ideas while he was at Virginia Polytechnic Institute as a Fulbright Visiting Professor. I have also benefited from everyday discussion with my colleagues Laurence J. Moore, Edward R. Clayton, and Anthony J. Lerro. Throughout the three years of preparing this book, I have received many helpful suggestions and criticism from my undergraduate and graduate students. I am especially grateful to Mr. Michael Mann and Mr. Warren Emerson, my graduate assistants, who helped me develop the FORTRAN program.

My colleague Laurence J. Moore read the second draft of the first seven chapters and improved them immensely. Of course any remaining errors are solely my responsibility. I am grateful to *Management Science* and *Policy Sciences* for granting permission to reproduce my co-authored articles that appeared in these journals. (Slight changes were made in these studies in order to maintain proper sequence of contents in the book. For example, discussion of the concept of goal programming was deleted from the articles, and the numbering of equations was changed. However, the content of the studies was not altered.) I would also like to thank Dean H. H. Mitchell and Robert L. King of the College of Business, Virginia Polytechnic Institute and State University, for their direct and indirect support.

The typing of drafts was done by Mrs. Joanna Evans and Mrs. Ginger Morrison. I am deeply grateful to my wife, Laura, and daughter, Tosca Moon, for their understanding endurance of lost evenings and weekends. Last, but not least, I thank Auerbach Publishers Inc., and Mr. William K. Fallon for their unfailing cooperation.

–Sang Moon Lee

Blacksburg, Virginia

Organization of Chapters

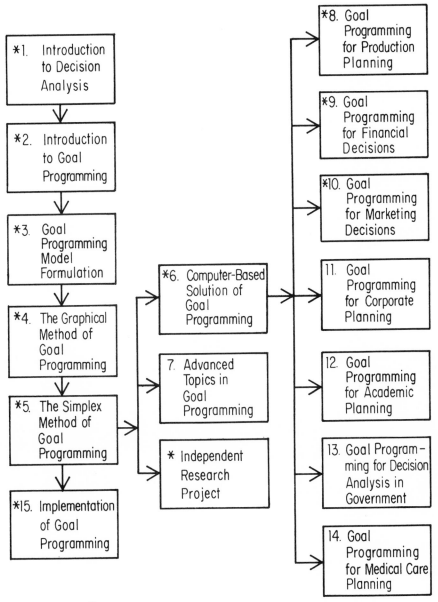

* Recommended for undergraduate courses

# PART 1

## DECISION ANALYSIS
## AND GOAL PROGRAMMING

# Chapter 1

## Introduction to Decision Analysis

Decision making has always been the primary task of man. The central subject to be presented in this book is decision analysis. Decision analysis is the analytical process by which one selects specific courses of action from a set of possible courses of action in order to achieve his goals. By way of introduction to decision analysis, we will first discuss the historical, social, and economic environment of man.

## MAN IN THE UNIVERSE

Since his earliest beginnings, man has demonstrated a unique capability to comprehend the meaning of his existence, to articulate his understanding via a system of communication, and to act on the basis of that understanding.[1] His capacity to learn about his environment and transmit such knowledge into action in order to improve it has been his primary characteristic of mankind.

Man's desire to take the most effective action has led to a continuous struggle to comprehend the norms and conditions under which the environment functions. The increase in man's body of knowledge has led to new discoveries, inventions, and innovations, and these have resulted in greater benefits, comforts, and challenges for man. The progress of a society has, therefore, been basically determined by the production and rate of increase in knowledge.[2]

Inasmuch as knowledge is sought for more effective action, this pursuit is the basis for a cognitive system that we shall call "science." Science is the process or organized body of knowledge conforming to established rules of inquiry, as well as to a system of propositions.[3]

Through knowledge, science, and specialization, man has recognized many of the relationships of his environmental systems. This new knowledge has provided man with the opportunity to manipulate environmental conditions to produce desired consequences. Man has thus acquired much control over nature, providing new horizons of civilization and growth. However, this is also the genesis of the problem of decision making, since different decisions may result in different consequences.[4] Man attempts to select the courses of action, from a set of alternative courses of action, that will achieve his objectives. Decision analysis is a formalized process for increasing man's understanding and control over environmental conditions. Thus, every new development in knowledge or science has a potentially practical implication for decision analysis.

## ECONOMIC MAN AND RATIONAL CHOICE

One of the primary incentives for man to pursue knowledge is the basic human and economic problem of satisfying unlimited human desires with limited resources. This has always been the most troublesome human problem.

There are two approaches man may employ to bridge the gap between desires and resources. First, he may try to increase his resources. This approach may take the form of hard work to make more money so that he can satisfy most of his desires. Or it may mean the pursuit of knowledge that will enable him to utilize existing limited resources in a more efficient manner. Finally, it may be a scientific breakthrough that yields new uses for relatively abundant resources, such as water and air.

The second approach is for man actually to limit his desires so that the existing resources become sufficient to satisfy them. The two approaches are quite in contrast, but both have found wide practice in the history of human society.

The first approach is more or less the accepted philosophy of Western culture. One who "makes it" through hard work or innovative ideas can expect great economic rewards. Such economic rewards also usually result in social and psychological rewards. In other words, "money talks" for the fulfillment of human desires. It is quite common to find in Western culture that millionaires are respected even more highly than statesmen, artists, scholars, or religious leaders. In a sense, it seems perfectly natural and appropriate for those who acquire greater control over resources to receive greater social respect, as we live in an environment of scarce resources.

The second approach has been a long-accepted practice in Eastern cultures. By exercising strong self-discipline, self-control, and sometimes even self-denial, one reduces his desires to the subsistence level. This philosophy is broadly defined as asceticism. Ascetics usually introduce some philosophic or religious accent into their daily life in order to enrich their "inner" happiness. Physiologically, the practice of asceticism is not an easy way of life. However, in Eastern cultures, ascetics are respected for their philosophy, knowledge, and courage to endure physical hardships. Such social recognition has made the practice of asceticism a meaningful way of life for many. If not completely ascetic, a simple, austere way of life has long been advocated in many Asiatic countries as the gentlemanly life style. It is interesting to note that traditionally the elite group in such societies comprised royalties, politicians, and scholars, all of whom were primarily engaged in the perpetuation of past norms and/or in literary or artistic endeavors. Scholars who got their hands dirty were never respected. Scholars sang the beauty of a simple and clean life with no materialistic desires. It now seems surprising upon reflection that the elite group that advocated such a way of life not only controlled those who were directly engaged in the actual production of goods and services for a long period of time but also received their respect.

The point, aside from the pros and cons of Eastern and Western culture, is that in a society where limitation of desires is a respected way of life, it is unlikely that decision analysis will be important or even necessary. Decision analysis becomes important in a cultural setting where the way of life is geared toward greater control over resources to fulfill human desires. It also appears that science becomes important only when the society departs from the philosophy that advocates limitation of human desires through self-control.

The traditional concept of "economic man" not only postulates that he is "economic" and "rational" but also suggests that he is the "optimizer," in the Western cultural sense. The economic man is assumed to be one who strives to allocate his scarce resources in the most economic manner and has the knowledge of the relevant aspects of his environment. He also possesses a stable system of preferences and the skill to analyze the alternative courses of action in order to achieve his desires. However, recent developments in the theory of the firm have cast considerable doubt on whether such assumptions of economic man can be applied to the decision maker in an organization in any realistic sense.[5]

To discuss the points above further, the process of rational choice will be evaluated. Rational choice involves the following aspects: (1) the set of alternative courses of action open to make a choice; (2) the relationships that determine the results (i.e., goal attainment, payoffs, satisfaction of desires, etc.); and (3) some measure of preference among the results.[6]

If an individual is responsible for the decision making of an organization, to be perfectly rational he must be capable of attaching definite preferences to each possible outcome of the alternative courses of action. Furthermore, he should be able to specify the exact outcomes by employing scientific analysis. In other words, the outcomes of alternatives must be specified in a certainty term or at least definite probabilities of occurrence must be attached.[7] According to broad empirical investigation, there is no evidence that any one individual is capable of performing such exact analysis for a complex decision problem. Moreover, there is a considerable doubt that the individual's value system is exactly identical to that of the firm for the determination of what is best for the organization.

There is an abundance of evidence that suggests that the practice of decision making is affected by the epistemological assumptions of the individual who makes the decision.[8] Indeed, the practices of scientific methodology and rational choice are not always directly applicable to decision analysis. The decision maker is constantly concerned with his environment, and he always relates possible decision outcomes and their consequences to the environment with its unique conditions. Stated differently, the decision maker is extremely conscious of the implications of the decision to his surroundings rather than simply considering isolated economic payoffs. This concern with the environmental context of the decision prompts a variety of modifications that further removes him from the classical concept of economic man.

Nevertheless, the decision maker attempts to employ an "approximate" rationality in order to maximize the attainment of organizational goals within the given set of constraints. He may fall far short of the economic man, but his decision-making process may at least be "intentionally" rational. If we define decision analysis as a rational choice process within the context of the decision maker's environmental concern and his limited knowledge, ability, and information, the paradox between the economic man and decision maker becomes increasingly vague.[9] There still remain discrepancies between the theory of rationality and realities of human life. These discrepancies, however, may provide valuable information for the analysis of human behavior in the organizational environment.

## MODERN DECISION ANALYSIS PROCESS

It is only in the past twenty to thirty years that scientific decision analysis has emerged as an important area of management and administration. Some leading authorities of management theory have gone so far as to define decision making as synonymous with management.[10]

Increased emphasis on decision analysis is the result of the aggregate effects of advances in management technology, the ever-increasing complexity of environment, and the improved capability of decision makers. In an attempt to improve the rationality in decision making, greater emphasis has been placed on techniques that would provide more concrete information about the decision environment and the outcomes of alternative courses of action. Hence, the trend of decision making has developed toward the quantitative and computer-oriented approaches. Decision analysis has become an analytical process that employs modern scientific method and systematic investigation in order to aid the decision maker in identifying the optimum course of action.

Modern decision technology does not completely replace the intuitive decision-making approach that has been widely practiced in the past and even today. The intuitive approach is based upon the experience of the decision maker who has become known for his decision-making abilities. He has a general awareness of the situation and he has some personal insights about future outcomes. With or without scientific decision analysis, a decision maker usually exercises some degree of personal judgment. Modern decision technology, therefore, should be intended to enrich and sharpen the judgment of the decision maker.

The process of decision analysis has the following characteristics: (1) consideration of organizational goals and environmental constraints; (2) an explicit analysis of relationships between component variables in the model; (3) verifiable outcome of the optimum course of action; (4) adaptation of the scientific method.[11]

The soundness or rationality of decision making is measured by the degree of organizational goals achieved by the decision. Therefore, the recognition of organizational goals provides the foundation for the need of a decision. Decision analysis is also constrained by environmental factors such as government regulations, union contracts, welfare of the public, etc. Organizational goals are generally established within these environmental constraints.

To derive the optimum solution, relationships between relevant variables must be explicitly analyzed. Such analysis enables the verification of the superiority of the optimum solution to other alternatives. Furthermore, the analysis of relationships between variables provides important managerial information such as shadow price, trade-off points, rate of substitution, etc.

The recommended course of action must be explicit and verifiable through the model. That is, the decision outcome will be identical no matter who the decision maker may be, so long as subjective judgment is not included in the process of analysis.

To reduce personal bias and subjectivity, an adaptation of the scientific approach is necessary. Adaptation of the scientific method implies systematic analysis of the decision system. Systematic investigation enables the decision maker to consider all pertinent factors related to the decision so that ultimate course of action can be identified from among a set of alternatives.

The process of decision analysis can be generally formulated as follows:

## 1. EXPLICIT RECOGNITION OF CONDITIONS NEEDING A DECISION

It is the responsibility of the decision maker to recognize the environmental and organizational conditions that call for a decision. For the decision to be made, objectives or goals to be achieved must be clearly identified and individual activities contributing to these objectives must be analyzed. Decision making requires identification of goals, and establishing goals always involves planning. The planning function determines what, how, and when action must be taken, and who should carry out the job.

## 2. SEARCH FOR ALTERNATIVE COURSES OF ACTION

When the goals of the decision have been determined, the next step is finding means to accomplish the goals. The manager's task in this process is the prediction and evaluation of consequences of various possible courses of action. Prediction of the future consequences of present decisions is a difficult task because there usually exists uncertainty in the decision environment. The predicted outcome of each course of action must be evaluated in terms of required inputs and the degree of uncertainty involved. In such situations, cost-benefit analysis is often utilized in evaluating various decision alternatives. Cost-benefit analysis is a management tool that evaluates alternatives quantitatively by relating benefits to associated costs. Many quantitative techniques must be combined in this process. Whatever techniques are used, it is always desirable to develop a model that allows forecasting the outcome of decisions according to the variation of inputs. It is important to design and generate additional alternatives when existing courses of action are not sufficient. A major deficiency of conventional judgment-based decision-making processes is the inadequate exploration of feasible alternatives.

## 3. MAKING THE CHOICE

The final stage of the decision process is making a choice among the alternative courses of action examined. This decision stage often requires a considerable degree of judgment. Due to the inadequacies of techniques that exist today and intangibles that are hard to measure quantitatively, qualitative analysis of the decision maker is generally an essential part of decision making.

In the three basic stages of the decision process, quantitative analysis plays a major role in the second stage, where alternative decision opportunities are identified and evaluated. Indeed, this systematic analysis of alternative opportunities is the crux of scientific decision analysis. However, in the first stage, where decision goals are identified, and in the third stage, where final choice is made, quantitative analysis also provides valuable information to assist the decision maker.

The new decision technology does not completely replace judgment in decision making. Instead, it relieves the decision maker from being involved in every minor well-structured decision problem. Further and even more important, new decision sciences aid his judgment in decision making for ill-structured major policy problems. Decision analysis will always remain a process of making rational choices assisted by new science and technology; however, the judgment of the decision maker will continue to play a major role.[12]

## GOAL PROGRAMMING FOR DECISION ANALYSIS

It has been shown that conceptually, economic man as a rational decision maker is a simple individual. He knows not only what he wants but he also knows exactly how to get it. As an optimizer, his primary economic role is the maximization of profits. If this were the situation in reality, there would be no need for decision analysis. Indeed, the decision maker may in reality have only a vague idea, if even that, what is the best outcome for the organization in a global sense. Furthermore, he is often quite incapable of identifying the optimum choice, because of either his lack of analytical ability or the complexity of the organizational environment. Modern decision analysis, therefore, becomes a useful tool to help the troubled decision maker.

Another problem is that there exists a considerable degree of varied opinion about the goals of an organization. One study that investigated the goal structures of 25 corporations as described by the top management revealed a wide range of political, social, economic, and ethical aspects that portrayed an image quite different from the economic man.[13] The items most frequently stated as the primary goal of the corporation were as follows:

| *Primary Goal* | *Frequency* |
| --- | --- |
| Personnel | 21 |
| Duties and responsibility to society in general | 19 |
| Consumers' needs | 19 |
| Stockholders' interests | 16 |
| Profit | 13 |
| Quality of product | 11 |
| Technological progress | 9 |
| Supplier relations | 9 |
| Corporate growth | 8 |
| Managerial efficiency | 7 |
| Duties to government | 4 |
| Distributor relations | 4 |
| Prestige | 2 |
| Religion as an explicit guide in business | 1 |

Some theorists may still claim that a variety of noneconomic goals are simply means to a long-run profit maximization. But a series of five-to-ten-year-span short-run noneconomic goals actually forms a long-run noneconomic goal for a period of fifty to one hundred years. Then, should we define the concept of long-run as a time period running well over the span of human life? A social scientist once observed that "once an organization becomes a going concern with many forces working to keep it alive the people who run it readily escape the task of defining its purposes."[14] Perhaps the multiplicity of vague organizational goals represents an attempt to escape from the task of clearly defining goals of the organization.

The decision maker is then, in reality, one who attempts to attain a set of goals to the fullest possible extent in an environment of conflicting interests, incomplete information, limited resources, and limited ability to analyze the complex environment. Such an attempt necessitates modern decision analysis so that the decision maker may strive toward the concept of economic man within the environmental constraints. However, the decision environment forces the decision maker to utilize some degree of subjective analysis. Thus, modern decision analysis attempts to introduce a systematic approach to the subjective decision-making process. The process of decision analysis is not a substitute for decision making, but it enables the decision maker to be more systematic in attaining his goals.

In addition to the realities discussed above, organizational structure based on the reward-punishment relationship and the bureaucratic process further complicates decision analysis. It is not only the welfare of the people who are to be affected by the decision that is important; the welfare of the decision maker himself and others within the organization also becomes a critical decision constraint. Especially in organizations of the public sector, decision making of an official has profound impact upon the lives of many people. The official must make a decision although he may not even know what is "good" for the people. In fact, quite possibly the people themselves may not know what is "good" for themselves. There is no protection against the enormous complexities of society. As a society, therefore, we have learned to respect statistics and numbers in order to analyze the preferences and desires of the mass.

The primary difficulty in modern decision analysis is the treatment of multiple conflicting objectives.[15] The question becomes one of value trades in the social structure of conflicting interests. It is indeed extremely difficult to answer questions such as what should be done now, what can be deferred, what alternatives are to be explored, and what should be the priority structure. Suppose a city is to formulate a long-range economic plan that is to evaluate long-range objectives and their resource requirements. The city needs new water tanks, extended transit systems, new streets, extended water, garbage, and sewage services, new welfare programs, job-training programs, etc. The resources required usually surpass resources available. There exists competition for resources among projects. If substantial resources are to be invested in water and sewage services, other projects must be reduced in scale, deferred, or abandoned completely. Therefore, the assignment of priorities to the competitive projects becomes the first decision problem. Next, the sources of funds should be analyzed. Should the city increase property tax, sales tax, city income tax, business tax, service fees, or use a special bond to acquire required resources for necessary projects? Obviously there are conflicting interests, and the decision maker must assign certain priorities according to his perceived equitability of value trades. The analysis of value trade may involve the consideration of utility functions. It is hard enough to construct a utility function for an individual, not to mention the difficulty of analyzing the aggregate utility for a complex problem that involves $n$ (a million, let's say) dimensional spaces. Utility analysis is therefore almost an impossible task.

A formal decision analysis that is capable of handling multiple conflicting goals through the use of priorities may be a new frontier of management science. The goal-programming approach appears to be an appropriate, powerful, and flexible technique for decision analysis of the troubled modern decision maker who is burdened with achieving multiple conflicting objectives under complex environmental constraints. Goal programming allows a simultaneous solution to a system of complex multiple objectives. Goal programming is capable of handling decision problems that deal with a single goal with multiple subgoals, as well as problems with multiple goals and multiple subgoals. The goal-programming approach utilizes an ordinal hierarchy among conflicting multiple goals so that the low-order goals are considered only after the higher-order goals are satisfied or have reached the desired limit. This book presents the concept, solution methods, and application examples of goal programming for complex modern decision analysis.

## STRUCTURE OF THE BOOK

Part I (Chapters 1 and 2) introduces the general background for goal programming. Chapter 1 has discussed the philosophic foundation of decision analysis and the function of goal programming in modern decision problems that involve multiple conflicting goals in complex social conditions. Chapter 2 introduces the concept of goal programming. First, a brief history of the development of goal programming is discussed. Chapter 2 also analyzes the limitations of linear programming and indicates the value of goal programming for decision analysis. The mathematical formulation of goal programming models is thoroughly examined under different goal structures and constraints through the use of simple examples. Finally, properties and limitations as well as application areas of goal programming are reviewed.

Part II (Chapters 3-7) introduces the methods and processes of goal programming. The most difficult problem in the application of management science to decision analysis is the formulation of the model for the real-world problems. Chapter 3 is therefore devoted to the detailed explanation of goal programming model formulation through a variety of examples. Chapter 4 presents a graphical solution method of goal programming. For a better explanation of the techniques some pertinent background materials are discussed, namely linear equations and inequalities and the graphical solution of linear programming. Chapter 5 is devoted to the simplex method of goal programming. To explain the detailed techniques involved, a simplex solution of linear programming is presented. By employing the same goal-programming problems discussed in Chapter 4, the simplex solution of the goal-programming problem is clearly illustrated. Chapter 6 presents the computer-based solution of goal programming problems. The chapter presents a FORTRAN computer program and exact card set-up procedure. In Chapter 7, advanced topics of goal programming are examined. The chapter explores several research areas of goal programming, such as the postoptimal sensitivity analysis, parametric programming, goal programming under uncertainty, and other related topics.

Part III (chapters 8-14) presents applications of goal programming. Chapter 8 discusses two case problems in which goal programming can be effectively utilized for production planning. First, a general production planning model is illustrated through a simple example. The second case illustrates a more complex aggregate production planning problem. Chapter 9 contains two studies that discuss applications of goal programming to broad financial planning and portfolio selection. In Chapter 10, goal programming is utilized for decision analysis in marketing. The chapter discusses two separate studies. The first study analyzes application of goal programming to the optimization of sales effort allocation. The second study explores advertising media scheduling through goal programming. In Chapter 11, overall organizational planning (or corporate planning) is examined for a real-world problem through the use of goal programming. Chapter 12 is devoted to contemporary decision problems in institutions of higher education. The study presents an academic resource allocation model in universities. Chapter 13 discusses an application of goal programming to economic planning in government agencies. The study presents an aggregative model for municipal economic planning. Chapter 14 examines decision problems in the health care service area. First, a goal programming model is developed for budget planning in health care clinics. The second study presents an aggregative resource allocation model for hospital administration.

Part IV of the book (chapter 15) presents concluding remarks. It includes discussion of the potential of goal programming, some philosophic implications of the technique, ideas concerning the analysis of multiple conflicting goals, steps for a successful implementation of goal programming, and an epilogue.

## REFERENCES

Anshen, Melvin. "The Manager and the Black Box." *Harvard Business Review,* November-December 1960, p. 60.

Astrom, Vincent. "Culture, Science, and Politics." In *The Making of Decisions: A Reader in Administrative Behavior,* ed. W. J. Gore and J. W. Dyson, pp. 85-92. London: The Free Press of Glencoe, 1964.

Boulding, Kenneth. "The Specialist with a Universal Mind." *Management Science,* vol. 14, no. 12 (August 1968), pp. 647-53.

McQuire, Joseph W. *Theories of Business Behavior.* Englewood Cliffs, N. J.: Prentice-Hall, 1964.

Schellenberger, Robert E. *Managerial Analysis.* Homewood, Illinois: Richard D. Irwin, Inc., 1969.

Schubik, Martin. "Approaches to the Study of Decision-Making Relevant to the Firm." In *The Making of Decisions: A Reader in Administrative Behavior,* ed. W. J. Gore and J. W. Dyson, pp. 31-50. London: The Free Press of Glencoe, 1964.

Selznik, Philip. *Leadership in Administration.* Evanston, Illinois: Row, Peterson & Co., 1957.

Simon, Herbert A. "A Behavioral Model of Rational Choice." *Quarterly Journal of Economics,* vol. 69, no. 1 (February 1955), pp. 99-118.

_____. *Models of Man.* New York: John Wiley & Sons, 1957.

_____. *The New Science of Management Decision.* New York: Harper & Brothers, 1960.

# Chapter 2

## Introduction to Goal Programming

### HISTORY OF GOAL PROGRAMMING

The origins of mathematical programming techniques go far back in mathematical history to the theories of linear and nonlinear equations and inequalities. However, George B. Dantzig is recognized as the father of linear programming. Dantzig's work was primarily in the search of techniques to solve logistics problems for military planning when he was employed by the United States Air Force in Washington, D.C., in the early 1940's. His research was encouraged by other scholars who were working on the same general subject: J. von Neumann, L. Hurwicz, and T. C. Koopmans. The original name given to the technique was "programming of interdependent activities in a linear structure," and that was later shortened to "linear programming."

From 1948 on, many scholars have joined Dantzig in refining the technique and exploring the application potential of linear programming. However, the team of A. Charnes and W. W. Cooper has played a key role in introducing and applying the technique to industrial problems. They have published excellent expository journal articles as well as textbooks in linear programming.

In their continuous research of linear programming, A. Charnes and W. W. Cooper developed the concept of goal programming.[1] They provided the name of goal programming in their well-known book on linear programming published in 1961.[2] The concept of goal programming first emerged as an issue for unsolvable linear programming problems. Charnes and Cooper explain:

> Closely related to the analysis of contradictions in unsolvable problems is the issue which will be called "goal attainment." Management sometimes sets such goals, even when they are unattainable within the limits of available resources, for a variety of reasons. For example, such goals may be established to provide incentives or to judge accomplishments, or they may be used as a safeguard to ensure that long-run considerations are not obliterated by immediately attainable objectives, etc. Any constraint incorporated in the functional will be called a "goal." Whether goals are attainable or not, an objective may then be stated in which optimization gives a result which comes "as close as possible" to the indicated goals.[3]

Charnes and Cooper provided the following illustration to introduce goal programming.

(2.1)  Maximize $Z = \$1x_1 + \$\frac{1}{2}x_2$
subject to
$$3x_1 + 2x_2 \leqslant 12$$
$$5x_1 \quad\quad \leqslant 10$$
$$x_1 + x_2 \geqslant 8$$
$$-x_1 + x_2 \geqslant 4$$
$$x_1, \quad x_2 \geqslant 0$$

, Figure 2.1 depicts the decision problem by indicating constraints on the graph. The two shaded areas indicate overlapping regions that can be considered as feasible solution areas in the sense that they satisfy some subsets of the constraints. However, since the two shaded areas do not intersect, there is no area of feasible solution. Thus, the above problem cannot be solved by the usual linear programming procedure.

Let us assume that the first two constraints in (2.1) represent available resources, i.e., machine capacities, and the third and fourth constraints represent management goals. Then, the objective function may be changed from profit maximization to goal attainment. As stated above, goals may not always be attainable. The objective of management may be rather a general approach of attaining the goals as closely as possible. Thus, the objective function can be

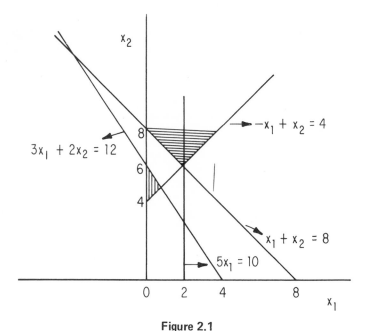

**Figure 2.1**

changed to: $\text{Min } Z = |x_1 + x_2 - 8| + |-x_1 + x_2 - 4|$. This is the basic idea of goal programming.

Y. Ijiri studied the detailed techniques of goal programming based on the general concepts developed by Charnes and Cooper. Ijiri's study presented the definition of "preemptive priority factors" to treat multiple goals according to their importance, assigning weights to goals of the same priority level, and suggesting the generalized inverse approach as the solution algorithm.[4] Ijiri reinforced and refined the concept of goal programming and developed it as a distinct mathematical programming technique. Ijiri's study was, however, primarily concerned with the development of the technique and its possible applications to accounting and managerial control.

Applications of goal programming have been relatively scarce. The first application of the technique was suggested by Charnes and others for advertising media planning.[5] This study presented a general goal programming model that can be utilized for any kind of advertising-media planning problems with some modifications. Charnes and Cooper, in cooperation with staff members of the Office of Civilian Manpower Management, Department of the Navy, also presented a goal programming model for manpower planning.[6]

Impressed by the potential of goal programming, B. Contini examined the goal programming method under conditions of uncertainty.[7] His study provided the mathematical feasibility for the stochastic approach to goal programming, based upon previous studies on chance constrained linear programming. V. Jaaskelainen presented a study in 1968 that applied goal programming to aggregate production planning.[8] This comprehensive study involves multiple-goal treatment for production, demand fluctuations, inventory control, and subcontracting, and it is based on the computer program developed by Jaaskelainen.

The author has explored applications of goal programming to various functional areas, such as academic planning, financial planning, economic planning, and hospital administration.[9] Some of these studies are presented in this book. The application of goal programming has been greatly enhanced in recent years with computer programs developed by the author and by V. Jaaskelainen.[10]

Since goal programming is a relatively new technique, its true potential is yet to be determined. However, it appears that the potential applicability of goal programming may be as wide as that of linear programming. Goal programming has a great deal of flexibility that is lacking in linear programming. Furthermore, the approach of multiple-goal attainment according to their priorities is readily suitable to most management decision problems. A further analysis of the concepts of goal programming will be discussed below.

## CONCEPTS OF GOAL PROGRAMMING

Organizational objectives vary according to the characteristics, types, philosophy of management, and particular environmental conditions of the organization. There is no single universal goal for all organizations. Profit maximization, which is regarded as the sole objective of the business firm in classical economic theory, is one of the most widely accepted goals of management. As reviewed in Chapter 1, in today's dynamic business environment, profit maximization is not always the only objective of management. In fact, business firms quite frequently place higher priorities on noneconomic goals than on profit maximization. Or, firms often seek profit maximization while pursuing other noneconomic goals. We have seen, for example, firms place great emphasis on social responsibilities, social contributions, public relations, industrial and labor relations, etc. Whether such goals are sought because of outside pressure or voluntary management decisions, noneconomic goals exist and are gaining a greater significance. The recent public awareness of the need for environmental control and ecology management has undoubtedly forced many firms to reevaluate their organizational objectives.[11]

If we grant that management has multiple conflicting objectives to achieve, the decision criteria should also be multidimensional. This implies that when a decision involves multiple goals, the quantitative technique used should be capable of handling multiple decision criteria. The linear programming technique that has been applied quite extensively to various decision problems has a limited value for problems involving multiple goals.

The primary difficulty with linear programming is not its inability to reflect complex reality. Rather, its difficulty lies in the unidimensionality of the objective function, which requires cost or profit information that is often almost impossible to obtain. For example, in a production scheduling problem, it is not easy to determine costs associated with the fluctuations in employment level, hiring, training, and laying off of employees, change in employee morale and in the public image of the firm, etc. This problem is more apparent in the public sector. Suppose the government is studying the feasibility of constructing a new airport in Washington, D.C. Here, there are many conflicting objectives and interests. The study must consider the capacity of the airport, accessibility of the location, traffic-flow planning, architectural style for the national prestige, noise level for the nearby residents, conservation of natural life in the area, and so on and so on. Obviously, linear programming is not generally suitable for such decision analysis.

To overcome the unidimensionality of the objective function required in linear programming, efforts have been made to convert various goals, costs, or value measures into one criterion, namely utility. If this process could be effectively performed, the limitation of linear programming would undoubtedly be reduced. However, exact measurement of utility is not a simple matter, and there is no effective methodology to develop a utility function today for either an individual or a group of people. Hence, decision making through linear programming via a utility function is only feasible in a theoretical sense.

To analyze the concept of goal programming, we should be more familiar with the concept of linear programming. First let us review the requirements of linear programming. The first basic requirement of linear programming is that the choice variables constituting the decision system be homogeneous and linear. In other words the choice variables (i.e., $x_1, x_2, \ldots$ etc.) may only be of the first order and their relative properties do not change if multiplied through by the same constant, i.e., $f(x) = f(cx) = cf(x)$. The second requirement is a set of constraints, or limited resources. To give this technique the flexibility required to achieve a solution, it is desirable to express the constraints in either "less than or equal to" or "equal to or greater than" form. Thus, a set of linear inequality constraints normally bound an area of feasible solution within which any point may be a solution, and external to the area there are no real solutions. A third requirement is that there should be an objective to achieve. This objective should be expressed as a linear function composed of the choice variables. The objective function must be homogeneous in the sense that types of measuring units represented by the variables (e.g., hours, manpower, pounds, units, etc.) will combine to provide a consistent unit of measure for the objective.

In the usual linear programming model, the objective is to maximize or minimize the objective criterion by meeting some constraint or goal. The constraint is expressed by some linear combination of choice variables that correspond to the inputs and/or outputs of the system under consideration. In addition, the individual choice variables are also usually constrained in some way. For example, in all linear programming problems, the variables considered must be positive or zero. Additional constraints may specify the maximum or minimum limits on the values of the choice variables.

The general form of linear programming model may be expressed as:

$$(2.2) \quad \text{Max} \quad Z = f(x) = \sum_{j=1}^{n} c_j x_j$$

$$\text{s.t.} \quad \sum_{j=1}^{n} a_{ij} x_j \leqslant b_i \ (i = 1,2, \ldots,m)$$

$$x_j \geqslant 0 \ (j = 1,2, \ldots,n)$$

or

$$(2.3) \quad \text{Min} \quad Z = f(x) = \sum_{j=1}^{n} c_j x_j$$

$$\text{s.t.} \quad \sum_{j=1}^{n} a_{ij} x_j \geqslant b_i \ (i = 1,2, \ldots,m)$$

$$x_j \geqslant 0 \ (j = 1,2, \ldots,n)$$

For complex linear programming problems, it is necessary to introduce slack variable, $s_i$, whenever inequality constraints exist in order to facilitate the iterative solutions. These slack variables represent slack activity or idle resources. In the optimization of any linear system, it is desirable to utilize any idle resources to the extent that they still contribute to the optimality of the objective function. In the simplex algorithm this is precisely the procedure utilized. Idle resources are used up until they no longer exist or no longer contribute to the achievement of the objective. At this point the optimum is achieved.

Linear programming basically requires consistency in the establishment of an objective function. The decision maker must strive to do one thing and one thing only, i.e., maximize profits, minimize costs, or some other objective criterion. Yet, many managerial decisions involve more or less conflicting goals. Even when the decision maker utilizes linear programming, implicitly he must weight the output of the solution against other objectives he could not include in the model. Furthermore, these other objectives may be in entirely different criteria from that of the objective function of the linear programming models. Therefore, even under normal circumstances the decision maker could end up with another more complex programming problem in his mind. Of course, such a situation could increase the complexity of the decision-making process, which in turn might actually reduce the decision maker's ability and objectivity. Consequently, this may lead to a local optimum rather than the global optimum sought for the organization. The goal programming approach is one way in which some of these difficulties can be alleviated.

The restrictive nature of linear programming for modern decision analysis is recognized by George Dantzig. He comments:

It is important to realize in trying to construct models of real-life situations, that life seldom, if ever, presents a clearly defined linear programming problem, and that simplification and neglect of certain characteristics of reality are as necessary in the application of linear programming as they are in the use of any scientific tool in problem solving.[12]

Goal programming is a modification and extension of linear programming. The goal programming approach allows a simultaneous solution of a system of complex objectives rather than a single objective. In other words, goal programming is a technique that is capable of handling decision problems that deal with a single goal with multiple subgoals, as well as problems with multiple goals with multiple subgoals. In addition, the objective function of a goal programming model may be composed of nonhomogeneous units of measure, such as pounds and dollars, rather than one type of unit.

Often, multiple goals of management are in conflict or are achievable only at the expense of other goals. Furthermore, these goals are incommensurable. Thus, the solution of the problem requires an establishment of a hierarachy of importance among these incompatible goals so that the low-order goals are considered only after the higher-order goals are satisfied or have reached the point beyond which no further improvements are desired. If management can provide an ordinal ranking of goals in terms of their contributions or importance to the organization and all goal constraints are in linear relationships, the problem can be solved by goal programming.

How is an objective function to be determined and expressed in algebraic form when there exist multiple, incommensurable, and conflicting goals? A simple answer might be to suggest an objective function composed of multiple goals. This is essentially what goal programming involves. In goal programming, instead of trying to maximize or minimize the objective criterion directly as in linear programming, deviations between goals and what can be achieved within the given set of constraints are to be minimized.

In the linear-programming solution procedure, the values of the choice variable dictated by the objective function criterion tend to "drive" the values of the slack variables. Unlike linear programming, the goal-programming objective function usually does not contain choice variables. Instead, it contains primarily the deviational variables that represent each type of goal or subgoal. The deviational variable is represented in two dimensions in the objective function, a positive and a negative deviation from each subgoal and/or constraint. Then, the objective function becomes the minimization of these deviations, based on the relative importance or priority assigned to them. The objective function, in effect, tends to cause the deviational variables to "drive" the values of the choice variables. It is, of course, possible to include the choice variables in the objective function if that is simpler or desired in the decision problem.

In a very simple goal programming problem where only one type of goal is involved, the model is not substantially different from a linear-programming model. The main difference arises when more than one goal, possibly conflicting and competitive, enters into the system.

The solution of linear programming is limited by quantification. Unless management can accurately quantify the relationship of the variables in cardinal numbers, the solution is only as good as the inputs. The distinguishing characteristic of goal programming is that it allows for an ordinal solution. Stated differently, management may be unable to specify the cost or utility of a goal or a subgoal, but often upper or lower limits may be stated for each subgoal. The manager is usually required to determine the priority of the desired attainment of each goal or subgoal and rank them in ordinal sequence for decision analysis. Economically speaking, the manager works with the problem of the allocation of scarce resources. Obviously, it is not always possible to achieve every goal to the extent desired by management. Thus, with or without goal programming the manager attaches a certain priority to the achievement of a certain goal. The true value of goal programming is, therefore, the solution of problems involving multiple, conflicting goals according to the manager's priority structure.

## MATHEMATICAL ANALYSIS OF GOAL PROGRAMMING

Goal programming is a linear mathematical model in which the optimum attainment of goals is achieved within the given decision environment. The decision environment determines the basic components of the model, namely the choice variables, constraints, and the objective function.

Choice variables are those real variables in the model whose values are arbitrarily assigned and changed in the search for the optimum set of values. The choice variables are related among themselves and to other variables whose values are specified according to the environmental or technological situation.

Constraints represent a set of relationships among variables, which restrict the values of choice variables.

An objective function is a mathematical expression, involving some variables in the model, whose value may be computed when the values of all other variables are determined.

Let us now consider the mathematical property of goal programming through a simple illustration.[13] First, goal programming involving a single goal with multiple subgoals will be discussed, followed by an analysis of multiple goals.

### A. Single Goal with Multiple Subgoals

Let us consider a case where a goal can be achieved by collectively achieving a set of subgoals, $x_1, x_2, \ldots, x_n$.

$$(2.4) \quad f(x_1, x_2, \ldots, x_n) = a_1 x_1 + a_2 x_2 + \ldots + a_n x_n = b$$

where $a_1, a_2, \ldots, a_n$ are real numbers. If we let x be a column vector with components $x_1, x_2, \ldots, x_n$ and let a be a row vector comprised of $a_1, a_2, \ldots, a_n$, then equation (2.4) can be expressed as:

$$(2.5) \quad ax = b$$

If the goal-programming formulation is used, the equation (2.5) can be rearranged as:

$$(2.6) \quad \text{Min} \quad Z = d^- + d^+$$
$$\text{s.t.} \quad ax + d^- - d^+ = b$$
$$x, d^-, d^+ \geqslant 0$$

where $d^+$ and $d^-$ represent deviational variables from the goal. In equation (2.6) it is assumed that the variable x is constrained to be non-negative. If there exists a solution for (2.6), the objective function always drives the values of $d^+$ and $d^-$ to zero. When $d^+$ and $d^-$ are minimized to zero, goal "b" will be achieved at a certain value of x. It should be noted that $d^+$ and $d^-$ are complementary to each other. If $d^+$ takes a nonzero value, $d^-$ will be zero, and vice versa. Since at least one of the variables will be zero, always $d^- \cdot d^+ = 0$.

## EXAMPLE 1

A furniture manufacturer produces two kinds of products, desks and tables. The gross margin from the sale of a desk is $80, and from the sale of a table $40. The goal of the plant manager is to earn a gross profit of $640 in the next week.

We can interpret the profit goal in terms of subgoals, which are sales volumes of desks and tables. Then, a goal programming model can be formulated as follows:

$$(2.7) \quad \text{Min} \quad Z = d^- + d^+$$
$$\text{s.t.} \quad \$80x_1 + \$40x_2 + d^- - d^+ = \$640$$
$$x_1, x_2, d^-, d^+ \geqslant 0$$

where $x_1$ is number of desks sold and $x_2$ is number of tables sold. If the profit goal of $640 is not completely achieved, then obviously the slack in the profit goal will be expressed by $d^-$, which represents the negative deviation from the goal. On the other hand, if the solution shows a profit in excess of $640, then $d^+$ will show some value. If the profit goal of $640 is exactly achieved, both $d^-$ and $d^+$ will be zero. In the above example, there is an infinite number of combinations of $x_1$ and $x_2$ that will achieve the goal. The solution will be any linear combination of $x_1$ and $x_2$ between two points ($x_1 = 8$, $x_2 = 0$) and ($x_1 = 0$, $x_2 = 16$).

## B. Subgoal Constraints

In equation (2.7), a solution was sought that would achieve the goal exactly so that ax = b, or if the goal was unachievable, approach it as closely as possible. The only constraint imposed on the subgoal was simply $x \geqslant 0$. However, quite often the actual environment of the organization imposes additional constraints on the subgoal such as:

(2.8)                 $Bx \leqslant h$

where B is an m $\times$ m matrix and h is an m-component column vector. Then, model (2.6) can now be expressed as follows:

(2.9)    Min      $Z = d^- + d^+$
         s.t.         $ax + d^- - d^+ = b$
                      $Bx \qquad \leqslant h$
                      $x, d^-, d^+ \geqslant 0$

## EXAMPLE 2

Consider the case presented in Example 1. Now let us suppose that in addition to the constraint considered in that example, the following two constraints are imposed. The marketing department reports that the maximum number of desks that can be sold in a week is six. The maximum number of tables that can be sold is eight.

Now the new goal-programming model can be presented in the following way:

(2.10)   Min      $Z = d^- + d^+$
         s.t.         $\$80x_1 + \$40x_2 + d^- - d^+ = \$640$
                      $x_1 \qquad\qquad\qquad \leqslant \quad 6$
                      $x_2 \qquad\qquad \leqslant \quad 8$
                      $x_1, x_2, d^-, d^+ \geqslant \quad 0$

The solution to (2.10) is $x_1 = 6$ and $x_2 = 4$. With that solution the deviational variables $d^+$ and $d^-$ will both be zero. The plant manager's profit goal can be achieved under the new constraints imposed on the subgoals.

## C. Analysis of Multiple Goals

The model illustrated in the two preceding examples can be extended to handle cases of multiple goals. It is assumed that these multiple goals are incompatible and incommensurable. Suppose there are m goals whose levels are expressed by an m-component column vector b, and that these multiple goals can be achieved by linear combinations of n subgoal variables represented by an n-component column vector x. If the relationship between goals and subgoals is expressed by A, which is an m $\times$ n matrix, then the model can be written as:

(2.11)                    $Ax = b$
                          $x \geqslant 0.$

Assuming that a solution exists for (2.11), the model can be transformed to:

(2.12)   Min        $Z = \sum\limits_{i=1}^{m} (d_i^- + d_i^+)$

         s.t.            $Ax + Id^- - Id^+ = b$
                         $x, d^-, d^+ \geqslant 0$

where $d^+$ and $d^-$ are m-component column vectors representing deviations from goals, and I is the m-dimensional identity matrix.

## EXAMPLE 3

Let us consider the furniture manufacturer case illustrated in Examples 1 and 2. Now the manager desires to achieve a weekly profit as close to $640 as possible. He also desires to achieve sales volume for desks and tables close to six and to four respectively. The manager's decision problem can be formulated as:

(2.13)   Min      $Z = d_1^- + d_2^- + d_3^- + d_1^+$
         s.t.      $\$80x_1 + \$40x_2 + d_1^- \qquad\qquad - d_1^+ = \$640$
                   $x_1 \qquad\qquad\qquad + d_2^- \qquad\qquad = \quad 6$
                   $x_2 \qquad\qquad + d_3 \qquad = \quad 4$
                   $x_1, x_2, d_1^-, d_2^-, d_3^-, d_1^+ \geqslant \quad 0$

The solution to this problem can be found by a simple examination of the problem: $x_1 = 6$, $x_2 = 4$, and all goals will be completely attained. Therefore, $d_1^- = d_2^- = d_3^- = d_1^+ = 0$.

## D.  Ranking and Weighting of Multiple Goals

In Example 3 we had a case in which all goals are achieved simultaneously within the given constraints. However, in the real decision environment this is rarely the case. Quite often, most goals are competitive in terms of need for scarce resources. In the presence of incompatible multiple goals the manager needs to exercise his judgment about the importance of the individual goals. Stated more simply, the most important goal must be achieved to the extent the manager desires before the next goal is considered.

Goals of the decision maker may simply be meeting a certain set of constraints. For example, the manager may set a goal concerning stable manpower level in the plant, which is simply a part of the production constraints. Or, the goal may be entirely a separate function from the constraints of the system. If that is the case, the goal constraint must be generated in the model. The decision maker must analyze the system and investigate whether all of his goals are expressed in the goal-programming model. When all constraints and goals are completely identified in the model, the decision maker must analyze each goal in terms of whether over or underachievement of the goal is satisfactory or not. Based on this analysis he can assign deviational variables to the regular and/or goal constraints. If overachievement is acceptable, positive deviation from the goal can be eliminated from the objective function. On the other hand, if underachievement of a certain goal is satisfactory, negative deviation should not be included in the objective function. If the exact achievement of the goal is desired, both negative and positive deviations must be represented in the objective function.

In order to achieve the ordinal solution—that is, to achieve the goals according to their importance—negative and/or positive deviations about the goal must be ranked according to the "preemptive" priority factors. In this way the low-order goals are considered only after higher-order goals are achieved as desired. If there are goals in k ranks, the "preemptive" priority factor $P_j$ ($j = 1, 2, \ldots, k$) should be assigned to the negative and/or positive deviational variables. The "preemptive" priority factors have the relationship of $P_j \ggg P_{j+1}$, which implies that the multiplication of n, however large it may be, cannot make $P_{j+1}$ greater than or equal to $P_j$. It is, of course, possible to refine goals even further by the means of decomposing the deviational variables. To do this, additional constraints and additional priority factors may be required.

One more step to be considered in the model formulation is the weighting of deviational variables at the same priority level. For example, if the sales goal involves two different products, there will be two deviational variables with the same priority factor. The criterion to be used in determining the differential weights of deviational variables is the minimization of the opportunity cost or regret. This implies that the coefficient of regret, which is always positive, should be assigned to the individual deviational variable with the identical $P_j$ factor. The coefficient of regret simply represents the relative amount of unsatisfactory deviation from the goal. Therefore, deviational variables on the same priority level must be commensurable, although deviations that are on different goal levels need not be commensurable.

The objective function of a goal programming problem consists of deviational variables with preemptive priority factors $P_j$'s for ordinal ranking and $\partial$'s for weighting at the same priority level. We let c be a 2m-component row vector whose elements are products of $P_j$ and $\partial$ such that:

$$(2.14) \qquad c = (\partial_1 P_{j1}, \partial_2 P_{j2}, \ldots, \partial_{2m} P_{j2m})^{14}$$

where $P_{ji}$ (i = 1, 2, . . ., 2m; j = 1, 2, . . ., k) are preemptive priority factors with the highest preemptive factor being $P_1$ and $\partial_i$'s (i = 1, 2, . . . 2m) are real numbers. We let d be a 2m-component column vector whose elements are $d^-$'s and $d^+$'s, such that:

$$(2.15) \qquad d = [d_1^-, d_2^-, \ldots d_m^-; d_1^+, d_2^+, \ldots, d_m^+)$$

Then, a goal-programming problem involving multiple conflicting goals can be formulated as:

$$
\begin{aligned}
(2.16) \quad \text{Min} \quad & cd \\
\text{s.t.} \quad & Ax + Rd = b \\
& x, d \geqslant 0
\end{aligned}
$$

where A and R are mxn and mx2m matrices respectively.

## EXAMPLE 4

Consider the following modified case of the illustration given in the previous examples. Production of either a desk or a table requires one hour of production capacity in the plant. The plant has a maximum production capacity of 10 hours a week. Because of the limited sales capacity, the maximum number of desks and tables that can be sold are six and eight per week, respectively. The gross margin from the sales of a desk is $80 and $40 for a table.

The plant manager has set the following goals, arranged in the order of importance.

1. First, he wants to avoid any underutilization of production capacity.
2. Second, he wants to sell as many desks and tables as possible. Since the gross margin from the sale of a desk is set at twice the amount of profit from a table, he has twice as much desire to achieve the sales goal for desks as for tables.
3. Third, he wants to minimize overtime operation of the plant as much as possible.

In the above example, the plant manager is to make a decision that will achieve his goals as closely as possible with the minimum sacrifice. Since overtime operation is allowed in this example, production of desks and tables may take more than the presently set production capacity of 10 hours. Therefore, the operational capacity can be expressed as:

$$(2.17) \qquad x_1 + x_2 + d_1^- - d_1^+ = 10$$

where $x_1$ is the number of desks to be produced and $x_2$ is the number of tables, $d_1^+$ is overtime operation, and $d_1^-$ is idle time when production of both types of products does not exhaust the capacity.

Accordingly, the sales-capacity constraints can be written as:

$$(2.18) \qquad \begin{aligned} x_1 + d_2^- &= 6 \\ x_2 + d_3^- &= 8 \end{aligned}$$

where $d_2^-$ is underachievement of sales goals for desks and $d_3^-$ represents the underachievement of sales goal for tables. It should be noted that $d_2^+$ and $d_3^+$ are not in the equations since the sales goals are given as the maximum possible sales volume.

In addition to variables and constraints stated above, the following "preemptive" priority factors are to be defined:

$P_1$: The highest priority is assigned by management to the underutilization of product capacity (i.e., $d_1^-$).

$P_2$: The second priority factor is assigned to the underutilization of sales capacity (i.e., $d_2^-$ and $d_3^-$). However, management puts twice the importance to $d_2^-$ as that assigned to $d_3^-$ in accordance with respective profit figures for desks and tables.

$P_3$: The lowest priority factor is assigned to the overtime in the production capacity (i.e., $d_1^+$).

Now the model can be formulated. The objective is the minimization of deviation from goals. The deviant variable associated with the highest preemptive priority must be minimized to the fullest possible extent. When no further imporvement is desirable or possible in the highest goal, then the deviations associated with the next highest priority factor will be minimized. The model can be expressed as:

$$
\begin{aligned}
(2.19) \quad \text{Min} \quad & Z = P_1 d_1^- + 2P_2 d_2^- + P_2 d_3^- + P_3 d_1^+ \\
\text{s.t.} \quad & x_1 + x_2 + d_1^- \qquad\qquad\quad - d_1^+ = 10 \\
& x_1 \qquad\quad + d_2^- \qquad\qquad\quad = 6 \\
& x_2 \qquad\qquad + d_3^- \qquad\quad = 8 \\
& x_1, x_2, d_1^-, d_2^-, d_3^-, d_1^+ \geqslant 0
\end{aligned}
$$

The optimal solution can be obtained through the simplex method of linear programming. From the simple investigation of the model we can derive the following optimal solution: $x_1 = 6$, $x_2 = 8$, $d_1^- = d_2^- = d_3^- = 0$, $d_1^+ = 4$. The first two goals are achieved, but the third goal is not completely attained since the overtime operation could not be minimized to zero. This kind of result reflects the everyday problem experienced in business when there are several conflicting goals.

## APPLICATION AREAS OF GOAL PROGRAMMING

An important property of goal programming is its capability to handle managerial problems that involve multiple incompatible goals according to their importance. If management is capable of establishing ordinal importance of goals in a linear decision system, the goal programming model provides management with the opportunity to analyze the soundness of their goal structure. In general, a goal programming model performs three types of analysis: (1) it determines the input requirements to achieve a set of goals; (2) it determines the degree of attainment of defined goals with given resources; and (3) it provides the optimum solution under the varying inputs and goal structures. The goal programming approach to be taken should be carefully examined by the decision maker before he employs the technique. The most important advantage of goal programming is its great flexibility, which allows model simulation with numerous variations of constraints and goal priorities.

Every quantitative technique has some limitations. Some of these are inherent to all quantitative tools and some are attributable to the particular characteristics of technique. The most important limitation of goal programming belongs to the first category. The goal programming model simply provides the best solution under the given set of constraints and priority structure. Therefore, if the decision maker's goal priorities are not in accordance with the organization objectives, the solution will not be the global optimum for the organization. For an effective application of goal programming—and for that matter of all mathematical techniques—a clear understanding of the assumptions and limitations of the technique is a prerequisite. The application of goal programming for managerial decision analysis forces the decision maker to think of goals and constraints in terms of their importance to the organization.

Goal programming can be applied to almost unlimited managerial and administrative decision areas. The following three are the most readily applicable areas of goal programming.

## 1. Allocation Problems

One of the basic decision problems is the optimum allocation of scarce resources. Let us assume that there are n different input resources that are limited to certain quantities and there are m different types of outputs that result from various combinations of the resources. The decision problem is to analyze the optimum combination of input resources to achieve certain goals set for outputs so that the total goal attainment can be maximized for the organization. A goal programming approach has been applied to the resource allocation problems in nonprofit institutions.[15] The study analyzes the resource requirements and actual allocation in order to achieve the administrative goals of the organization. Also, goal programming has been applied to the sales effort allocation problem in the marketing area.

## 2. Planning and Scheduling Problems

Many decision problems involve some degree of planning and/or scheduling. In order to achieve certain goals in the future, decisions must be made concerning present and future actions to be taken. To accomplish desired outputs, the optimum combination of inputs in certain time periods must be identified. These inputs may include manpower, materials, time, production capacity, technology, etc. Many problems, such as production scheduling, location determination, financial planning, personnel planning, marketing strategy planning, etc., can be analyzed by goal programming.

Goal programming has also been applied to media planning, manpower planning, aggregate production scheduling, financial planning, location determination, etc.

## 3. Policy Analysis

For government agencies and nonprofit organizations, the basic decision problem involves the assignment of priorities to various goals and development of programs to achieve these goals. Such decision process constitutes the policy analysis of the organization. Through the application of goal programming the organization is able to ascertain the soundness of its policies, the input requirements for achievement of set goals, and the degree of goal attainment with the given resources. This review and evaluation process is an integral part of policy analysis. Therefore, goal programming is particularly well suited for decision analysis in public and nonprofit organizations.

Here we have discussed three apparent application areas of goal programming. No single list could possibly exhaust all the potential application fields of goal programming. Goal programming can be utilized in those areas where linear programming has been extensively applied. The real value of goal programming is in complex decision problems that involve multiple incompatible goals in multiple dimensions.

## LIMITATIONS OF GOAL PROGRAMMING

Now that we have had a chance to understand the basic concept, mathematical formulation, and application areas of goal programming, it seems appropriate to discuss limitations of the technique. It has been pointed out earlier that every quantitative technique has limitations. Some limitations are inherent to all quantitative tools, and some are attributable to the particular characteristics of individual techniques. Here we will limit our discussion to the limitations of goal programming that are attributable to the underlying assumptions of linear mathematical programming technique.[16]

## 1. Proportionality

It has been pointed out clearly that goal programming is an extension of linear programming. This implies that the objective function, constraints, and goal relationships must be linear. This means that the measure of goal attainment and resource utilization must be proportional to the level of each activity conducted individually. Decision problems that involve some nonlinear relationships because of the lack of proportionality are extremely difficult to solve by goal programming at this time. In fact, this is one of the future research needs in goal programming. It is possible to formulate a goal programming model for a nonlinear programming problem by employing the piecewise linear approximation. However, such a case is indeed an infrequent exception.

## 2. Additivity

The condition that goal attainment and resource utilization be proportional to the level of each activity conducted individually does not ensure linearity. A nonlinearity may occur if there exist joint interactions among some activities of the goal attainment or the total utilization of resources. To ensure linearity, therefore, the activities must be additive in the objective function and constraints.

## 3. Divisibility

Another limitation of goal programming is that fractions of decision variables must be acceptable in the solution. In other words, the optimum solution of a goal programming problem often yields noninteger values for the decision variables. For many decision problems this limitation imposes no actual limitation. For example, if the unit used for the decision variables in a product mix problem is number of boxes and a box contains 100 pieces of the product, a fractional solution value is perfectly satisfactory. There are cases, however, where the decision variables must be integers to have physical significance. There have been no studies made thus far in the integer goal programming area.

## 4. Deterministic

In the normal goal-programming problem, all of the model coefficients ($a_{ij}$, $b_i$, and $P_j$) must be constants. In other words, the problem requires a solution in a static decision environment. However, in reality the decision environment is usually dynamic rather than static. Therefore, the model coefficients are neither known nor constant. This limitation is a most severe one, as goal programming models are usually formulated for future decision making. The model coefficients are based on forecasts of future conditions. Information and forecasting methods available are generally inadequate for the precise determination of coefficients. It is also possible that model coefficients are random variables that have unique probability distributions for the value they take on when the solution is implemented. There have been some studies made concerning stochastic approach to goal programming. In Chapter 7, some of these studies will be reviewed.

In concluding this chapter, several points must be discussed. First, some may question whether the decision maker's priority structure is a good basis for decision analysis. In today's organizational structure, decision making is the responsibility of managerial personnel. It is not by principle but by necessity that decisions must be made by certain individuals. The decision maker's judgment and priority structure for organizational objectives may not be the most effective in a global sense. However, in the reward-punishment type of organizational environment and bureaucratic decision process, the decision maker tends to work toward the most efficient achievement of organizational objectives. Therefore, the decision maker's judgment concerning the organization's multiple conflicting goals appears to be the most appropriate basis for decision analysis.

The second question that may arise is whether the preemptive priority structure on an ordinal ranking is better than utility analysis. If there existed a systematic methodology that could effectively measure the decision maker's utility for multiple goals, obviously the utility approach would be most convenient, as it could utilize regular linear programming techniques. However, there is no effective utility analysis method at present. Furthermore, under normal circumstances the decision maker is not capable of measuring precisely how much more important the first goal is than the second in cardinal value. Often the only knowledge the decision maker has is that in his judgment, a first goal is more important for the organization than a second goal. Hence, a preemptive priority structure in terms of ordinal ranking of goals is the more appropriate vehicle for decision analysis. Furthermore, a continuous refinement of goals through a decomposition procedure can result in a greatly improved preemptive priority structure.[17]

The third point to examine is that a practical decision problem that completely satisfies all the assumptions of goal programming is indeed very rare. Often, however, the goal programming model is the most applicable technique available to derive a reasonable recommendation for action. As long as the goal programming solution provides a better recommendation for action than other techniques, including the decision maker's intuition, it justifies the application of the technique. Nevertheless, it should be clearly pointed out that the user of goal programming must be fully aware of the limitations and approximations involved before proceeding with the goal programming approach for a decision analysis.

## REFERENCES

Charnes, A., and Cooper, W. W. *Management Models and Industrial Applications of Linear Programming.* New York: John Wiley & Sons, Inc., 1961.

_____, *et al.* "A Goal Programming Model for Media Planning." *Management Science,* vol. 14, no. 8 (April 1968), pp. 423-30.

_____, *et al.* "Note on an Application of a Goal Programming Model for Media Planning." *Management Science,* vol. 14, no. 8 (April 1968), pp. 431-36.

_____, and Ferguson, R. "Optimal Estimation of Executive Compensation by Linear Programming." *Management Science,* vol. 1, no. 2 (January 1955), pp. 138-51.

_____, and Nilhaus, R. J. "A Goal Programming Model for Manpower Planning." Management Science Research Report No. 115 (also see No. 188). Carnegie-Mellon University, August 1968.

Contini, B. "A Stochastic Approach to Goal Programming." *Operations Research* (May-June 1968), pp. 576-86.

Dantzig, George B. *Linear Programming and Extensions.* Princeton, N. J.: Princeton University Press, 1963.

Hillier, Frederick S. and Lieberman, Gerald J. *Introduction to Operations Research.* San Francisco: Holden-Day, Inc., 1967.

Ijiri, Y. *Management Goals and Accounting for Control.* Chicago: Rand-McNally, 1965.

Jaaskelainen, Veikko. *Accounting and Mathematical Programming.* Helsinki: 1969.

_____. "A Goal Programming Model of Aggregate Production Planning." *Ekonomisk Tidskrift* (Swedish Journal of Economics), no. 2 (1969), pp. 14-29.

Lee, Sang M. "Decision Analysis Through Goal Programming." *Decision Sciences,* vol. 2, no. 2 (April 1971), pp. 172-80.

_____. "An Aggregative Resource Allocation Model for Hospital Administration." A paper presented at the Third Annual Meeting of the American Institute for Decision Science, October 1971.

_____. "An Aggregative Budget Planning Model for Hospital Administration." A paper presented at the 12th American Meeting of the Institute of Management Science, September 1971.

_____, and Clayton, E. "A Goal Programming Model for Academic Resource Allocation." *Management Science,* vol. 18, no. 8 (April 1972), pp. 395-408.

_____, and Jaaskelainen, V. "Goal Programming for Financial Planning." *Liiketaloudellinen Aikakauskirja* (The Finnish Journal of Business Economics), vol. 3 (1971), pp. 291-303.

_____, Lerro, A., and McGinnis, B. "Optimization of Tax Switching for Commercial Banks." *Journal of Money, Credit, and Banking,* vol. 3, no. 2 (May 1971), pp. 293-303.

_____, and Sevebeck, William. "An Aggregative Model for Municipal Economic Planning." *Policy Sciences,* vol. 2, no. 2 (June 1971), pp. 99-115.

Ruefli, T., "A Generalized Goal Decomposition Model." *Management Science,* vol. 17, no. 8 (April 1971), pp. 505-18.

# PART 2

## METHODS AND PROCESSES
## OF GOAL PROGRAMMING

# Chapter 3

## Goal Programming Model Formulation

It is important to consider formulation of the model before getting into the details of goal programming solution or discussing advanced topics of goal programming. The most difficult problem in the application of management science to decision analysis is, in fact, the formulation of the model for the practical problem in question. Model formulation is the process of transforming a real-world decision problem into a management science model. With the great advances in the use of computers, solution is not generally as difficult a problem as the formulation of the model.

In order to provide some experience and insight into formulating and analyzing a goal programming problem, a wide variety of examples will be presented. One key to successful application of goal programming is the ability to recognize when a problem can be solved by goal programming and to formulate the corresponding model.

## VARIATIONS OF THE OBJECTIVE FUNCTION

The general goal programming model was presented in Chapter 2 as:

(3.1)   Min      cd
        s.t.     $Ax + Rd = b$
                 $x, d \geqslant 0.$

The objective function (Min cd) is simply a minimization function of deviational variables with certain priority factors and weights assigned to them. A number of variations in the objective function may be achieved according to the goal structure of the decision analysis. Let us examine the following five practical variations:[1]

## 1. MINIMIZATION OF $(d^- + d^+)$

Given that the goal constraint is expressed by $Ax + d^- - d^+ = b$, the minimization of $d^- + d^+$ will minimize the absolute value of $Ax - b$. In other words, minimization of both negative and positive deviations will tend to search for the x which achieves the goal $Ax = b$ exactly. As discussed in Chapter 2, at least one deviational variable will be zero, depending upon the level of goals and technical feasibility of the system. For example, if $Ax > b$, then $d^- = 0$ and $d^+ = Ax - b$, whereas if $Ax < b$, then $d^+ = 0$ and $d^- = b - Ax$. If $Ax = b$ (the goal is achieved exactly as desired), of course $d^- = d^+ = 0$. Regardless of the condition of the goal constraint, the minimization of $d^- + d^+$ searches for the value of x which minimizes $d^-$ or $d^+$ whichever is larger, i.e., a positive value.

## 2. MINIMIZATION OF $d^-$

If the objective function is constructed to minimize the negative deviation $d^-$ from the goal, the solution set will consist of all x's such that $Ax \geqslant b$ by minimizing $d^-$ to zero, if such solutions are possible in the model. If it is not possible to minimize $d^-$ to zero, the solution set will consist of all x's that minimize $(b - Ax)$ to the extent possible.

## 3. MINIMIZATION OF $d^+$

If the objective function is to minimize the positive deviation from the goal, the solution will identify all x's which satisfy $Ax \leqslant b$, provided such solutions are possible. If the model cannot minimize $d^+$ to zero, the solution consists of all x's which minimize $(Ax - b)$ to the fullest possible extent.

## 4. MINIMIZATION OF $(d^- - d^+)$

The minimization of $(d^- - d^+)$ has the same effect of maximizing Ax. If we denote $d = (d^- - d^+)$, the deviational variable d is unrestricted in sign. Then the goal programming model can be written as:

$$(3.2) \quad \text{Min} \quad d$$
$$\text{s.t.} \quad Ax + d = b$$
$$x,d \geqslant 0$$

Since $d = b - Ax$, we can transform the objective function to minimize $(b - Ax)$. Because b is a constant, the function is equivalent to the maximization of Ax. In practice, however, the maximization of Ax can also be achieved if we set b to a very large number and minimize $d^-$. Therefore, in most cases the function of "minimize $d^- - d^+$" is only rarely used.

## 5. MINIMIZATION OF $(d^+ - d^-)$

The effect of the function to minimize $(d^+ - d^-)$ is equivalent to the minimization of Ax. Again, if we denote $d = d^+ - d^-$, the goal programming model can be written as:

$$(3.3) \quad \text{Min} \quad d$$
$$\text{s.t.} \quad Ax - d = b$$
$$x,d \geqslant 0$$

Since we know that $d = Ax - b$ and b is a constant, the objective function is the same as to minimize Ax. In most real-world problems, we rarely minimize $(d^+ - d^-)$, because we can get the identical result by minimizing $d^+$ if we set b substantially small. This will be discussed further at a later point in the chapter.

Model formulation for the examples presented in this section will be based on the assumption that the solution procedure to be employed will be either the graphical or the simplex technique. Although there are other solution techniques for goal programming, such as the generalized inverse procedure suggested by Ijiri, only the graphical and simplex methods will be discussed in this book.

## EXAMPLES OF MODEL FORMULATION

### A. Example 1

The manufacturing plant of an electronics firm produces two types of television sets, both color and black-and-white. According to past experience, production of either a color or a black-and-white set requires an average of one hour in the plant. The plant has a normal production capacity of 40 hours a week. The marketing department reports that, because of limited sales opportunity, the maximum number of color and black-and-white sets that can be sold are 24 and 30 respectively for the week. The gross margin from the sale of a color set is $80, whereas it is $40 from a black-and-white set.

In this example, if the president had only the single goal of profit maximization under the normal production and marketing conditions, the decision problem could be easily formulated as a linear programming problem, as illustrated below:

$$
\begin{aligned}
(3.4) \quad \text{Max} \quad & Z = \$80_1 + \$40x_2 \\
\text{s.t.} \quad & x_1 + x_2 \leqslant 40 \\
& x_1 \quad\quad \leqslant 24 \\
& \quad\quad x_2 \leqslant 30 \\
& x_1, x_2 \geqslant 0
\end{aligned}
$$

where $x_1$ represents the number of color television sets and $x_2$ the number of black-and-white sets. The optimum solution thus derived would be $x_1 = 24$, $x_2 = 16$, and the total profit $Z = \$2,560$.

The above profit maximization problem can also be solved by the goal programming approach:

$$
\begin{aligned}
(3.5) \quad \text{Min} \quad & Z = p_1(d_1^+ + d_2^+ + d_3^+) + p_2 d_4^- \\
\text{s.t.} \quad & x_1 + x_2 + d_1^- - d_1^+ = 40 \\
& x_1 \quad\quad + d_2^- - d_2^+ = 24 \\
& \quad\quad x_2 + d_3^- - d_3^+ = 30 \\
& \$80x_1 + \$40x_2 + d_4^- - d_4^+ = \$10,000
\end{aligned}
$$

The objective function in the goal programming formulation indicates that the highest priority, $p_1$, must be assigned to the minimization of $d_i^+$. Since in the linear programming model the first three constraints are "less than or equal to" inequalities, the solution must be within the region that satisfies these three constraints. Thus, in the goal programming problem, assigning the highest preemptive priority factor to the minimization of positive deviations, $d_i^+$, in the first three constraints, the same restrictions are met.

The second priority factor of the goal programming is assigned to the minimization of $d_4^-$—that is, minimize the underachievement of some arbitrarily high profit goal. Arbitrarily, $10,000 is set, knowing that we will never be able to achieve such a high profit. The minimization of underachievement of the profit goal will drive the values of $x_1$ and $x_2$, within the area of feasible solution, as closely to $10,000 as possible. The solution of the goal programming problem is the same as the linear programming solution: $x_1 = 24$, $x_2 = 16$, and $Z = \$2,560$.

The above goal programming formulation indicates that goal programming is also effective in deriving the optimum solution for linear programming problems. However, it may be easier to solve the problem by linear programming. Actually, the profit maximization (or cost minimization) is only one simple case in which there is only a single goal, in the variety of goal programming problems.

In the above problem, the rest of the case reads as follows:

The president of the company has set the following goals as arranged in the order of their importance to the organization.

1.  First, he wants to avoid any underutilization of normal production capacity (no layoffs of production workers).
2.  Second, he wants to sell as many television sets as possible. Since the gross margin from the sale of a color television set is twice the amount from a black-and-white set, he has twice as much desire to achieve sales for color sets as for black-and-white sets.
3.  Third, the president wants to minimize the overtime operation of the plant as much as possible.

Now, let us formulate the constraints of the problem. In the above example, the president has a set of multiple goals that he desires to achieve as closely as possible. Let us formulate the constraints first.

## 1. PRODUCTION CAPACITY

Since the goal regarding overtime operation was given the lowest priority, it is quite possible that production of both color and black-and-white sets may take more than the presently set production capacity of 40 hours. The operational capacity restriction can be expressed as:

(3.6) $$x_1 + x_2 + d_1^- - d_1^+ = 40$$

where $x_1$ is the number of color sets, $x_2$ is the number of black-and-white sets, $d_1^-$ is underutilization of normal operation hours (40), and $d_1^+$ is overtime operation of the plant. It should be remembered, from the discussions in Chapter 2, that at least one of the deviational variables between $d_1^-$ and $d_1^+$ must be zero in the solution. In other words $d_1^- \cdot d_1^+ = 0$. Suppose $x_1 + x_2 = 30$ in the solution, then there is an underutilization of normal capacity $(d_1^-)$ of 10 hours and there will be no overtime operation $(d_1^+ = 0)$. On the other hand, if $x_1 + x_2 = 50$, the amount of overtime operation $(d_1^+)$ will be 10 hours and the underutilization $(d_1^-)$ will be zero. If total production consumes exactly the plant capacity, i.e., $x_1 + x_2 = 40$, then there will be neither underutilization nor the overtime operation. Consequently, both $d_1^-$ and $d_1^+$ will be zero.

## 2. SALES CAPACITY

The marketing department reported that the *maximum* number of color and black-and-white television sets that can be sold are 24 and 30 respectively for the week. The word *maximum* implies that the sales department cannot possibly move more than the stated numbers. In this case, we can actually leave out positive deviations from the goals, i.e., $d_i^+$'s. However, if one wishes to include those variables, it is perfectly permissible. Then, the sales constraints will be:

(3.7) $$x_1 + d_2^- - d_2^+ = 24$$
$$x_2 + d_3^- - d_3^+ = 30$$

where $d_1^-$ and $d_2^-$ represent the underachievement of sales goals for color and black-and-white sets respectively and $d_1^+$ and $d_2^+$ are overachievement of sales goals.

In addition to the variables and constraints stated above, the following preemptive priority factors are defined in order to pursue the various stated goals:

$p_1$: The highest priority is assigned to minimizing the underutilization of production capacity (i.e., $d_1^-$).

$p_2$: The second priority factor is assigned to minimizing the underachievement of sales goals (i.e., $d_2^-$ and $d_3^-$). However, the president also wants to assign differential weights (not preemptive but ordinary numerical weights) to the achievement of sales goal according to the gross margin ratio between color and black-and-white television sets. Hence, the president assigns twice the weights to $d_2^-$ as he assigns $d_3^-$.

$p_3$: The lowest priority factor is assigned to minimizing the overtime operation of the plant (i.e., $d_1^+$).

Thus, the objective function can now be formulated. The objective is to minimize deviations from the goals. The value of the deviational variable associated with the highest preemptive priority must first be minimized to the fullest possible extent. When no further improvement is possible or desired in the highest goal, then we attempt to minimize the value of the deviational variables associated with the next highest priority factor, and so forth. Thus, the objective function can be formulated as:

$$(3.8) \quad \text{Min} \quad Z = p_1 d_1^- + 2p_2 d_2^- + p_2 d_3^- + p_3 d_1^+$$

Therefore, the complete model is:

$$(3.9) \quad \begin{aligned} \text{Min} \quad & Z = p_1 d_1^- + 2p_2 d_2^- + p_2 d_3^- + p_3 d_1^+ \\ \text{s.t.} \quad & x_1 + x_2 + d_1^- \qquad\qquad - d_1^+ \qquad\qquad = 40 \\ & x_1 \qquad\quad + d_2^- \qquad\qquad - d_2^+ \qquad = 24 \\ & x_2 \qquad\qquad + d_3^- \qquad\qquad - d_3^+ = 30 \\ & x_1, x_2, d_1^-, d_2^-, d_3^-, d_1^+, d_2^+, d_3^+ \geqslant 0 \end{aligned}$$

In the above model $Z$ in the objective function can be interpreted as the total of the unattained portions of management goals expressed in terms of priority factors.

## B. Example 2

The manager of the only record shop in a college town has a decision problem that involves multiple goals. The record shop employs five full-time and four part-time salesmen. The normal working hours per month for a full-time salesman are 160, and 80 hours per month for part-time salesmen. According to performance records of the salesmen, the average sales has been five records per hour for full-time salesmen and two records per hour for part-time salesmen.

The average hourly wage rates are $3 for full-time salesmen and $2 for part-time salesmen. Average profit from the sale of a record is $1.50.

In view of past sales records and increased enrollment at the local college, the manager feels that the sales goal for the next month should be 5,500 records. Since the shop is open six days a week, overtime is often required of salesmen (not necessarily overtime but extra hours for the part-time salesmen). The manager believes that a good employer-employee relationship is an essential factor of business success. Therefore, he feels that a stable employment level with occasional overtime requirement is a better practice than an unstable employment level with no overtime. However, he feels that overtime of more than 100 hours among the full-time salesmen should be avoided because of the declining sales effectiveness caused by fatigue.

The manager has set the following goals:

1.  The first goal is to achieve a sales goal of 5,500 records for the next month.
2.  The second goal is to limit the overtime of full-time salesmen to 100 hours.
3.  The third goal is to provide job security to salesmen. The manager feels that full utilization of employees' regular working hours (no layoffs) is an important factor for a good employer-employee relationship. However, he is twice as concerned with the full utilization of full-time salesmen as with the full utilization of part-time salesmen.
4.  The last goal is to minimize the sum of overtime for both full-time and part-time salesmen. The manager desires to assign differential weights to the minimization of overtime according to the net marginal profit ratio between the full-time and part-time salesmen.

Based on the problem stated above, the following constraints can be formulated:

## 1. SALES GOAL

Achievement of the sales goal, which is set at 5,500, is a function of total working hours of the full-time and part-time salesmen and their productivity (sales per hour) rates.

$$(3.10) \qquad 5x_1 + 2x_2 + d_1^- - d_1^+ = 5,500$$

where $x_1$ = total full-time salesmen hours/month

$x_2$ = total part-time salesmen hours/month

$d_1^-$ = underachievement of sales goal

$d_1^+$ = overachievement of sales goal

5  = sales per hour for full-time salesmen

2  = sales per hour for part-time salesmen

5,500 = sales goal for month.

## 2. REGULAR WORKING HOURS

Salesmen hours are determined by the regular working hours for each type of salesman and the number of full-time and part-time salesmen employed. Since we denoted $x_1$ as the total full-time salesmen hours/month, with five full-time salesmen the total regular working hours per month will be 5 x 160 = 800. For the part-time salesmen, the total salesmen hours/month will be 4 x 80 = 320. Thus we have:

$$(3.11) \qquad x_1 + d_2^- - d_2^+ = 800$$
$$x_2 + d_3^- - d_3^+ = 320$$

where $d_2^-$ = underutilization of the total regular full-time salesmen hours/month

$d_2^+$ = overtime given to full-time salesmen/month

$d_3^-$ = underutilization of the total regular part-time salesmen hours/month

$d_3^+$ = extra working hours given to part-time salesmen.

## 3. OVERTIME

In the goal programming solution, to achieve a certain goal we must have a deviational variable to minimize.[2] If there is no deviational variable to minimize to achieve the goal, one must create one by introducing a new constraint. In this record shop example, the manager's second goal is to limit the overtime of full-time salesmen to 100 hours. We do not have a deviational variable to minimize in order to achieve this goal in the above-formulated constraints. Therefore, we must introduce a new constraint. Now it becomes apparent that this is a case of decomposition of a certain goal. Note that the manager listed the minimization of overtime for full-time salesmen as a part of the fourth goal. In essence, he has two separate goals regarding the overtime work of full-time salesmen. To limit the overtime of full-time salesmen to 100 hours, we should introduce the following constraint:

(3.12)                    $d_2^+ + d_{21}^- - d_{21}^+ = 100$

where $d_2^+$ = actual overtime of full-time salesmen

$d_{21}^-$ = difference between the actual overtime of full-time salesmen and desired 100 hours of overtime

$d_{21}^+$ = overtime in excess of desired 100 hours.

We introduced both the negative and positive deviation from the allowed 100 hours of overtime because the actual overtime can be less than, equal to, or even more than 100 hours. Now, we have a deviational variable to minimize to achieve the second goal, i.e., $d_{21}^+$. It should be noted that the above constraint can also be expressed in a different way by adding 100 to the right-hand side of the regular working-hour constraint of the full-time salesmen as below:

(3.13)                    $x_1 + d_4^- - d_4^+ = 900$

In this problem either of the above two constraints can be used to solve the problem.

The goal programming model for the above problem is thus formulated below:

(3.14)  Min      $Z = p_1 d_1^- + p_2 d_{21}^+ + 2p_3 d_2^- + p_3 d_3^- + 3p_4 d_3^+ + p_4 d_2^+$

s.t.

$$
\begin{array}{llllll}
5x_1 + 2x_2 + d_1^- & & -d_1^+ & & & = 5{,}500 \\
x_1 & + d_2^- & & -d_2^+ & & = 800 \\
x_2 & + d_3^- & & -d_3^+ & & = 320 \\
& & d_{21}^- + d_2^+ & & -d_{21}^+ & = 100 \\
\end{array}
$$

$x_1, x_2, d_1^-, d_2^-, d_3^-, d_{21}^-, d_1^+, d_2^+, d_3^+, d_{21}^+ \geqslant 0$

In the above model, the differential weight of 3 is assigned to $d_3^+$ at the $p_4$ level on the basis of net marginal profit ratio per hour between the full-time and part-time salesmen. The productivity ratio (sales per hour) between the full-time and part-time salesmen is 5 to 2, while the hourly wage rate for overtime is $4.50 and $2.00. The marginal profit per hour of overtime is $3.00 for the full-time salesmen and $1.00 for the part-time salesmen. The relative cost of an hour of overtime for the part-time salesmen is three times that of the full-time salesmen. Therefore, $3p_4$ is assigned to $d_3^+$, whereas $p_4$ is assigned to $d_2^+$.

## C. Example 3

A government agency produces two broadly classified economic goods, consumption goods and investment goods.[3] The inputs utilized for the production are material resources and labor. It is assumed that material resources are required to produce both consumption and investment goods. On the other hand, labor is required only for the production of investment goods. All other activities are neglected for the sake of simplicity.

The production process is presented in Figure 3.1. It is assumed that the requirement of material resources is exactly proportional to the output.

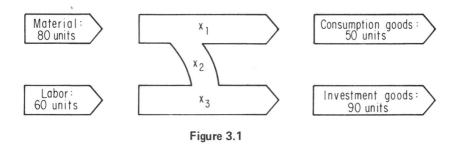

**Figure 3.1**

In other words, one unit of material produces one unit of output, either consumption or investment goods. On the other hand, two units of labor are required to produce a unit of investment goods. Further, let us assume that both kinds of input or input combinations cost one monetary unit.

The administrator is faced with the problem of identifying the required inputs $x_1$, $x_2$, and $x_3$ in order to achieve a set of government goals. His goals are listed in an ordinal ranking:

1. Produce at least 50 units of consumption goods
2. Produce exactly 90 units of investment goods
3. Boost the local economy by utilizing at least 80 units of material resources and 60 units of labor
4. Limit the budget requirement to 120 monetary units for the system input
5. Minimize the input requirements as much as possible.

The above government decision problem can be formulated as a goal programming problem as below:

## 1. INPUTS

The input goals for material and labor are stated as 80 and 60 units respectively.

$$(3.15) \qquad \begin{aligned} x_1 + x_2 + d_1^- - d_1^+ &= 80 \\ x_3 \qquad + d_2^- - d_2^+ &= 60 \end{aligned}$$

where $x_1$ = number of units of material resources required to produce consumption goods

$x_2$ = number of units of material resources required to produce investment goods

$x_3$ = number of units of labor required to produce investment goods

$d_1^-$ = underutilization of available material input

$d_1^+$ = overutilization of material input

$d_2^-$ = underutilization of available labor input

$d_2^+$ = overutilization of labor input.

## 2. OUTPUT REQUIREMENTS

The administrator stipulated that the required outputs for the consumption and investment goods are 50 and 90 units, respectively.

$$(3.16) \qquad \begin{aligned} x_1 + d_3^- - d_3^+ &= 50 \\ x_2 + 0.5x_3 + d_4^- - d_4^+ &= 90 \end{aligned}$$

## 3. BUDGET CONSTRAINT

The administrator desires to limit the budget requirement to 120 monetary units for the input.

$$(3.17) \qquad x_1 + x_2 + x_3 + d_5^- - d_5^+ = 120$$

The complete goal programming model for the above problem can be formulated as below:

$$(3.18) \quad \text{Min } Z = p_1 d_3^- + p_2(d_4^- + d_4^+) + p_3(d_1^- + d_2^-) + p_4 d_5^+ + p_5(d_1^+ + d_2^+)$$

$$
\begin{aligned}
\text{s.t.} \quad x_1 \; + x_2 \qquad\qquad + d_1^- \qquad\qquad\quad - d_1^+ \qquad\qquad &= 80 \\
x_3 \; + d_2^- \qquad\qquad\quad - d_2^+ \qquad\qquad &= 60 \\
x_1 \qquad\qquad\quad + d_3^- \qquad\qquad\qquad\quad - d_3^+ \qquad\quad &= 50 \\
x_2 + 0.5x_3 \; + d_4^- \qquad\qquad\qquad\qquad - d_4^+ \;\; &= 90 \\
x_1 + x_2 + x_3 \qquad\qquad + d_5^- \qquad\qquad\qquad\qquad\quad - d_5^+ &= 120 \\
x_1, x_2, x_3, d_1^-, d_2^-, d_3^-, d_4^-, d_5^-, d_1^+, d_2^+, d_3^+, d_4^+, d_5^+ &\geqslant \;\; 0
\end{aligned}
$$

## D. Example 4

American Computer Hardwares, Inc. produces three different types of computers: Epic, Galaxie, and Utopia. The production of all computers is conducted in a complex and modern assembly line. The production of an Epic requires five hours in the assembly line, a Galaxie requires eight hours, and a Utopia requires 12 hours. The normal operation hours of the assembly line are 170 per month. The marketing and accounting departments have estimated that profits per unit for the three types of computers are \$100,000 for the Epic, \$144,000 for the Galaxie, and \$252,000 for the Utopia. The marketing department further reports that the demand is such that the firm can expect to sell all the computers it produces in the month.

The president of the firm has established the following goals according to their importance:

1. Avoid underutilization of capacity in terms of regular hours of operation of the assembly line.
2. Meet the demand of the northeastern sales district for five Epics, five Galaxies, and eight Utopias (differential weights should be assigned according to the net profit ratios among the three types of computers).
3. Limit overtime operation of the assembly line to 20 hours.
4. Meet the sales goal for each type of computer: Epic, 10; Galaxie, 12; and Utopia, 10 (again assign weights according to the relative profit function for each computer).
5. Minimize the total overtime operation of the assembly line.

With the practice and experience gained from the previous three examples, we can easily set up a goal programming model for the above problem.

## 1. NORMAL OPERATION CAPACITY OF THE ASSEMBLY LINE

The normal capacity of the assembly line for the month is 170 hours. With this capacity, the firm produces three types of computers. The total operation hours required to produce the computers will be simply a function of production rate (in number of hours) for a unit of each type of computer. So we can formulate the normal operation capacity of the assembly line as:

$$(3.19) \qquad 5x_1 + 8x_2 + 12x_3 + d_1^- - d_1^+ = 170$$

where $x_1$ = number of Epic computers

$x_2$ = number of Galaxie computers

$x_3$ = number of Utopia computers

$d_1^-$ = underutilization of normal operation hours of the assembly line

$d_1^+$ = overtime operation of the assembly line.

## 2. SALES CONSTRAINT

First, the firm has outstanding orders from the northeastern sales district as follows:

$$(3.20) \qquad \begin{aligned} x_1 + d_2^- - d_2^+ &= 5 \\ x_2 + d_3^- - d_3^+ &= 5 \\ x_3 + d_4^- - d_4^+ &= 8 \end{aligned}$$

By now, we should be able to define all the variables without writing them down. The firm also has sales goals for the month, which are:

$$(3.21) \qquad \begin{aligned} x_1 + d_5^- - d_5^+ &= 10 \\ x_2 + d_6^- - d_6^+ &= 12 \\ x_3 + d_7^- - d_7^+ &= 10 \end{aligned}$$

It should be pointed out that the above constraint can be written in a different manner. In (3.19), $d_1^+$ represents the production of computers in excess of the demand by the northeastern sales district. Remembering that the sales goal for the month includes the sales of the northeastern district, we can formulate goals for the $d_i^+$ variables. For example, the firm tries to sell 10 Epic computers in the month. Since it will deliver five computers to the northeastern sales district, the firm can attain the sales goal for the Epic if $d_5^+$ comes out to 5 or greater. Thus, the above constraint can also be formulated as (3.22). It is up to the model formulator as to which constraint to use. It appears, however, that constraint (3.21) is a better constraint since we don't have to be confused with a greater number of deviational variables and their decompositions.

$$
\begin{array}{l}
(3.22) \qquad d_2^+ + d_{21}^- - d_{21}^+ = 5 \\
\qquad\qquad d_3^+ + d_{31}^- - d_{31}^+ = 7 \\
\qquad\qquad d_4^+ + d_{41}^- - d_{41}^+ = 2
\end{array}
$$

## 3. OVERTIME OPERATION OF THE ASSEMBLY LINE

We have already mentioned that frequently we have to introduce new constraints in order to define deviational variables that we must minimize to achieve certain goals. The president listed the minimization of overtime operation of the assembly line to 20 hours. As we do not have a deviational variable to minimize in order to achieve this goal, we introduce the following constraint:

$$
(3.23) \qquad d_1^+ + d_{11}^- - d_{11}^+ = 20
$$

where $d_{11}^-$ = the difference between the actual overtime operation of the assembly line and the allowed 20 hours of overtime operation.

$d_{11}^+$ = overtime operation of the plant in excess of 20 hours.

Now, the above problem can be formulated as a goal programming model.

$$
\begin{aligned}
(3.24)\ \ \text{Min } Z = {} & p_1 d_1^- + 20p_2 d_2^- + 18p_2 d_3^- + 21p_2 d_4^- + p_3 d_{11}^+ + 20p_4 d_5^- + \\
& 18p_4 d_6^- + 21p_4 d_7^- + p_5 d_1^+
\end{aligned}
$$

| s.t. | | | | | | |
|---|---|---|---|---|---|---|
| $5x_1 + 8x_2 + 12x_3 + d_1^-$ | | | $- d_1^+$ | | $= 170$ |
| $x_1$ | $+ d_2^+$ | | $- d_2^+$ | | $= 5$ |
| $x_2$ | $+ d_3^-$ | | $- d_3^+$ | | $= 5$ |
| $x_3$ | $+ d_4^-$ | | $- d_4^+$ | | $= 8$ |
| $x_1$ | $+ d_5^-$ | | $- d_5^+$ | | $= 10$ |
| $x_2$ | $+ d_6^-$ | | $- d_6^+$ | | $= 12$ |
| $x_3$ | $+ d_7^-$ | | $- d_7^+$ | | $= 10$ |
| $d_{11}^-$ | $+ d_1^+$ | | $- d_{11}^+$ | | $= 20$ |

$$x_1, x_2, x_3, d_1^-, d_2^-, d_3^-, d_4^-, d_5^-, d_6^-, d_7^-, d_{11}^-, d_1^+, d_2^+, d_3^+, d_4^+, d_5^+, d_6^+, d_7^+, d_{11}^+ \geqslant 0$$

In the objective function of the above model we note differential weights assigned to the second and fourth priority factors. We remember that the criterion used for the weights was the net profit ratio among the three types of computers. Here, we have an assumption that the cost of operating the assembly line is proportional to which computer the line is producing. The net profit ratio, therefore, will simply be determined by dividing the profit by operation hours required to produce each type of computer. For the Epic, the profit is $100,000 per unit and it requires five hours of assembly operation. Hence, the profit per hour of assembly operation for Epic is $20,000. Similarly, profits per hour of assembly operation for the Galaxie and Utopia will be $18,000 and $21,000 respectively. The differential weights are based on these figures.[4]

## E. Example 5

Midwestern Foods, Inc., specializes in the sale of wheat. The firm has definite information concerning the cost at which it can buy and the price at which it can sell the wheat in the next four months. The sale of wheat is restricted by the storage capacity of the firm. The normal capacity of the firm's storage facility is 3,000 bushels (overloading of 2,000 is allowed in emergencies). The estimated cost $c_i$ and the price $p_i$ during the next four months are given as follows:

| Months | 1 | 2 | 3 | 4 |
|---|---|---|---|---|
| Cost ($c_i$) | $4 | $4 | $4 | $7 |
| Price ($p_i$) | 6 | 7 | 5 | 6 |

The quantity of the purchase is assumed to be entirely based upon the revenue generated from sales. It is also assumed that sales are made at the beginning of the month, followed by purchases. At the beginning of the first month there are 2,000 bushels of wheat in the warehouse.

The president of the firm has the following multiple goals, listed in ordinal importance with respect to what he desires to achieve in the next four months:

1. In the first month, only the normal capacity of the warehouse be used.
2. The firm should have at least $20,000 at the beginning of the fourth month for purchases.
3. The firm should reserve at least $2,000 for emergency purposes in each month.
4. The firm should maximize total profit during the entire four-month period.

Before the above problem is formulated as a goal programming model, there are several things we should consider. First, purchase is based upon the cumulative amount of sales in the preceding months. The amount of sales is also restricted by the storage capacity and the purchase in the preceding months. Profit is a combined function of purchases, sales, prices, and costs. If we define sales in month i as $y_i$ and purchase in month i as $x_i$, we can formulate the following constraints:

## 1. SALES

The quantity of sales in the first month will be based on the wheat on hand, since sales precedes purchases in a month. Sales in the subsequent months will be based on the purchase in the preceding months and what is left in the warehouse in the preceding month. Thus we have:

$$
\begin{array}{lll}
(3.25) & y_1 & + d_1^- - d_1^+ = 2{,}000 \\
& y_1 + y_2 + \qquad\quad - x_1 & + d_2^- - d_2^+ = 2{,}000 \\
& y_1 + y_2 + y_3 \quad - x_1 - x_2 & + d_3^- - d_3^+ = 2{,}000 \\
& y_1 + y_2 + y_3 + y_4 - x_1 - x_2 - x_3 & + d_4^- - d_4^+ = 2{,}000
\end{array}
$$

In the above constraints, the sales constraint in the first month is easy to follow. From the second month on, it might be a little confusing. The quantity of sales at the beginning of the second month will be a function of whatever is left in the warehouse at the end of the first month, which will be the sum of the purchase in the first month and what is left from the original 2,000 bushels after the first month's sales. In other words, $y_2 \leqslant (2{,}000 - y_1) + x_1$. Rearranging the constraint, it can be written as: $y_1 + y_2 - x_1 \leqslant 2{,}000$. By introducing deviational variables we derive the sales constraint for the second month.

## 2. PURCHASES

The quantity of purchases depends upon the warehouse space and also the cash availability. At the beginning of the first month, it was assumed that there was no cash on hand. Therefore, sales in the first month will be the sole source of funds for the purchase. However, we also have the cash reserve requirement at the end of month of $2,000. The cash reserve constraints will be introduced later. The warehouse space available for purchase in the first month will be the original space of 1,000 bushels plus space provided by sales in the first month. Hence, we can derive $x_1 \leqslant 1{,}000 + y_1$. Furthermore, it was assumed that the quantity of purchase is entirely based on the revenue generated from sales. Hence, we can formulate purchases in the next four months as follows:

$$
\begin{aligned}
(3.26) \quad -y_1 &&&+ x_1 &&&+ d_5^- - d_5^+ &= 1{,}000 \\
-y_1 - y_2 &&&+ x_1 + x_2 &&&+ d_6^- - d_6^+ &= 3{,}000 \\
-y_1 - y_2 - y_3 &&&+ x_1 + x_2 + x_3 &&&+ d_7^- - d_7^+ &= 3{,}000 \\
-y_1 - y_2 - y_3 - y_4 &&&+ x_1 + x_2 + x_3 + x_4 &&&+ d_8^- - d_8^+ &= 3{,}000 \\
-6y_1 &&&+ 4x_1 &&&+ d_9^- - d_9^+ &= 0 \\
-6y_1 - 7y_2 &&&+ 4x_1 + 4x_2 &&&+ d_{10}^- + d_{10}^+ &= 0 \\
-6y_1 - 7y_2 - 5y_3 &&&+ 4x_1 + 4x_2 + 4x_3 &&&+ d_{11}^- + d_{11}^+ &= 0 \\
-6y_1 - 7y_2 - 5y_3 - 6y_4 &&&+ 4x_1 + 4x_2 + 4x_3 + 7x_4 &&&+ d_{12}^- + d_{12}^+ &= 0
\end{aligned}
$$

## 3. CASH RESERVE OF $2,000 AT END OF EACH MONTH

It is assumed that the initial cash balance at the beginning of the first month is zero. The source of the cash, therefore, is simply the difference between the sales and purchases.

$$
\begin{aligned}
(3.27) \quad 6y_1 &&&- 4x_1 &&&+ d_{13}^- - d_{13}^+ &= 2{,}000 \\
6y_1 + 7y_2 &&&- 4x_1 - 4x_2 &&&+ d_{14}^- - d_{14}^+ &= 2{,}000 \\
6y_1 + 7y_2 + 5y_3 &&&- 4x_1 - 4x_2 - 4x_3 &&&+ d_{15}^- - d_{15}^+ &= 2{,}000 \\
6y_1 + 7y_2 + 5y_3 + 6y_4 &&&- 4x_1 - 4x_2 - 4x_3 - 7x_4 &&&+ d_{16}^- - d_{16}^+ &= 2{,}000
\end{aligned}
$$

## 4. PROFIT OF $20,000 AT THE BEGINNING OF THE FOURTH MONTH

The total profit at the beginning of the fourth month is a cumulative function of sales in the four-month period and purchases in the first three months.

$$
\begin{aligned}
(3.28) \qquad & 6y_1 + 7y_2 + 5y_3 + 6y_4 - 4x_1 - 4x_2 - 4x_3 + \\
& d_{17}^- - d_{17}^+ = 20{,}000
\end{aligned}
$$

## 5. PROFIT MAXIMIZATION FOR THE ENTIRE FOUR-MONTH PERIOD

The last equation of (3.26) represents the total profit function for the entire four-month period. To maximize total profit we can assign a large right-hand side (rhs) value of, say, \$100,000. By minimizing the underachievement of this profit goal ($d_{17}^-$), we will attempt to maximize profit as below:

$$(3.29) \qquad 6y_1 + 7y_2 + 5y_3 + 6y_4 - 4x_1 - 4x_2 - 4x_3 - 7x_4 + d_{18}^- - d_{18}^+ = 100,000$$

It should be noted that we can also achieve the same result by using a decomposed subgoal. For example, in the equation below we would minimize $d_{19}^-$.

$$(3.30) \qquad d_{17}^+ + d_{19}^- - d_{19}^+ = 98,000$$

The complete model can now be formulated as:

$$(3.31) \quad \text{Min } Z = p_1(d_1^+ + d_2^+ + d_3^+ + d_4^+ + d_6^+ + d_7^+ + d_8^+ + d_9^+ + d_{10}^+ + d_{11}^+ + d_{12}^+) + p_2 d_5^+ + p_3 d_{17}^- + p_4(d_{13}^- + d_{14}^- + d_{15}^- + d_{16}^-) + p_5 d_{18}^-$$

s.t.

$$
\begin{aligned}
y_1 &&&&&& + d_1^- - d_1^+ &= 2,000 \\
y_1 + y_2 &&- x_1 &&&& + d_2^- - d_2^+ &= 2,000 \\
y_1 + y_2 + y_3 &&- x_1 - x_2 &&&& + d_3^- - d_3^+ &= 2,000 \\
y_1 + y_2 + y_3 + y_4 &&- x_1 - x_2 - x_3 &&&& + d_4^- - d_4^+ &= 2,000 \\
- y_1 &&+ x_1 &&&& + d_5^- - d_5^+ &= 1,000 \\
- y_1 - y_2 &&+ x_1 + x_2 &&&& + d_6^- - d_6^+ &= 3,000 \\
- y_1 - y_2 - y_3 &&+ x_1 + x_2 + x_3 &&&& + d_7^- - d_7^+ &= 3,000 \\
- y_1 - y_2 - y_3 - y_4 &&+ x_1 + x_2 + x_3 + x_4 &&&& + d_8^- - d_8^+ &= 3,000 \\
- 6y_1 &&+ 4x_1 &&&& + d_9^- - d_9^+ &= 0 \\
- 6y_1 - 7y_2 &&+ 4x_1 + 4x_2 &&&& + d_{10}^- - d_{10}^+ &= 0 \\
- 6y_1 - 7y_2 - 5y_3 &&+ 4x_1 + 4x_2 + 4x_3 &&&& + d_{11}^- - d_{11}^+ &= 0 \\
- 6y_1 - 7y_2 - 5y_3 - 6y_4 &&+ 4x_1 + 4x_2 + 4x_3 + 7x_4 &&&& + d_{12}^- - d_{12}^+ &= 0 \\
6y_1 &&- 4x_1 &&&& + d_{13}^- - d_{13}^+ &= 2,000 \\
6y_1 + 7y_2 &&- 4x_1 - 4x_2 &&&& + d_{14}^- - d_{14}^+ &= 2,000 \\
6y_1 + 7y_2 + 5y_3 &&- 4x_1 - 4x_2 - 4x_3 &&&& + d_{15}^- - d_{15}^+ &= 2,000 \\
6y_1 + 7y_2 + 5y_3 + 6y_4 &&- 4x_1 - 4x_2 + 4x_3 - 7x_4 &&&& + d_{16}^- - d_{16}^+ &= 2,000 \\
6y_1 + 7y_2 + 5y_3 + 6y_4 &&- 4x_1 - 4x_2 - 4x_3 &&&& + d_{17}^- - d_{17}^+ &= 20,000 \\
6y_1 + 7y_2 + 5y_3 + 6y_4 &&- 4x_1 - 4x_2 - 4x_3 - 7x_4 &&&& + d_{18}^- - d_{18}^+ &= 100,000
\end{aligned}
$$

In the objective function of the model, the highest priority is assigned to the deviational variables representing the facts of life. For example, the minimization of $d_1^+$ to $d_4^+$ simply shows that sales cannot be more than what the firm has on hand. Minimizing $d_6^+$ through $d_8^+$ assures that purchases should be limited to the given capacity of the firm. In many real-world problems, often there are occasions where high priorities are assigned to satisfy the given set of restrictions imposed by the decision environment.

## F. Example 6

The Blacksburg Men's Store, which is a branch of a larger store in a nearby city, specializes in sales of quality men's clothing. Presently the store is operating with the full-time manager, who works on salary, and eight part-time salesmen, who earn an hourly wage of $1.40 plus a 20 percent discount on any clothes they purchase in the store. Among the part-time salesmen, four are experienced in selling men's clothing and the others are new to the job.

Each month the store receives a sales quota. The manager breaks it down into his quota and the quota for his part-time salesmen as a group. For the next month the store has received a quota of $25,000. The manager allotted $13,000 for himself and $12,000 for the part-time team. From records of past experience, the manager sells an average of $54.40 worth of clothing per hour. The group of four experienced part-time salesmen sell an average of $32.25 per hour and the group of inexperienced salesmen sell an average of $26.25 per hour.

The manager's regular working schedule per month has been 188 hours, and for each part-time salesmen it has been 50 hours. The manager realizes, however, that he has to put in many extra hours to meet the monthly quotas. He would like to limit his overtime hours to 44, so that his part-time salesmen get enough hours to meet their quota and to earn the money they desire.

As an incentive to the manager and other employees, the main store offers them bonus and commission plans. The manager receives 3 percent bonus on the total sales volume that the store achieves above its sales quota for each month. The manager's objective is to earn an average $50 per month from this bonus plan. The part-time salesmen receive a 5 percent commission on all sales that they make over their quota. The bonus is then split equally among the salesmen. The manager feels that if part-time salesmen put forth some effort, they should be able to earn about $10 each in commission per month.

The manager's goals are listed below in ordinal ranking of importance:

1. The store must meet its sales quota for the month of $25,000.
2. The manager desires to meet his sales quota for the month of $13,000.
3. The manager would like to limit his overtime to 44 hours for the month.

4. The part-time salesmen must meet their group sales quota for the month of $12,000.
5. The manager wants the part-time salesmen to work at least 400 hours for the month.
6. The manager would like to earn $50 in bonus. He would also like to see his part-time salesmen earn a commission of $10 each.
7. If possible, the manager would like to work no more than 188 hours for the month.
8. He wants to minimize the total extra hours that part-time salesmen work in the month.

## 1. SALES QUOTAS

Meeting the sales quota of $25,000 for the store depends on the average sales per hour of the manager and the part-time salesmen and the number of hours they work during the month.

$$(3.32) \qquad 54.50x_1 + 32.25x_2 + 26.25x_3 + d_1^- - d_1^+ = 25,000$$

where $x_1$ = number of hours worked by the manager

$x_2$ = number of hours worked by experienced part-time salesmen

$x_3$ = number of hours worked by new part-time salesmen

$d_1^-$ = underachievement of the store's sales quota

$d_1^+$ = overachievement of the store's sales quota.

Meeting the manager's sales quota of $13,000 for the month may be expressed as:

$$(33.3) \qquad 54.50x_1 + d_2^- - d_2^+ = 13,000$$

The part-time salesmen's quota of $12,000 for the month will be:

$$(3.34) \qquad 32.25x_2 + 26.25x_3 + d_3^- - d_3^+ = 12,000$$

## 2. NORMAL WORKING HOURS

The working hours for the manager and part-time salesmen are based on what is desired to be the normal number of working hours for the month.

(3.35)
$$x_1 + d_4^- - d_4^+ = 188$$
$$x_2 + d_5^- - d_5^+ = 200$$
$$x_3 + d_6^- - d_6^+ = 200$$

## 3. THE MANAGER'S OVERTIME

The manager desires to limit his overtime to 44 hours in the month.

(3.36)
$$d_4^+ + d_{41}^- - d_{41}^+ = 44$$

## 4. BONUS AND COMMISSION

The manager wants to earn at least \$50 bonus for the month, and he would like to see each of his salesmen receive \$10 in commission.

(3.37)
$$.03[(54.50x_1 + 32.25x_2 + 26.25x_3) - 25{,}000] + d_7^- - d_7^+ = 50$$
$$.05[(32.25x_2 + 26.25x_3) - 12{,}000] + d_8^- - d_8^+ = 80$$

Now that the constraints, variables and constraints needed for the model have been defined, it is possible to formulate the complete model.

(3.38)

$$\text{Min } Z = p_1 d_1^- + p_2 d_2^- + p_3 d_{41}^+ + p_4 d_3^- + 13 p_5 d_5^- + 10 p_5 d_6^- +$$
$$p_6(d_7^- + d_8^-) + p_7 d_4^+ + 10 p_8 d_5^+ + 13 p_8 d_6^+$$

s.t.
$$54.50x_1 + 32.25x_2 + 26.25x_3 + d_1^- - d_1^+ = 25{,}000$$
$$54.40x_1 \qquad\qquad\qquad + d_2^- - d_2^+ = 13{,}000$$
$$32.25x_2 + 26.25x_3 + d_3^- - d_3^+ = 12{,}000$$
$$x_1 \qquad\qquad\qquad\qquad + d_4^- - d_4^+ = \quad 188$$
$$x_2 \qquad\qquad\qquad + d_5^- - d_5^+ = \quad 200$$
$$x_3 + d_6^- - d_6^+ = \quad 200$$
$$d_4^+ + d_{41}^- - d_{41}^+ = \quad 44$$
$$1.635x_1 + .9675x_2 + .7875x_3 + d_7^- - d_7^+ = \quad 800$$
$$1.6125x_2 + 1.3125x_3 + d_8^- - d_8^+ = \quad 680$$

In the objective function of the above model we assign differential weights to the fifth and eighth goals. The ratio of sales per hour between experienced and new salesmen is 13 to 10, therefore this ratio is assigned to the underutilization of regular working hours. In the eighth goal the weight is reversed to the minimization of extra hours, since the ratio of relative cost of an extra working hour between the two salesmen teams becomes 10 to 13.

## STEPS OF GOAL PROGRAMMING MODEL FORMULATION

Thus far we have illustrated model formulation of goal programming problems with relatively simple examples. The steps we have taken in the model formulation can be briefly summarized as follows:

### 1. Define Variables and Constants

The first step in model formulation is the definition of choice variables and the right-hand side. The right-hand-side constants may be either available resources or specified goal levels. It requires a careful analysis of the problem in order to identify all relevant variables that have some effect on the set of goals stated by the decision maker.

### 2. Formulate Constraints

Through an analysis of the relationships among choice variables and their relationships to the goals, a set of constraints should be formulated. A constraint may be either a system constraint representing a relationship among the variables or a goal constraint that defines the relationship between choice variables and the goals. It should be remembered that if there is no deviational variable to minimize in order to achieve a certain goal, a new constraint must be created. Also, if further refinement of goals and priorities is required, it may be facilitated by decomposing certain deviational variables.

### 3. Develop the Objective Function

Through the analysis of the decision maker's goal structure, the objective function must be developed. First, the preemptive priority factors should be assigned to certain deviational variables that are relevant to goal attainment. Second, if necessary, differential weights must be assigned to deviational variables at the same priority level. It is imperative that goals at the same priority level be commensurable.

## PROBLEMS

1.  A production manager faces the problem of job allocation between his
    two teams. The processing rate of the first team is 5 units per hour and
    the processing rate of the second team is 6 units per hour. The normal
    working hours for both teams are 8 hours per day. The production mana-
    ger has the following goals for the next day as arranged in order of im-
    portance:
    A.  The manager wants to avoid any underachievement of production
        level, which is set at 120 units of product.
    B.  Any overtime operation of team 2 beyond 3 hours should be
        avoided.
    C.  The sum of overtime should be minimized (assign differential
        weights according to the relative cost of overtime hour—assume
        that the operating cost of the two teams is identical).
    D.  Any underutilization of regular working hours should be avoided
        (again assign weights according to the relative productivity of the
        two teams).

    Formulate the above problem as a goal programming model.

2.  Modern Fashions, Inc., produces two types of ladies' bathing suits: regular
    and bikini. The production of all bathing suits is done in a modern sewing
    center. A regular bathing suit requires an average of five minutes and a
    bikini requires an average eight minutes in the sewing center. The normal
    operation hours of the center are, by running two shifts, 80 hours per week.
    The president wishes to achieve the following goals, listed in ordinal rank-
    ing of importance:
    A.  Achieve the profit goal of $2,000 for the week.
    B.  Limit the overtime operation of the sewing center to eight hours.
    C.  Meet the sales goal for each type of bathing suit: regular, 500; bikini,
        400.
    D.  Avoid any underutilization of regular operation hours of the sewing
        center.

    1.  Formulate a goal programming model from the above problem.
    2.  What will be the change in the objective function if the president
        decides to achieve the sales goal *exactly* as stated?

3.  American Electronics, Inc. produces the most sophisticated color television
    sets. The company has two production lines. The production rate of line 1
    is 2 sets per hour, whereas it is 1 1/2 sets per hour in line 2. The regular
    production capacity is 40 hours a week for both lines. The expected

profit from an average color television set is $100. The top management of the firm has the following goals for the week in ordinal ranking:

A. Meet the production goal of 180 sets for the week.

B. Limit the overtime of line 1 to 10 hours.

C. Avoid the underutilization of regular working hours for both lines (assign differential weights according to the production rate of each line).

D. Limit the sum of overtime operation for both lines. (Assign weights according to the relative cost of overtime hour. Let us assume that the cost of operation is identical for the two production lines.)

1. Formulate the above problem as a goal programming model.

2. If top management desires to put the profit goal of $19,000 for the week as the top priority goal above the stated four goals, how would the model be changed?

3. If top management has only one goal for profit maximization subject to the regular production capacity for both lines, how would the goal programming model be formulated?

4. Colonial Furnitures, Inc., produces three products—desks, tables, and chairs. All furniture is produced in the central plant. To produce a desk requires three hours in the plant, a table takes two hours, and a chair requires only an hour. The regular plant capacity is 40 hours a week. According to the marketing department, the maximum number of desks, tables, and chairs that can be sold are 10, 10, and 12 respectively. The president of the firm has established the following goals according to their importance:

A. Avoid any underutilization of production capacity.

B. Meet the order of Gatewood store for 7 desks and 5 chairs.

C. Avoid the overtime operation of the plant beyond 10 hours.

D. Achieve the sales goals of 10 desks, 10 tables, and 12 chairs.

E. Minimize the overtime operation as much as possible.

Formulate the above problem as a goal programming model.

# REFERENCES

Ijiri, Yuji. *Management Goals and Accounting for Control.* Chicago: Rand-McNally, 1965.

Pitkanen, Eero. "Goal Programming and Operational Objectives in Public Administration." *Swedish Journal of Economics,* vol. 72, no. 3 (1970), pp. 207-214.

# Chapter 4

# The Graphical Method
of
Goal Programming

The objective of a goal programming solution is the achievement of a given set of goals to the fullest possible extent. The solution must be achieved within the given decision environment and according to the decision maker's priority structure for the goals. In Chapter 3 we discussed model formulation procedures for goal programming problems. We will next discuss methods for solving goal programming problems. Two basic solution methods are presented in this book: the graphical method and the simplex method. The same illustrative examples are used to introduce each method so that the reader can grasp both the similarities and the differences between the two methods.

This chapter presents the graphical method of goal programming. Since we can effectively depict problems that involve only two dimensions on the graph, the graphical method is illustrated through a problem that involves only two choice variables. It is possible to depict three dimensional problems graphically; however the procedure becomes so tedious when there are a number of constraints that it is simpler to solve it by the simplex method. For most complex real-world problems, the graphical technique is never used. Instead, the simplex or the generalized inverse technique is employed. However, the graphical method provides a conceptual framework for understanding the solution process of goal programming problems.

In order to facilitate an understanding of the graphical solution method, it is useful to review some of the properties of linear equations and linear inequalities. Further, the graphical method of linear programming will be briefly reviewed. Then we will discuss the graphical method of goal programming through several simple examples.

## LINEAR EQUATIONS AND INEQUALITIES

The linear equation is a mathematical expression in which the two sides of the equation must be equal. All variables, $x_i$, in linear equations are of the first degree. Therefore, all linear equations involving two variables will appear geometrically as straight lines in two-dimensional space. Linear equations involving n variables will portray hyperplanes in n-dimensional space.

Let us examine two linear equations involving two variables.

(4.1)          $y = 6 - 2x$
(4.2)          $y = -2 + x$

For the first equation, the y intercept (or constant) is 6 and the slope (or coefficient) is $-2$. Similarly, the second equation has the y intercept of $-2$ and the slope of $+1$. The y intercept is the point on the y-axis that the straight line passes through when the value of x is zero. The slope of the linear function is the unit change in y associated with one unit change in x, or $\frac{\Delta y}{\Delta x}$.

Now, the above two linear equations can be plotted on the graph as shown in Figure 4.1. The first equation passes through (0,6) and (0,3), and the second equation passes through (0,−2) and (2,0). The intersecting point for the two equations can easily be found by solving them simultaneously. For example, the point on the x-axis can be found by solving $6-2x = 2+x$, which shows the value of $x = 2^{2/3}$. The value of $y = 2/3$ can be found by substituting $x = 2^{2/3}$ in $y = 6-2x$.

Many real-world problems cannot be expressed in the exact form of equations. Therefore, the identification and use of inequalities become inevitable. Inequalities are also mathematical statements that specify certain conditions. For example, let us consider the following two inequalities:

(4.3)          $y \leqslant 6 - 2x$
(4.4)          $y \geqslant -2 + x$

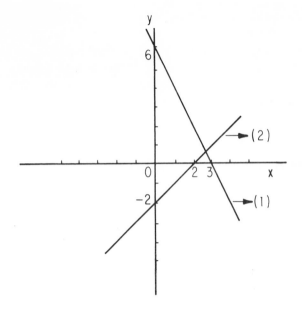

**Figure 4.1**

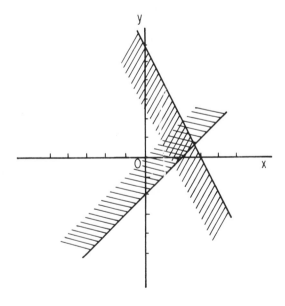

**Figure 4.2**

Whereas the graphical presentation of a linear equation is a straight line, the graphical representation of a linear inequality is a closed half-space. The two inequalities presented above are depicted in Figure 4.2. For the first inequality, any point on the straight line $y = 6-2x$ and to the left of the line, shown as a shaded area, satisfies the inequality statement. Similarly, for the second inequality, any point on the straight line $y = -2+x$ and in the upper shaded area satisfies the condition.

The discussion thus far has been limited to examples involving two variables. However, the concept of equations and inequalities can also be treated for higher-dimensional problems. For example, let us examine the equation (4.5) involving three variables, which is illustrated in Figure 4.3.

(4.5) $\qquad x + y + z = 10$

Since there are three variables the graphical representation of the equation is a plane in three-dimensional space.

For an inequality involving three variables the closed shadow area will be bounded by a plane instead of a straight line as is the case in the two-dimensional problems.

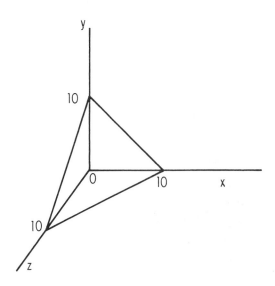

**Figure 4.3**

## THE GRAPHICAL METHOD OF LINEAR PROGRAMMING

In this section the graphical method of linear programming will be discussed to provide the basic background knowledge needed for graphical analysis of goal programming.

Let us first examine a maximization problem:

(4.6)    $$\text{Max } Z = \$10x_1 + \$8x_2$$
$$\text{s.t.} \quad 2x_1 + 4x_2 \leqslant 80$$
$$3x_1 + x_2 \leqslant 60$$
$$x_1, x_2 \geqslant 0$$

The standard restriction for all linear programming problems is that all choice variables must be non-negative, shown by the last constraint.

The first step of the graphical method is to depict the straight lines and the shaded areas of the constraints as expressed by inequalities. The Figure 4.4 shows the shaded areas and the area of feasible solution where two shaded areas overlap. This area represents a region that satisfies both conditions specified by the two constraints. This area of feasibility is often referred to as a convex set in

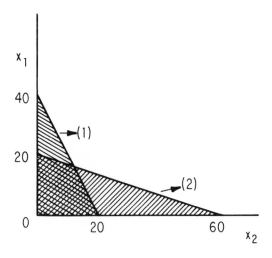

**Figure 4.4**

linear programming. Any point outside the area does not satisfy all the restrictions, and therefore the solution is infeasible. The straight lines to be graphed can easily be derived by solving the constraints in terms of one variable, which will be treated as the y-axis. In this case let us take $x_1$. Then the two constraints can be transformed as $2x_1 + 4x_2 \leqslant 80 \rightarrow x_1 \leqslant 40 - 2x_2$ and $3x_1 + x_2 \leqslant 60 \rightarrow x_1 \leqslant 20 - 1/3x_2$.

The second step is to determine the optimum solution. One easy way to identify the optimum solution point is by analyzing the total profits of all points that form the area of feasible solution. For example, in our example we know that the total profit is zero at the origin. The profit increases as we move out from the origin as far as possible within the area of feasible solution. That implies that the optimum point will be one of the three points that form the boundaries of the area: $(x_1 = 20, x_2 = 0)$, $(x_1 = 16, x_2 = 12)$, and $(x_1 = 0, x_2 = 20)$. The intersecting point for the two straight lines can easily be identified by solving two equations simultaneously as explained already. The profit at these points can be determined by substituting the above values in the objective function, yielding: $200, $256, and $160. Obviously, the maximum profit is $256, resulting at the point where the two straight lines intersect.

Another method for finding the optimum point is by making use of the iso-profit function. The iso-profit function can be derived by solving the objective function in terms of the variable that we treat as the y-axis for constraints. For example, $Z = $10x_1 + $8x_2$ can be expressed as $x_1 = Z/10 - 4/5x_2$ where Z represents the total profits. As the term *iso* implies, the iso-profit function is a straight line on which every point has an identical total profit.

Since the y intercept is unknown, the best we can do is simply plot the iso-profit function according to its slope. As explained earlier, the profit is zero at the origin and increases as we move out from the origin. Now, we plot the iso-profit function on the graph and move out from the origin as far as possible within the area of feasible solution. The total profit increases as long as the iso-profit line with the same slope, $-4/5$, moves farther from the origin. This implies that the last point we touch with the iso-profit line within the area of feasible solution will be the optimum point.

The first iso-profit line we plotted on the graph, as shown in Figure 4.5, is between the two points $(x_1 = 4, x_2 = 0)$ and $(x_1 = 0, x_2 = 5)$. Any point on this iso-profit line has an identical total profit of $40. As we move further out with the slope of the iso-profit function, the profit increases. The second iso-profit line is drawn connecting the points $(x_1 = 8, x_2 = 0)$ and $(x_1 = 0, x_2 = 10)$. The total profit at any point on this line is $80. The last point we will touch with the iso-profit line within the area of feasible solution connects the points $(x_1 = 25.6,$

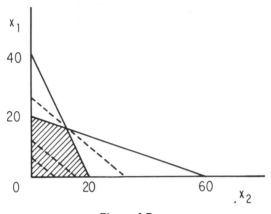

**Figure 4.5**

$x_2 = 0$) and ($x_1 = 0$, $x_2 = 32$), and also passes through the intersecting point of the two straight lines. At this point, ($x_1 = 16$, $x_2 = 12$), the iso-profit line satisfies both constraints and maximizes profit at \$256.

The easiest way to identify the optimum point is comparing slopes of those straight lines that form the area of feasible solution with the slope of the iso-profit function. The two slopes of the straight lines that form the boundaries of the feasible solution area are $-1/3$ and $-2$ respectively. As long as the slope of the iso-profit line ($-4/5$) lies between these two slopes, the last point the iso-profit line will touch within the area of feasible solution is the intersecting point of the two straight lines. This is clearly evident in Figure 4.6. The slope of

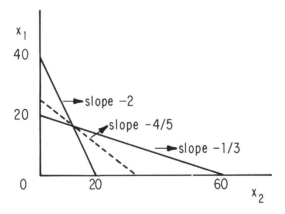

**Figure 4.6**

$-4/5$ is not so steep as $-2$ but steeper than $-1/3$, and therefore the iso-profit line will pass through the intersecting point of the two lines at the tip of the area of feasible solution. This is the case in the preceding problem, and consequently the profit is maximized at the intersection point of the two constraint lines.

If the iso-profit function had a slope of $-3$, which is steeper than either of the two slopes of the constraint functions, the last point the iso-profit line would touch within the area of feasible solution would be the extreme point on the $x_2$-axis. Therefore, the optimum point will of course be $(x_1 = 0, x_2 = 20)$. By the same token, if the iso-profit function had a slope of $-1/5$, which is flatter than either of the constraint slopes, the optimum point would be $(x_1 = 20, x_2 = 0)$, as shown in Figure 4.7.

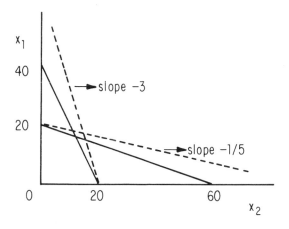

**Figure 4.7**

If the iso-profit function reveals a slope that is identical to one of the slopes of the constraint functions, the last points touched by the iso-profit line within the feasible solution area will be exactly the same as the straight line itself. For example, if the slope of the iso-profit function is $-1/3$, any point on the line between points of $(x_1 = 20, x_2 = 0)$ and $(x_1 = 16, x_2 = 12)$ will be the maximizing point. This reasoning is clearly shown in Figure 4.8 by the heavy portion of the line connecting $(x_1 = 20, x_2 = 0)$ and $(x_1 = 0, x_2 = 60)$.

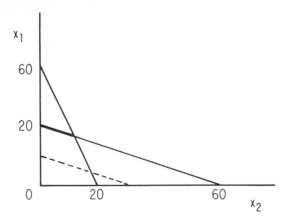

**Figure 4.8**

## THE GRAPHICAL METHOD OF GOAL PROGRAMMING

In this section, building on what we have learned thus far in this chapter, a graphical solution of goal programming problems will be discussed. Let us recall that the objective of goal programming is not the maximization or minimization of a single objective criterion. Instead, the objective is to attain a certain set of goals, which are rated by priority factors, as closely as possible. As discussed in Chapter 2, the procedure we will follow is the minimization of deviation between the goal and what we can achieve within the given decision environment. Deviation from the goal with the highest priority factor will be minimized to the fullest possible extent, and the deviation for the second goal will be minimized after considering the first goal, and so on.

The goal programming problem is, therefore, always a minimization problem. To explain the graphical solution of goal programming, let us consider the following three examples.

### A. Example 1

A textile company produces two types of linen materials, a strong upholstery material and a regular dress material. The upholstery material is produced according to direct orders from furniture manufacturers. The dress material, on the other hand, is distributed to retail fabric stores. The average production rates for the upholstery material and for the dress material are identical: 1,000 yards per hour. By running two shifts, the operational capacity of the plant is 80 hours per week.

The marketing department reports that the maximum estimates sales for the following week is 70,000 yards of the upholstery material and 45,000 yards of the dress material. According to the accounting department, the approximate profit from a yard of upholstery material is $2.50, and from a yard of dress material $1.50.

If the president of the firm has only one goal — profit maximization — this problem can be formulated as a linear programming problem as shown below:

(4.7)    $\text{Max } Z = \$2500x_1 + \$1500x_2$

$$\text{s.t.} \quad \begin{array}{rl} x_1 + & x_2 \leqslant 80 \\ x_1 & \leqslant 70 \\ & x_2 \leqslant 45 \\ x_1, x_2 \geqslant & 0 \end{array}$$

The $x_1$ and $x_2$ variables are expressed in terms of 1,000's of yards in the model.

The problem is presented on the graph in Figure 4.9 and the area of feasible solution is indicated by the shaded area. Three constraints form the boundaries

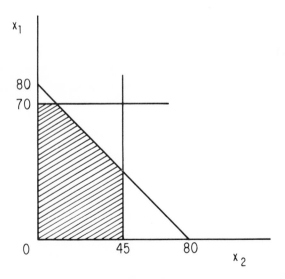

**Figure 4.9**

of the feasible solution area, and their slopes are $-1, 0$, and $\infty$. The slope of the iso-profit function is $-3/5$. Since this slope lies between $0$ and $-1$, the optimum point will be the intersection point between the production capacity constraint, $x_1 + x_2 \leqslant 80$, and sales constraint for the upholstery material, $x_1 \leqslant 70$. The optimum point is identified as $(x_1 = 70, x_2 = 10)$ and the total profit at that point is $\$190,000$.

If the president of the firm has only the one objective of profit maximization, goal programming is not even required. However, the rest of the case reads as follows:

The president of the company believes that a good employer-employee relationship is an important factor for business success. Hence, he decides that a stable employment level is a primary goal for the firm. Therefore, whenever there is demand exceeding normal production capacity, he simply expands production capacity by providing overtime. However, he also feels that overtime operation of the plant of more than 10 hours per week should be avoided because of the accelerating costs. The president has the following four goals:

1. The first goal is to avoid any underutilization of production capacity (i.e., to maintain stable employment at normal capacity).
2. The second goal is to limit the overtime operation of the plant to 10 hours.
3. The third goal is to achieve the sales goals of 70,000 yards of upholstery material and 45,000 yards of dress material.
4. The last goal of the president is to minimize the overtime operation of the plant as much as possible.

Since the above problem involves multiple incompatible goals, linear programming is not an effective approach to the case. A goal program model should be developed for the solution. The case presents three basic constraints: production, sales, and overtime operation of the plant.

## 1. PRODUCTION CAPACITY

The production capacity at present is limited to 80 hours by running two shifts. However, since overtime operation of the plant is allowed to a certain extent the constraint can be expressed as

$$(4.8) \qquad x_1 + x_2 \gtrless 80.$$

By introducing deviational variables to the constraint it can be expressed as

(4.9)                           $x_1 + x_2 + d_1^- - d_1^+ = 80$

where $x_1$ = number of hours used for producing the upholstery material
  $x_2$ = number of hours used for producing the regular dress material
  $d_1^-$ = underutilization of production capacity as set at 80 hours of opera-
      tion
  $d_1^+$ = overutilization of normal production capacity beyond 80 hours.
  It should be noted, as discussed in Chapter 2, that $d_1^-$ takes on a value only
when $d_1^+$ is zero and vice-versa. Therefore, the product of $d_1^- \times d_1^+$ is always zero.

## 2. SALES CONSTRAINTS

In this case, the *maximum* sales for upholstery and dress materials are set at
70,000 and 45,000 yards, respectively. Hence, it is assumed that overachieve-
ments of sales beyond the maximum limits are impossible. Then, the sales
constraints will be (as before, $x_1$ and $x_2$ are expressed in thousands):

(4.10)                          $x_1 \leqslant 70$
                                $x_2 \leqslant 45$

and these can be converted as below with certain deviational variables:

(4.11)                          $x_1 + d_2^- = 70$
                                $x_2 + d_3^- = 45$

where $d_2^-$ = underachievement of sales goal of upholstery material
  $d_3^-$ = underachievement of sales goal of dress material.

## 3. OVERTIME OPERATION CONSTRAINT

From the case itself, only production and sales constraints can be
formulated. However, the analysis of goals indicates that overtime operation of
the plant is to be minimized to 10 hours or less. To solve the problem by goal

programming, we need a deviational variable that represents the overtime operation of the plant beyond 10 hours. By minimizing this deviational variable to zero we can achieve the goal. Since there is no such deviational variable in the three constraints presented above, we must create a new constraint.

The overtime operation of the plant, $d_1^+$, should be limited to 10 hours or less. However, it may not be possible to limit the overtime operation to 10 hours or less in order to meet higher order goals. Therefore, $d_1^+$ can be smaller than, equal to, or even greater than 10 hours. By introducing some new deviational variables, a constraint regarding overtime can be expressed as:

$$(4.12) \qquad d_1^+ + d_{12}^- - d_{12}^+ = 10$$

where $d_{12}^-$ = negative deviation of overtime operation from 10 hours

$\qquad d_{12}^+$ = overtime operation beyond 10 hours.

We recall from the discussion in Chapter 3 that the above goal decomposition constraint can also be expressed as $x_1 + x_2 + d_4^- - d_4^+ = 90$. The expression of the goal decomposition with the choice variables is more effective for the graphical solution. If we use the constraint based only on deviational variables, we have to convert it in such a way that it is expressed by choice variables.

Now the complete model can be formulated. The objective is the minimization of deviations from goals with certain assigned priorities. The deviant variable with the highest priority must be minimized to the fullest possible extent. When no further improvement is possible for the highest goal, the other deviational variables are to be minimized according to their assigned priority factors. One thing to be noted here is the priority factor $P_3$ that is assigned to the underachievement of sales goals for two types of materials. Sales goals for both materials are considered equally important. However, the profit contribution rate of each material differs somewhat. A yard of upholstery material contributes $2.50 profit, and a yard of dress material only $1.50 profit. Therefore, differential weights must be assigned to sales goals of these materials, even though they are on the same priority level. The profit contribution ratio between the upholstery and dress materials is 5 to 3. Hence, these are assigned as differential weights. The differential weights imply that management is relatively more concerned with the achievement of the sales goal for the upholstery material than that for the dress material. Now, the model can be formulated:

(4.13)  $\begin{aligned} \text{Min } Z &= P_1 d_1^- + P_2 d_4^+ + 5P_3 d_2^- + 3P_3 d_3^- + P_4 d_1^+ \end{aligned}$

$$\begin{aligned} \text{s.t.} \quad x_1 + x_2 + d_1^- \qquad\qquad\quad - d_1^+ \quad &= 80 \\ x_1 \qquad\quad + d_2^- \qquad\qquad\qquad &= 70 \\ x_2 \qquad\quad + d_3^- \qquad\qquad &= 45 \\ x_1 + x_2 \qquad\qquad\quad + d_4^- \quad - d_4^+ &= 90 \\ x_1, x_2, d_1^-, d_2^-, d_3^-, d_4^-, d_1^+, d_4^+ &\geqslant 0 \end{aligned}$$

The choice variables were expressed in terms of 1,000's of yards in the model.

In order to solve this goal programming problem by the graphical method, the constraints must first be plotted on a graph as shown in Figure 4.10. Since production capacity may be less than or equal to or even greater than 80 hours, the shaded areas can be on either side of the straight line, as noted by the arrow signs. Now, if goals presented by the president are to be met within the sales-constraints areas, the area of feasible solution will be ABDO.

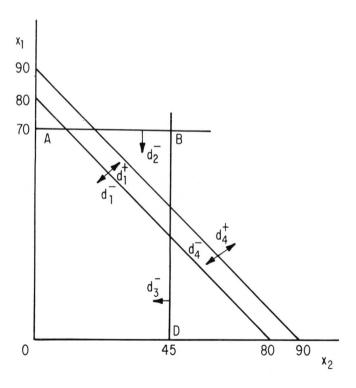

**Figure 4.10**

If we use (4.12) as the overtime constraint, in order to introduce the constraint on the graph we have to modify the production capacity. From the overtime operation constraint we can obtain $d_1^+ = 10 + d_{12}^+ - d_{12}^-$. By substituting this value into the production capacity an additional production constraint can be obtained as $x_1 + x_2 + d_{12}^- - d_{12}^+ = 90$. This constraint indicates that operation of the plant could be equal to, less than, or greater than 90 hours. If we are to achieve our goals within the sales constraints, the area of feasible solution should remain as ABDO.

We have plotted all constraints on the graph. The next step is the analysis of the objective function. The first goal is to avoid the underutilization of production capacity or the minimization of $d_1^-$ to zero. In order to achieve this goal, the production capacity constraints of $x_1 + x_2 - d_1^+ = 80$ must be analyzed. The arrow pointing toward the origin from the above function must be minimized to zero. The area of feasible solution is now the upper part from the straight line as shown in Figure 4.11.

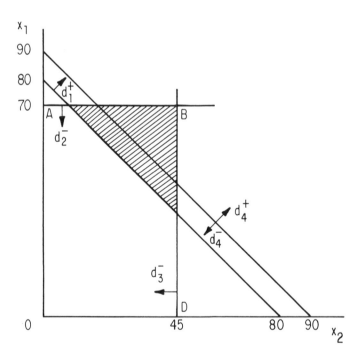

**Figure 4.11**

The second goal of the president is to limit the overtime operation of the plant to 10 hours. To achieve this goal the area of feasible solution must be limited to the shaded area shown by Figure 4.12. Now, the first two most important goals will be attained as long as production takes place within the shaded area.

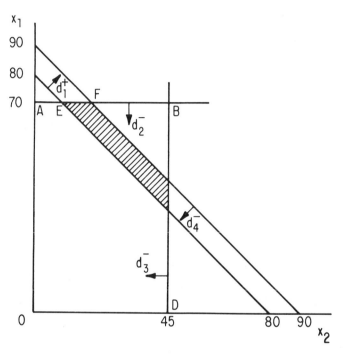

**Figure 4.12**

The third goal presented by the president is to achieve maximum sales. Since profit ratio between the upholstery material ($x_1$) and dress material ($x_2$) is 5 to 3, we should try to sell as much upholstery material as possible before trying to sell dress material. The maximum quantity of upholstery material we can sell is, of course, 70,000 yards. This sales goal can be met on the line EF within the shaded area. Next, we should try to achieve the maximum sales for the dress material, 45,000 yards. However, in order to achieve this goal we have to reach point B. This point is outside the shaded area, and therefore our first two goals could not be attained at that point. We have no desire to achieve the third goal at the expense of the first two goals. The maximum sales goal for dress material obviously cannot be achieved within the given decision environment. The maximum possible sales goal for dress material must be found on the straight line segment EF.

At point E, $x_1$ is of course 70,000 and $x_2$ will be 10,000. At point F, $x_1$ remains 70,000 but $x_2$ will be 20,000. It is evident that point F will be the optimum point to meet the firm's three goals.

The last goal of the president is to minimize the total overtime operation of the plant. This goal does not actually change our optimum point, since the overtime operation of the plant has been limited to 10 hours. If we eliminate the overtime operation of 10 hours, this will achieve the fourth goal, but only by sacrificing 10,000 yards of the dress material. In other words, we are going back to point E from F. Of course, we do not wish to attain the fourth goal at the expense of the third goal.

At the optimum point F, the production will be 70,000 yards of upholstery material and 20,000 yards of dress material, and the profit will be $205,000. At this point, our first two goals are attained completely but the last two goals could not be achieved completely since there exists the underachievement of sales goal for the dress material by 25,000 yards and the overtime operation of 10 hours. However, the solution achieves all the goals as closely as possible according to the stated priorities and within the given decision environment.

The reader may think that linear programming could solve the problem if we treated the first two goals as constraints and maximized the profit within the constraints. In other words, formulate the linear programming problem as:

$$(4.14) \quad \text{Max } Z = 2500x_1 + 1500x_2$$
$$\text{s.t.} \quad x_1 \qquad \leqslant 70$$
$$x_2 \leqslant 45$$
$$x_1 + \quad x_2 \leqslant 90$$
$$x_1, x_2 \qquad \geqslant 0$$

The solution to the problem is to produce 70,000 yards of upholstery material and 20,000 yards of dress material for a profit of $205,000. As expected, the optimum solution is identical in this case. Does this mean that linear programming would yield the identical answer if we converted some of the management goals to constraints? The answer is absolutely *not*. First of all, it may be quite possible that none of the goals set by the decision maker involve financial goals, i.e., profit maximization or cost minimization. Second, the constraints set by certain goals will not form a convex set or a single area of feasible solution, so that there might be no solution to the linear programming model.

To illustrate the above statement, let us change the case we have discussed very slightly. Suppose the best customer of the textile company ordered 100,000 yards of upholstery material. For various reasons the president does not wish to fail to meet this order; therefore, the highest priority is assigned to this goal. $P_1$ is then assigned to $d_2^-$, the underachievement of this sales goal. The second goal of the president is to avoid any underutilization of production capacity; therefore, $P_2$ is assigned to $d_1^-$. The third goal is to minimize the overtime operation of the plant as much as possible. The fourth goal is assigned to the achievement of the sales goal for dress material, so $P_4$ is assigned to $d_3^-$.

If we attempt to utilize linear programming to maximize profit, the model would be developed as:

$$(4.15) \quad \text{Max } Z = 2500x_1 + 1500x_2$$
$$x_1 \quad\quad\quad \geqslant 100$$
$$x_2 \leqslant \ \ 45$$
$$x_1 + \quad x_2 \leqslant \ \ 90$$
$$x_1, x_2 \quad\quad \geqslant \ \ \ 0$$

The graphical presentation of the model is shown in Figure 4.13. Obviously, there is no area of feasible solution and consequently the above problem is unsolvable by linear programming.

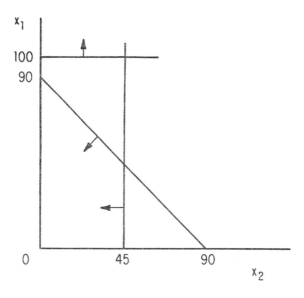

**Figure 4.13**

If we utilize goal programming, the above problem can easily be solved. The goal programming model is formulated as below:

(4.16)  $\text{Min } Z = P_1 d_2^- + P_2 d_1^- + P_3 d_1^+ + P_4 d_3^-$

$$\begin{array}{ll}
\text{s.t.} & x_1 + x_2 + d_1^- \quad\quad - d_1^+ \quad\quad = 80 \\
& x_1 \quad\quad + d_2^- \quad\quad - d_2^+ = 100 \\
& x_2 \quad\quad + d_3^- \quad\quad = 45 \\
& x_1, x_2, d_1^-, d_2^-, d_3^-, d_1^+, d_2^+ \geqslant 0
\end{array}$$

The graphical presentation of the model is shown in Figure 4.14.

Now, the first goal is to meet the sales goal of 100,000 yards of upholstery material ordered by the best customer. Therefore, the area of feasible solution will be on and above the straight line of $x_1 = 100$. The second goal is to minimize the underutilization of production capacity, which is set at 80 hours. By satisfying the first goal the second goal is automatically attained. The third goal is to minimize the overtime operation of the plant as much as possible. It is clear that in order to satisfy the first goal we have to run the plant 100 hours, an overtime operation of 20 hours. Hence, the third goal cannot be completely attained, but may be partially attained by moving $x_2$ to zero. Our last goal is to

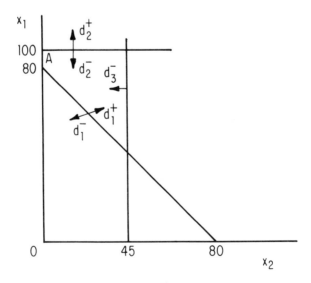

**Figure 4.14**

meet the sales goal for the dress material (i.e., $x_2 = 45$). The only way to meet this goal, however, is by providing more overtime hours for the plant. This implies that trying to attain the fourth goal sacrifices the third goal. Therefore this goal has to be ignored.

The goal programming solution of the problem is $x_1 = 100$ and $x_2 = 0$. At this point A, our first two important goals are satisfied; however, the third goal is not completely attained, and the last goal is not achieved at all. From this example, it should be clear that goal programming attempts to solve the problem by identifying the optimum point according to specified environmental conditions and stated priority structures. However, linear programming could not be used even to attempt to solve the problem when there is no single area of feasible solution.

Some questions may be raised as to why the decision maker sets goals that are beyond the normal operating constraints. However, in reality management often sets such goals even when they are unattainable within the limits of normally available resources, for a variety of reasons. For example, such goals may be established to analyze additional resource requirements and improve the current position of the organization. Or they may be established in order to ensure that long-run goals are not obliterated by short-run objectives. It is also possible that such goals are established to provide incentive to the employees or to judge organizational performance. The point is that management does quite frequently establish goals that may not be attainable within the normal organizational capacity.

Thus far, we have discussed the graphical analysis of goal programming. The goal programming solution is not usually a one-shot procedure. Quite often the decision maker would like to see the change in the output according to various combinations of inputs. Goal programming is a very suitable technique for such sensitivity analysis. In fact, there are three basic functions of a goal programming model. First, the model identifies required inputs to achieve stated goals under the given conditions. Second, the goal programming model reveals the degree of goal attainment with the given inputs under the specified conditions. Third, but not the least important, the model facilitates a simulation analysis of various combinations of inputs, constraints, and priority structures of goals.

## B. Example 2

In order to provide more insight into the graphical solution method of goal programming let us examine a second problem.

Oriental Rugs, Inc., produces the world's finest oriental rugs. The company's production facility consists of two production lines. Production line 1 is staffed with skilled workers who can produce an average of 2 rugs per hour. Line 2 is capable of producing an average of only 1½ rugs per hour, as it is staffed with relatively new employees. The regular working hours for the next week are 40 for each line. The profit from an average rug is $100. It is estimated that the operating costs of the two lines are virtually identical. The president of the firm has listed the following multiple goals to achieve in the coming week in ordinal ranking of importance:

1.  Meet the production goal of 180 rugs for the week.

2.  Limit the overtime operation of team 1 to five hours.

3.  Avoid the underutilization of regular working hours for both teams (assign differential weights according to the productivity of each team).

4.  Limit the sum of overtime operation for both teams (apply differential weights according to the relative cost of overtime hour. Assume that the cost of running each assembly line is identical).

From the above case the following constraints can be formulated.


## 1. PRODUCTION CAPACITY

The total production is a function of the sum of production rate for each production line multiplied by the number of hours each line is in operation. The production goal of 180 rugs will be based on this production function:

$$(4.17) \qquad 2x_1 + 1\tfrac{1}{2}x_2 + d_1^- - d_1^+ = 180$$

where $x_1$ = number of hours line 1 is in operation

$x_2$ = number of hours line 2 is in operation

$d_1^-$ = underachievement of production goal which is set at 180 rugs

$d_1^+$ = production in excess of 180 rugs.


## 2. REGULAR OPERATION HOURS

The regular working hours are usually limited to 40 hours for both production lines. However, if overtime is required to meet the production goal, working hours have to be increased accordingly.

(4.18)  $x_1 + d_2^- - d_2^+ = 40$
        $x_2 + d_3^- - d_3^+ = 40$

where $d_2^-$ = underutilization of regular working hours of 40 in line 1
      $d_2^+$ = overtime operation in line 1
      $d_3^-$ = underutilization of regular working hours of 40 in line 2
      $d_3^+$ = overtime operation in line 2.

## 3. OVERTIME OPERATION FOR LINE 1

According to goal 2 the overtime operation of line 1 should be limited to five hours. The reason for this may be the accelerating wage costs involved with the skilled personnel in line 1. Obviously, there are more experienced personnel in line 1 than line 2 if we assume that their production rates differ despite the same production facilities. The constraint for the overtime operation for line 1 is necessary in order to create a deviational variable that represents overtime operation of line 1 beyond five hours. By minimizing this deviational variable to zero we will be able to achieve our second goal.

(4.19)  $x_1 + d_4^- - d_4^+ = 45$

where $d_4^-$ = operation hours of line 1 that total less than 45 hours
      $d_4^+$ = overtime operation of line 1 beyond five hours.

Before the complete model is formulated, there are a couple of points to be considered. First, our third goal is to avoid the underutilization of regular working hours for both teams. This goal implies that we do not wish to lay off any employees. This goal may be based on the fact that the plant is unionized and the contract prohibits any involuntary layoff of employees. Or it may be the policy of the organization. Many firms believe that favorable employee relations, public image of the firm, and union relations come from a stable employment policy. At any rate, the president wishes to assign differential weights to the third goal according to the productivity of each line. Since the productivity of line 1 is 2 units per hour and that of line 2 only 1½ units per hour, the president wishes to avoid the underutilization of regular working hours of line 1 more than line 2. The differential weights, then, will be 2 for $d_2^-$ and 1½ for $d_3^-$. Or, if we wish to assign the whole numbers, we can assign 4 to $d_2^-$ and 3 to $d_3^-$.

Another point to be considered is our fourth goal, which is to limit the sum of overtime operation for both lines. Again, we have to assign differential weights to deviational variables that represent overtime operation for two lines. The criterion to use is the relative cost of overtime. The production rates ratio for the two teams is 2 to 1½. Therefore, the relative cost of an hour of overtime is greater for line 2 than for line 1. The relative costs of overtime ratio for two lines will be 1½ to 2. Hence, we should assign 1½ to $d_2^+$ and 2 to $d_3^+$. If we wish to handle whole numbers the differential weights could be 3 and 4 respectively.

Now, the goal programming model can be formulated:

$$(4.20) \quad \text{Min } Z = P_1 d_1^- + P_2 d_4^+ + 4P_3 d_2^- + 3P_3 d_3^- + 4P_4 d_3^+ + 3P_4 d_2^+$$

$$\begin{array}{llll}
\text{s.t.} & 2x_1 + 1\tfrac{1}{2}x_2 + d_1^- & -d_1^+ & = 180 \\
& x_1 \quad\quad + d_2^- & -d_2^+ & = 40 \\
& x_2 + d_3^- & -d_3^+ & = 40 \\
& x_1 \quad\quad\quad\quad + d_4^- & -d_4^+ & = 45 \\
\end{array}$$

$$x_1, x_2, d_1^-, d_2^-, d_3^-, d_4^-, d_1^+, d_2^+, d_3^+, d_4^+ \geqslant 0$$

The above problem is shown on the graph in Figure 4.15. Now, our first goal is to achieve the production goal of 180 rugs. Therefore, the area of feasible

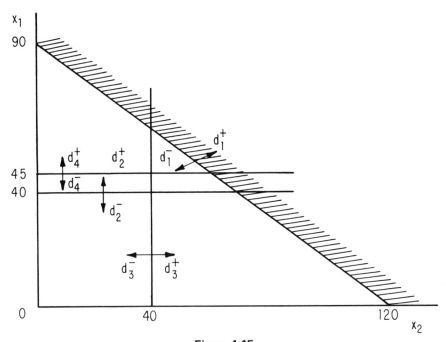

**Figure 4.15**

solution must be upper part from the straight line $x_1 = 90 - 3/4x_2$ shown by
the shaded area. Our second goal calls for the limitation of overtime operation
for line 2 to five hours. This can be plotted as $x_1 \leqslant 45$, as shown in Figure 4.16.
Now, it is obvious that the double shaded area from point A to the right satisfies
the first two goals.

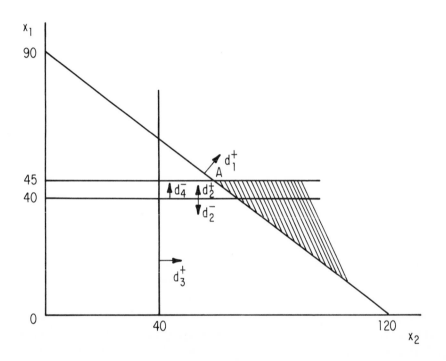

**Figure 4.16**

The third goal is to avoid underutilization of regular working hours for both
lines. Since a greater weight is assigned to the underutilization of line 1, we
should examine the working hours of line 1 first. The constraint calls for
$x_1 \geqslant 40$. Plotting this on the graph we obtained the area of feasible solution
shown in Figure 4.17. Any point in the shadowed area satisfies the first two
goals as well as the first part of the third goal.

Now, we are also to avoid any underutilization of regular working hours for
line 2. It is clearly evident in the above figure, however, that the shadowed area
already satisfies this constraint. For example, point A or B or any point on the
line AB actually indicates overtime operation for line 2.

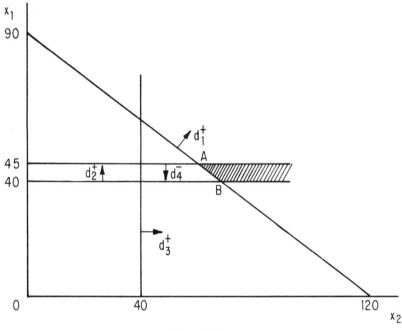

**Figure 4.17**

Our last goal is to minimize the sum of overtime for both lines. However, the limitation of overtime for line 2 is given a greater weight. Therefore, the overtime operation of line 2 must first be minimized to the fullest extent. Obviously, at point A the overtime operation of line 2 will be minimized while meeting the first three goals. Next, we try to minimize the overtime operation for line 1. Since we cannot decrease the overtime operation of line 1 any further from point A without providing more overtime to line 2, point A has to be the optimum point.

Now, let us analyze the optimum point. At point A, $x_1$ will be 45 hours. Substituting this value in the production function $x_1 = 90 - \frac{3}{4}x_2$, we obtain the value 60 for $x_2$. With these values ($x_1 = 45, x_2 = 60$) we could attain our first three goals completely, but could not attain the last goal. Although we could not minimize overtime operation of both lines to zero, it has been minimized to the greatest possible extent. At the optimum point, the total production will be 180 rugs, which yields a profit of $18,000.

## C.  Example 3

The last example we will discuss in this chapter is the record-shop problem described in Chapter 3 as Example 2. The model we formulated was:

(4.21)   $\text{Min } Z = P_1 d_1^- + P_2 d_{21}^+ + 2P_3 d_2^- + P_3 d_3^- + P_4 d_2^+ + 3P_4 d_3^+$

$$\begin{aligned}
\text{s.t.} \quad 5x_1 + 2x_2 + d_1^- \qquad\quad - d_1^+ \qquad\qquad\qquad &= 5,500 \\
x_1 \qquad + d_2^- \qquad\quad - d_2^+ \qquad\qquad &= \phantom{5,}800 \\
x_2 \quad + d_3^- \qquad\quad - d_3^- \qquad &= \phantom{5,}320 \\
d_{21}^- \qquad\qquad + d_2^+ - d_{21}^+ &= \phantom{5,}100 \\
x_1, x_2, d_1^-, d_2^-, d_3^-, d_{21}^-, d_1^+, d_2^+, d_3^+, d_{21}^+ &\geqslant \qquad 0
\end{aligned}$$

For convenience, we convert the overtime goal constraint to $x_1 + d_4^- - d_4^+$ $= 900$. Consequently, the second goal in the objective function will also be changed to $P_2 d_4^+$.

The model constraints are plotted on the graph as shown in Figure 4.18. The first goal of the manager is to achieve the sales goal by minimizing $d_1^-$ to zero. Therefore, the solution should be on or above the line $x_1 = 1,100 - 2/5x_2$. The second goal calls for the limitation of overtime for the full-time salesmen to 100 hours. By minimizing $d_4^+$, the solution area becomes the shaded area shown in Figure 4.19. It is evident that the double shaded area satisfies the first two goals.

The third goal is to avoid underutilization of regular working hours for the full-time and part-time salesmen. Since we assigned twice the weight to the minimization of $d_4^-$, we must restrict the solution area on or above the line $x_1 = 800$. Now the solution area has become a very thin strip, shown in Figure 4.20. The fourth goal is to minimize the sum of overtime given to the full-time and part-time salesmen. Since we assigned three times the weight to minimization of part-time salesmen's overtime as to minimization of full-time salesmen's overtime, $d_3^+$ must be minimized first. The closest we can come to the

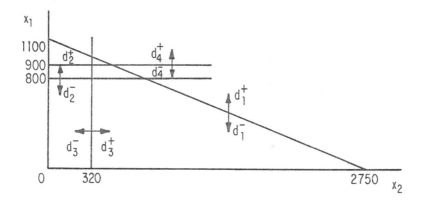

Figure 4.18

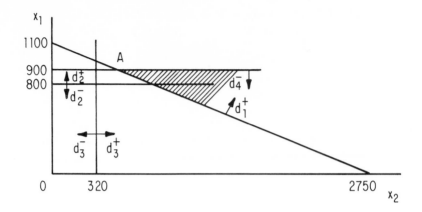

**Figure 4.19**

straight line $x_2 = 320$ is point A. Now we would like to minimize $d_2^+$. However, if we move down to B in order to minimize $d_2^+$, we are increasing $d_3^+$. Therefore, the optimum solution of the problem will be point A, where $x_1 = 900$, $x_2 = 500$, $d_2^+ = 100$, $d_3^+ = 180$. With this solution the manager of the record shop is able to attain the first three most important goals completely, but the last goal is not achieved, as there are overtime hours in the solution.

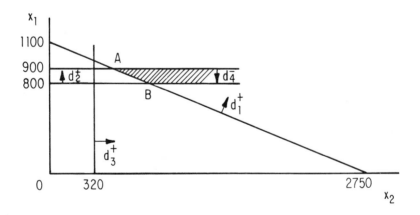

**Figure 4.20**

## SUMMARY

In this chapter we have examined the graphical method of goal programming solution. The step-by-step procedure to follow in order to identify the optimum point can be summarized as follows:

1. Identify and develop the model constraints from the problem.
2. Create new constraints if the deviational variable required in order to attain a certain stated goal is not present in the constraints already identified.
3. Develop the objective function by assigning appropriate priority and weight factors to deviational variables.
4. Plot the straight lines for all constraints involving choice variables.
5. Plot the straight lines for additional constraints that involve only deviational variables by converting it into a constraint involving choice variables.
6. Identify the point or area that satisfies the first goal, second goal, and so forth until the last goal is achieved to the fullest possible extent.
7. Identify the optimum point and analyze it in terms of degrees of goal attainment.

In this chapter we have discussed goal programming problems which involve only two choice variables. The graphical analysis is simple when we deal with such two-dimensional problems. However, the purpose of the graphical analysis was primarily for the purpose of illustrating the concept of goal programming solutions rather than as a general technique for solving goal programming problems. Therefore, the graphical method is recommended only when the problem involves two structural variables. A modified simplex method will be demonstrated in the following chapter as a general technique for solving goal programming problems.

## PROBLEMS

1. Solve the following problem by the graphical method of goal programming.

$$\text{Min } Z = P_1 d_1^- + 2P_2 d_3^+ + P_2 d_2^+ + P_3 d_1^+$$

$$
\begin{array}{llll}
\text{s. t.} & 2x_1 + x_2 + d_1^- & -d_1^+ & = 150 \\
& x_1 \quad\quad +d_2^- & -d_2^+ & = 40 \\
& x_2 \quad +d_3^- & -d_3^+ & = 40 \\
& x_1, x_2, d_1^-, d_2^-, d_1^+, d_2^+, d_3^+ \geq & 0
\end{array}
$$

2.  Solve the following problem by the graphical method of goal programming.

$$\text{Min } Z = P_1 d_2^- + 3P_2 d_4^+ + 2Pd_3^+ + P_3 d_1^-$$

$$
\begin{array}{llll}
\text{s. t.} & 5x_1 + 6x_2 + d_1^- & -d_1^+ & = 2{,}000 \\
 & x_1 + x_2 + d_2^- & -d_2^+ & = 375 \\
 & x_1 \quad\quad + d_3^- & -d_3^+ & = 180 \\
 & x_2 \quad + d_4^- & -d_4^+ & = 180 \\
 & x_1, x_2, d_1^-, d_2^-, d_3^-, d_4^-, d_1^+, d_2^+, d_3^+, d_4^+ \geq & & 0
\end{array}
$$

3.  Solve Example 1 presented in Chapter 3 by using the graphical method of goal programming.

4.  Solve Problem 1 at the end of Chapter 3 by using the graphical method of goal programming.

5.  Solve Problem 2 at the end of Chapter 3 by using the graphical method of goal programming. What will be the change in the optimum solution if the president decides to achieve the sales goal exactly as stated in the problem?

6.  Solve Problem 3 at the end of Chapter 3 by using the graphical method of goal programming. (a) If top management desires to put the profit goal of $19,000 for the week as the top priority goal above the stated four goals how would the model be changed? (b) If top management has only one goal for profit maximization subject to the regular production capacity for both lines, how would the goal programming solution be achieved through the graphical method?

# Chapter 5

## The Simplex Method of Goal Programming

In this chapter a modified simplex procedure is introduced to solve goal programming problems. The simplex method is not, however, the only available solution technique for complex problems. The generalized inverse technique has also been introduced by Ijiri.[1] The solution of goal programming problems by the simplex method, however, has not been thoroughly discussed in the literature. Furthermore, the computer program introduced in Chapter 6 is based primarily on the simplex procedure. We shall, therefore, examine the simplex solution of goal programming in great detail.

The simplex method is an algorithmic procedure that employs an iterative process so that the optimum solution is achieved through progressive operations. Theoretically, the simplex method is capable of solving any linear mathematical programming problem. In practice, only problems with a limited number of variables and constraints are solved by hand through the simplex procedure. If the magnitude of the problem is such that it may require more than a reasonable time, one should solve the problem through the computer if such a facility is available. According to the author's experience, students tend to utilize the computer if the problem requires more than an hour of solution time by hand, if the professor does not specify solution by hand.

The simplex solution procedure for goal programming problems is very similar to the simplex solution of linear programming problems; however, there are several distinct differences. It is helpful, therefore, to review the simplex procedure of linear programming problems before we discuss the goal programming solution.

## THE SIMPLEX METHOD OF LINEAR PROGRAMMING

Since goal programming is based on a minimization procedure, it is appropriate to examine the following minimization problem of linear programming:

$$(5.1) \quad \text{Min } Z = \$2x_1 + \$1x_2$$
$$\text{s.t.} \quad 2x_1 + 3x_2 \geqslant 6$$
$$4x_1 + 2x_2 \geqslant 8$$
$$x_2 \leqslant 2$$

It should be remembered that the first step of the simplex procedure is to formulate equations of the model constraints by introducing slack or artificial variables. It is assumed that the reader is familiar with this process from previous exposure to linear programming. We can transform the above model to equations as follows:[2]

$$(5.2) \quad \text{Min } Z = \$2x_1 + \$1x_2 + OS_1 + OS_2 + OS_3 + MA_1 + MA_2$$
$$\text{s.t.} \quad 2x_1 + 3x_2 - S_1 + A_1 = 6$$
$$4x_1 + 2x_2 - S_2 + A_2 = 8$$
$$x_2 + S_3 = 2$$

The initial step of the simplex procedure assumes that the first solution is at the origin. In other words, choice variable $x_1$ and $x_2$ are both zero. Therefore, from the constraints we can easily identify the variables that will be in the solution base. Table 5.1 indicates the initial simplex tableau for the problem. Since it is assumed that the reader has a basic knowledge of the simplex solution procedure for linear programming, only a brief explanation of the table shall be presented here.[3]

Table 5.1

| $C_j$ | | | 2 | 1 | 0 | 0 | 0 | M | M |
|---|---|---|---|---|---|---|---|---|---|
| | $V$ | $C$ | $x_1$ | $x_2$ | $S_1$ | $S_2$ | $S_3$ | $A_1$ | $A_2$ |
| M | $A_1$ | 6 | 2 | 3 | −1 | | | 1 | |
| M | $A_2$ | 8 | ④ | 2 | | −1 | | | 1 |
| O | $S_3$ | 2 | | 1 | | | 1 | | |
| | $Z_j$ | 14M | 6M | 5M | −M | −M | 0 | M | M |
| | $Z_j$−$C_j$ | | 6M−2 | 5M−1 | −M | −M | 0 | 0 | 0 |

The $C_j$ column and row simply indicate the cost associated with each variable as defined in the objective function. Recall that we are utilizing the "large M" method, which assigns a large cost "M" to artificial variables. The column $V$ represents the variables in the solution base and column $C$ indicates the right-hand-side (rhs) values of the constraints. There are many variations in notation and procedures of the simplex method in various linear programming books according to the preference of the author. However, the basic methodology is universal. In our example, we will define $Z_j$ and $Z_j − C_j$. $Z_j$, which represents the resource requirements, is simply the sum of each $C_j$ times the constant or coefficient in the column of interest. The optimum column— that is, the column containing the variable that will enter the solution base in the second stage—is identified by finding the largest positive $Z_j − C_j$ value. The key row or replaced row is the row containing the variable to be removed from the solution base. It is selected by finding the minimum positive or zero value obtained by dividing $C$ constants by the coefficients in the optimum column.

Of course the reason why the $x_1$ column is selected as the optimum column is that the variable $x_1$ can reduce the total cost at the fastest rate, as shown by its $Z_j − C_j$. The maximum number of units of $x_1$ that can be introduced in the solution is identified as 2 when the column $C$ constant 8 is divided by coefficient 4 of column $x_1$. Thus, row $A_2$ is identified as the outgoing row. Table 5.2 illustrates the second-stage tableau and the final solution. Again, remembering that the simplex procedure is based on the matrix inverse technique, let us review how new values are derived in the second tableau. There are two procedures we have to employ to find new values in the next interaction:

1.  For the key row ($x_1$ in Table 5.2), the new values are determined by dividing values in the first table in row $A_2$ by coefficient 4 in the $x_2$ column.
2.  For the remaining rows, the new values are determined by subtracting the product of (the coefficient at the intersection of the optimum column and each row times the new value derived in the key row for each column) from the original value. For example, the new rhs constant of $A_1$ in Table 5.2 is derived by the calculation procedure of $6 - 2 \times 2 = 2$.

Table 5.2

| $C_j$ | | | 2 | 1 | 0 | 0 | 0 | M | M |
|---|---|---|---|---|---|---|---|---|---|
| | V | C | $x_1$ | $x_2$ | $S_1$ | $S_2$ | $S_3$ | $A_1$ | $A_2$ |
| M | $A_1$ | 2 | | (2) | −1 | 1/2 | | 1 | |
| 2 | $x_1$ | 2 | 1 | 1/2 | | −1/4 | | | |
| 0 | $S_3$ | 2 | | 1 | | | 1 | | |
| | $Z_j$ | 2M+4 | 2 | 2M+1 | −M | M/2−1/2 | 0 | M | |
| | $Z_j$−$C_j$ | | 0 | 2M | −M | M/2−1/2 | 0 | 0 | |
| 1 | $x_2$ | 1 | | 1 | −1/2 | 1/4 | | | |
| 2 | $x_1$ | 3/2 | 1 | | 1/4 | −3/8 | | | |
| 0 | $S_3$ | 1 | | | 1/2 | −1/4 | 1 | | |
| | $Z_j$ | 4 | 2 | 1 | 0 | −1/2 | 0 | | |
| | $Z_j$−$C_j$ | | 0 | 0 | 0 | −1/2 | 0 | | |

Now, it should be apparent in Table 5.2 that variable $A_2$ is eliminated from the table because once an artificial variable is removed from the solution base it is no longer needed for further analysis. By employing the same procedure used in the first table, the optimum column ($x_2$) and the key row ($A_1$) are identified. The third table indicates that the solution base has a total cost of $4 ($Z_j$ value under column C), and it cannot be reduced any further as shown by the zero or negative $Z_j - C_j$ values. The optimum solution is, therefore, $x_1 = 1$, $x_2 = 1\frac{1}{2}$, $S_3 = 1$, and $Z_j = 4$.

## THE SIMPLEX METHOD OF GOAL PROGRAMMING

### Example 1

The best way to explain the simplex method of goal programming is through an example. So, let us examine the problem presented as Example 1 in Chapter 4. The president of a textile company has the following goal programming problem:

$$(5.3) \quad \text{Min } Z = P_1 d_d^- + P_2 d_4^+ + 5P_3 d_2^- + 3P_3 d_3^- + P_4 d_1^+$$

$$\begin{array}{llll}
\text{s.t.} & x_1 + x_2 + d_1^- & - d_1^+ & = 80 \\
& x_1 & + d_2^- & = 70 \\
& x_2 & + d_3^- & = 45 \\
& x_1 + x_2 & + d_4^- - d_4^+ & = 90 \\
\end{array}$$

$$x_1, x_2, d_1^-, d_2^-, d_3^-, d_4^-, d_1^+, d_4^+ \geqslant 0$$

The reason we formulated the last constraint (minimization of overtime operation of the plant in excess of 10 hours) by using a choice variable was that we must have all constraints expressed in terms of choice variables in order to solve the problem by the graphical solution method. However, since we now utilize the simplex procedure, we can also solve the problem if the constraint is formulated as below:

$$(5.4) \quad d_1^+ + d_{11}^- - d_{11}^+ = 10$$

where $d_1^+$ = overtime operation of the plant

$d_{11}^-$ = difference between the actual overtime operation of the plant and 10 hours of overtime

$d_{11}^+$ = overtime operation of the plant in excess of 10 hours.

If we use the above-formulated constraint, of course the second goal in the objective function should read $P_2 d_{11}^+$. Either model is perfectly acceptable for the simplex solution of the problem.

Before the first simplex table is presented for the above problem, there are several things we must consider.

First, in goal programming, the purpose of the objective function is to minimize the total unattained goals. This is achieved by minimizing the deviational variables through the use of certain preemptive priority factors or differential weight. There is no profit maximization or cost minimization per se in the objective function. Therefore, the preemptive factors and differential weights take the place of the $C_j$ as used in linear programming.

Second, the objective function is expressed by assigning priority factors to certain variables. These preemptive priority factors are multidimensional as they are ordinal rather than cardinal values. In other words, priority factors at different levels are not commensurable. This implies that the simplex criterion ($Z_j$ or $Z_j - C_j$) cannot be expressed by a single row as is done in the case of linear programming. Rather, the simplex criterion becomes a matrix of m x n size, where m represents the number of preemptive priority levels and n is the number of variables including both choice and deviational variables.

Third, since the simplex criterion is expressed as a matrix rather than a row, we must design a new procedure in identifying the optimum column. The relationship between the preemptive priority factors is $P_j >>> P_{j+1}$, which means $P_j$ always takes priority over $P_{j+1}$. It is, therefore, clear that the selection procedure of the optimum column must consider the level of priorities.

Table 5.3 presents the initial tableau of this goal programming problem. The basic assumption in formulating the initial table of goal programming is identical to that of linear programming. We assume that the initial solution is at the origin, where values of all the choice variables are zero. In the first constraint, therefore, the total operation hours of the plant are of course zero, since $x_1 = x_2 = 0$. Naturally, there cannot be any overtime operation of the plant

Table 5.3

| $C_j$ | | | $P_1$ | | 5$P_3$ | 3$P_3$ | | $P_4$ | $P_2$ |
|---|---|---|---|---|---|---|---|---|---|
| | $V$ | $C$ | $x_1$ | $x_2$ | $d_1^-$ | $d_2^-$ | $d_3^-$ | $d_{11}^-$ | $d_1^+$ | $d_{11}^+$ |
| $P_1$ | $d_1^-$ | 80 | 1 | 1 | 1 | | | | $-1$ | |
| 5$P_3$ | $d_2^-$ | 70 | (1) | | | 1 | | | | |
| 3$P_3$ | $d_3^-$ | 45 | | 1 | | | 1 | | | |
| | $d_{11}^-$ | 10 | | | | | | 1 | 1 | $-1$ |
| | $P_4$ | 0 | | | | | | | $-1$ | |
| $Z_j - C_j$ | $P_3$ | 485 | 5 | 3 | | | | | | |
| | $P_2$ | 0 | | | | | | | | $-1$ |
| | $P_1$ | 80 | 1 | 1 | | | | | $-1$ | |

($d_1^+ = 0$). Therefore, underutilization of the normal plant capacity ($d_1^-$) will be 80 hours. Hence, the variable $d_1^-$ is entered in the solution base and the constant (rhs) becomes 80. By the same token, $d_2^-$ and $d_3^-$ are also in the solution base. In the last constraint ($d_1^+ + d_{11}^- - d_{11}^+ = 10$), since the plant is not in operation $d_1^+$ has to be zero. Then, overtime operation of the plant in excess of 10 hours ($d_{11}^+$) must also be zero. Consequently, $d_{11}^-$ takes the constant of 10 as shown in Table 5.3. It is always the rule that in the initial table of goal programming negative deviational variables ($d_i^-$) will appear in the solution base.

Now, let us examine $C_j$. In goal programming $C_j$ is represented by the preemptive priority factors and the differential weights as shown in the goal programming objective function. Most goal programming problems involve a large number of variables. For that reason, in order to make the table easier to read empty spaces are left in the table where zero should appear.

The simplex criterion ($Z_j - C_j$) is a $4 \times 8$ matrix because we have four priority levels and eight variables (2 choice, 6 deviational) in the model. The goal programming procedure first achieves the most important goal to the fullest possible extent and then considers the next-order goal, and so forth. It should be readily apparent that the selection of the optimum column should be based on the per-unit contribution rate of each variable in achieving the most important goal. When the first goal is completely attained, then the optimum column selection criterion will be based on the achievement rate for the second goal, and so on. That is why we listed the preemptive priority factors from the lowest to the highest so that the optimum column can be easily identified at the bottom of the table. To make the simplex table relatively simple, we have omitted the matrix of $Z_j$ all together. It requires a little more calculation this way but the simplified tableau is well worth that inconvenience.

The goal programming problem is a minimization problem. In the minimization problem of linear programming, the values in the constant (rhs) column of the simplex criterion represent the total cost of the solution. In goal programming those values ($P_4 = 0$, $P_3 = 485$, $P_2 = 0$, and $P_1 = 80$) in the constant column represent the unattained portion of each goal. For example, in the initial tableau where the textile plant is not even in operation, the second and the fourth goals are already completely attained. How could this be possible? Examining the objective function, we can find that the second goal is to minimize the overtime operation of the plant in excess of 10 hours and the fourth goal is to minimize the total overtime operation of the plant. It should be evident by now to the reader that since we are not even operating the plant at this point (at the origin), naturally there is no overtime operation. Consequently, we have already attained the second and fourth goals. The underachievement of the first goal is 80 because the underutilization of the normal plant operation capacity is 80 hours. For the third goal, the underachievement of the goal is 485. We remember that differential weights of 5 and 3 are assigned to the underachievement of sales goals for the upholstery and dress fabric materials. Since these two goals are commensurable and are at the same preemptive priority level, this procedure is absolutely appropriate. However, it is not as easy to interpret the underachievement of 485 for the third goal as other unattained goals where no differential weights are assigned.

Now, let us examine the calculation of $Z_j - C_j$ in Table 5.3. We have already discussed that $C_j$ values represent the priority factors assigned to deviational variables and $Z_j$ values are products of the sum of $C_j$ times constants or coefficients. Thus, $Z_j$ value in the $x_1$ column will be $(P_1 \times 1 + 5P_3 \times 1)$, or $P_1 + 5P_3$. $C_j$ value in the $x_1$ column is zero as shown by the blank in the $C_j$ row. Therefore, $Z_j - C_j$ for the $x_1$ column is $P_1 + 5P_3$. Since $P_1$ and $P_3$ are not commensurable, we must list them separately in the $P_1$ and $P_3$ rows in the simplex criterion $(Z_j - C_j)$. Consequently, $Z_j - C_j$ value will be 1 at the $P_1$ row and 5 at the $P_3$ row in the $x_1$ column. By employing the same procedure, $Z_j - C_j$ of the $x_2$ column can be derived. It will be $(P_1 \times 1 + 3P_3 \times 1) - 0$, or $P_1 + 3P_3$. For the following three columns, $d_1^-$, $d_2^-$, and $d_3^-$, $Z_j - C_j$ will be zero since $Z_j$ values are identical to the respective $C_j$ values.

For the $d_{11}^-$ column $Z_j - C_j$ is zero because $Z_j$ and $C_j$ are both zero. For the $d_1^+$ column, we can easily calculate the $Z_j$ value of $-P_1$ from the table. Since $C_j$ value of the column is $P_4$, $Z_j - C_j$ will be $-P_1 - P_4$. Therefore, $-1$ is listed at the $P_1$ row and also at the $P_3$ row in the column. The last column, $d_{11}^+$, indicates zero $Z_j$. However, its $C_j$ value is $P_2$. Thus, $Z_j - C_j$ for the column becomes $-P_2$. Accordingly, we list $-1$ at the $P_2$ row for the $d_{11}^+$ column.

As mentioned earlier, we have combined calculational procedure for identifying $Z_j$ and $Z_j - C_j$ values in the modified simplex tableau. The procedure requires more calculations in our heads, but it certainly makes the tableau simpler to handle. It is especially true if the problem under consideration is a very complex one. For example, if a problem containing 5 preemptive priorities and 25 variables is being analyzed, we can save 5x25 matrix by calculating $Z_j - C_j$ in one shot.

Now, let us move on to select the optimum column and key row. The criterion used to determine the optimum column is the rate of contribution of each variable in achieving the most important goal $(P_1)$. In other words, the column with the largest positive value at the $P_1$ level in $Z_j - C_j$ will be selected as the optimum column. In Table 5.3, there are two identical positive values in $x_1$ and $x_2$ columns. In order to break this tie, we check the next-lower priority levels. Since there is a greater value in $x_1$ column at the $P_3$ level as compared to the $x_2$ column, we select $x_1$ as the optimum column. The key row is the row which has the minimum value when we divide the rhs constants by the coefficients in the optimum column. The coefficient of 1 is circled in Table 5.3 to indicate that it is at the intersection of the optimum column and the key row. By entering $x_1$ into the solution base, the underutilization of the regular plant capacity and the underachievement of sales goal for the upholstery material will be affected. This is clear from an observation of coefficients existing in the $d_1^-$ and $d_2^-$ rows.

By utilizing the regular simplex procedure of linear programming, the first tableau is revised to obtain the second tableau as shown in Table 5.4. The plant is in operation for 70 hours in order to produce 70,000 yards of upholstery material. Therefore, the underutilization of regular plant capacity is now 10 hours as shown by the constant in the $d_1^-$ row. We have also completely achieved the sales goal for upholstery material, and therefore $d_2^-$ has been removed from the solution base. The calculation of new coefficients in goal programming is usually easier than in linear programming because in goal programming there are many coefficients with unit (1) value. In the $d_3^-$ and $d_{11}^-$ rows in Table 5.4, the coefficients will remain exactly as they were in Table 5.3. This is because the intersectional elements, i.e., the element at the intersection of the optimum column and each row in question, are zero. Calculating the new values for the key row $(x_1)$ is achieved by dividing old row values by the pivot element of 1.

Table 5.4

| $C_j$ | | | $P_1$ | | $5P_3$ | $3P_3$ | | $P_4$ | $P_2$ |
|---|---|---|---|---|---|---|---|---|---|
| | $V$ | $C$ | $x_1$ | $x_2$ | $d_1^-$ | $d_2^-$ | $d_3^-$ | $d_{11}^-$ | $d_1^+$ | $d_{11}^+$ |
| $P_1$ | $d_1^-$ | 10 | ① | 1 | −1 | | | | −1 | |
| | $x_1$ | 70 | 1 | | | 1 | | | | |
| $3P_3$ | $d_3^-$ | 45 | | 1 | | | 1 | | | |
| | $d_{11}^-$ | 10 | | | | | | 1 | 1 | −1 |
| $Z_j - C_j$ | $P_4$ | 0 | | | | | | | −1 | |
| | $P_3$ | 135 | | 3 | −5 | | | | | |
| | $P_2$ | 0 | | | | | | | | −1 |
| | $P_1$ | 10 | | 1 | −1 | | | | −1 | |

The only row where some additional calculation is required is the $d_1^-$ row, but this is also relatively easy since there are no complex fractions involved.

Let us examine Table 5.4 more closely. The $Z_j$ values in the constant column ($P_4 = 0$, $P_3 = 135$, $P_2 = 0$, $P_1 = 10$) indicate that the unattained portion of the first goal has decreased considerably, 70 to be exact. This is a good sign because the goal programming model is a minimization problem and the value of $Z_j$ should decrease at each step toward the optimum point. As our immediate concern is the achievement of the most important goal, we should simply examine whether $Z_j$ has decreased at the $P_1$ level at the end of each step. When $Z_j$ at the $P_1$ level is completely minimized to zero, our attention should then be focused on the $Z_j$ value at the $P_2$ level, and so on. In Table 5.4, $Z_j$ at the $P_3$ level has also decreased by the amount of 350, as the production of 70,000 yards of the upholstery material automatically enables the achievement of the sales goal for the upholstery material.

The optimum column is identified as $x_2$ in Table 5.4. The key row of $d_1^-$ is determined by the usual procedure. The best way to achieve the most important goal completely is by producing 10,000 yards of regular fabric material. The production of 70,000 yards of upholstery and 10,000 yards of dress materials will require 80 hours of plant operation.

Table 5.5 presents the third-stage solution. The solution indicates that production of 70,000 yards of upholstery and 10,000 yards of dress materials is

Table 5.5

| $C_j$ | | | | | | $P_1$ | $5P_3$ | $3P_3$ | | $P_4$ | $P_2$ |
|---|---|---|---|---|---|---|---|---|---|---|---|
| | $V$ | $C$ | $x_1$ | $x_2$ | $d_1^-$ | $d_2^-$ | $d_3^-$ | $d_{11}^-$ | $d_1^+$ | $d_{11}^+$ |
| | $x_2$ | 10 | | 1 | 1 | −1 | | | −1 | | |
| | $x_1$ | 70 | 1 | | | 1 | | | | | |
| $3P_3$ | $d_3^-$ | 35 | | | −1 | 1 | 1 | | 1 | | |
| | $d_{11}^-$ | 10 | | | | | | 1 | ① | −1 | |
| | $P_4$ | 0 | | | | | | | | −1 | |
| $Z_j - C_j$ | $P_3$ | 100 | | | −3 | −2 | | | 3 | | |
| | $P_2$ | 0 | | | | | | | | | −1 |
| | $P_1$ | 0 | | | −1 | | | | | | |

sufficient to achieve the first, second, and fourth goals. However, the third goal is not completely attained since the sales goal of dress material is still 35,000 yards short of complete attainment; $d_3^-$ of 35 shown in the solution base indicates this. As there is no further goal attainment required at $P_1$ and $P_2$ levels, all coefficients in the $Z_j-C_j$ are either zero or negative as shown in the table.

The selection of the optimum column should be determined at the $P_3$ level. The $d_1^+$ column is obviously the optimum column as the only positive value at the $P_3$ level of $Z_j-C_j$ is in this column. The key row is $d_{11}^-$. The procedure is both rational and sensible. It employs overtime operation of the plant to attain the third goal to a greater extent. Since we assigned the fourth priority factor to the minimization of overtime operation of the plant, in essence we are attaining the third goal at the expense of the fourth goal.

Table 5.6 presents the optimum solution to the problem. It is optimum in the sense that this solution enables the decision maker to attain his goals as closely as possible within the given decision constraints and priority structure. Note the decrease of $Z_j$ value at the $P_3$ level from 100 to 75. In order to decrease the underachievement of the third goal we sacrificed the complete attainment of the fourth goal by 10 units as shown at the $P_4$ level. The optimum solution is $x_1 = 70$, $x_2 = 20$, $d_1^+ = 10$, $d_3^- = 25$. In other words, the company should produce 70,000 yards of upholstery material and 20,000 yards of dress material with 10 hours of overtime operation of the plant, resulting in 25,000

Table 5.6

| $C_j$ | | | | | $P_1$ | $5P_3$ | $3P_3$ | | $P_4$ | $P_2$ |
|---|---|---|---|---|---|---|---|---|---|---|
| | $V$ | $C$ | $x_1$ | $x_2$ | $d_1^-$ | $d_2^-$ | $d_3^-$ | $d_{11}^-$ | $d_1^+$ | $d_{11}^+$ |
| | $x_2$ | 20 | | 1 | 1 | −1 | | | 1 | −1 |
| | $x_1$ | 70 | 1 | | | 1 | | | | |
| $3P_3$ | $d_3^-$ | 25 | | | −1 | 1 | 1 | −1 | | 1 |
| $P_4$ | $d_1^+$ | 10 | | | | | | 1 | 1 | −1 |
| | $P_4$ | 10 | | | | | | 1 | | −1 |
| $Z_j - C_j$ | $P_3$ | 75 | | | −3 | −2 | | −3 | | 3 |
| | $P_2$ | 0 | | | | | | | | −1 |
| | $P_1$ | 0 | | | −1 | | | | | |

yards of underachievement in the sales goal of the fabric material. With this solution the president of the firm is able to attain his two most important goals completely, and the next two goals as completely as possible under the given constraints.

In Table 5.6, since the third goal is not completely attained, there is a positive value in $Z_j - C_j$ at the $P_3$ level. We find it (3) in the $d_{11}^+$ column. Obviously, we can attain the third goal to a greater extent if we introduce $d_{11}^+$ in the solution. We find, however, a negative value $(-1)$ at a higher priority level, i.e., at the $P_2$ level. This implies that if we introduce $d_{11}^+$, we would improve achievement of the third goal at the expense of achieving the second goal. Thus, we cannot introduce $d_{11}^+$ into the solution. The same logic applies to the $d_{11}^-$ column, where we find a positive value at the $P_4$ level. The rule is that if there is a positive element at a lower priority level in $Z_j - C_j$, the variable in that column cannot be introduced into the solution as long as there is a negative element at a higher priority level.

There is one more bit of analysis that we can derive from Table 5.6. From an analysis of the $Z_j - C_j$ values we can point out where conflict exists among the goals. Conflict exists between the second and third goals in column $d_{11}^+$, and between the third and fourth goals in column $d_{11}^-$. Now the decision maker can precisely determine how he must rearrange the priority structure if the underachieved goals at the lower levels are to be completely attained. This process provides the decision maker an opportunity to evaluate the soundness of his priority structure for the goals. Furthermore, from an analysis of the coefficients in the main body of the table the decision maker can identify the exact trade-offs between goals. For example, in Table 5.6 we can see that if we introduce 25 units of $d_{11}^+$ into the solution, the third goal will be completely attained. However, this procedure will "undo" the second goal by the same quantity. The marginal substitution rate in this case is one to one. The same is true between the third and fourth goals, as the coefficients and $d_3^-$ of $d_1^+$ rows in the $d_{11}^-$ column indicate. An analysis of the final solution table provides a great deal of information and insight about the decision environment and the decision maker's priority structure of goals. The reader can now compare the solution derived by the simplex method with the one derived by the graphical approach in Chapter 4.

## STEPS OF THE SIMPLEX METHOD OF GOAL PROGRAMMING

Now that we have illustrated how to solve a goal programming problem by the modified simplex method, we can summarize the solution steps to aid in future solutions:

## 1. SET UP THE INITIAL TABLEAU FROM THE GOAL PROGRAMMING MODEL

We assume that the initial solution is at the origin. Therefore, all the negative deviational variables in the model constraints should enter the solution base initially. List the constants (rhs) and the coefficients of all variables in the main body of the table. Also list the preemptive priority factors and differential weights to the appropriate variables by examining the objective function. In the simplex criterion $(Z_j-C_j)$, list priority levels in the V column from the lowest at the top to the highest at the bottom. $Z_j$ values must be calculated and recorded in the C column. The last step is to calculate $Z_j-C_j$ values for each column starting from the first choice variable to the last positive deviational variable.

## 2. DETERMINE THE NEW ENTERING VARIABLE

This step is identical to the identification of the optimum column. First, we find the highest priority level that has not been completely attained by examining the $Z_j-C_j$ values in the constant (rhs) column. When the priority level is determined, we proceed to identify the variable column that has the largest positive $Z_j-C_j$ value. The variable in that column will enter the solution base in the next iteration. If there is a tie between the largest positive values in $Z_j - C_j$ at the highest priority level, check the next lower priority levels and select the column that has a greater value at the lower priority level. If the tie cannot be broken, choose one on an arbitrary basis. The other column will be chosen in subsequent iterations.

## 3. DETERMINE THE EXITING VARIABLE FROM THE SOLUTION BASE

This process is identical to finding the key row. Calculate the value of the constant divided by the coefficients in the optimum column. Select the row that has the minimum positive or zero value. The variable in that row will be replaced by the variable in the optimum column in the next interation. If there exists a tie when constants are divided by coefficients, find the row that has the variable with the higher priority factor. This procedure enables the attainment of higher order goals first, and thereby reduces the number of iterations.

## 4. DETERMINE THE NEW BASIC FEASIBLE SOLUTION

First, find the new constant and coefficients of the key row by dividing old values by the intersectional element, i.e., the element at the intersection of the key row and the optimum column. Second, find the new values for all other rows by using the calculation procedure [of old value-intersectional element of that row x new value in the key row in the same column]. Now, complete the table by finding $Z_j$ and $Z_j-C_j$ values for the priority rows.

## 5. DETERMINE WHETHER THE SOLUTION IS OPTIMAL

First, analyze the goal attainment level of each goal by checking the $Z_j$ value for each priority row. If the $Z_j$ values are all zero, this is the optimal solution. Second, if there exists a positive value of $Z_j$, examine the $Z_j-C_j$ coefficients for that row. If there are positive $Z_j-C_j$ values in the row, determine whether there are negative $Z_j-C_j$ values at a higher priority level in the same column. If there are negative $Z_j-C_j$ values at a higher priority value for the positive $Z_j-C_j$ values in the row of interest, the solution is optimum. Third, if there exists a positive $Z_j-C_j$ value at a certain priority level and there is no negative $Z_j-C_j$ value at a higher priority level in the same column, this is not the optimum solution. Therefore, return to Step 2 and continue. Figure 5.1 illustrates the simplex solution procedure for goal programming problems.

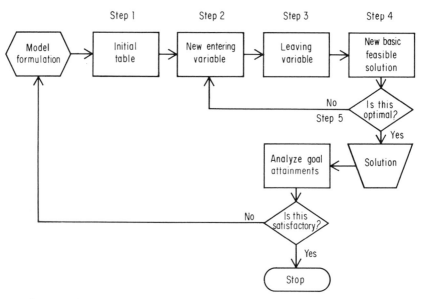

**Figure 5.1. Flowchart of the simplex procedure of goal programming**

## Example 2

Now, it would seem appropriate to solve a somewhat more complex goal programming problem by the simplex procedure. Let us examine the record-shop case discussed in Chapter 3, Example 2, and solved by the graphic method in Chapter 4, Example 3. The problem involves a decision analysis by the manager of a record shop in a small college town. The model formulated was:

$$
(5.5) \quad \text{Min } Z = P_1 d_1^- + P_2 d_{21}^+ + 2P_3 d_2^- + P_3 d_3^- + P_4 d_2^+ + 3P_4 d_3^+
$$

$$
\begin{aligned}
\text{s.t.} \quad 5x_1 + 2x_2 + d_1^- \qquad\qquad\quad - d_1^+ \qquad\qquad\qquad &= 5{,}500 \\
x_1 \qquad\quad + d_2^- \qquad\quad - d_2^+ \qquad\qquad\qquad &= 800 \\
x_2 \quad\quad + d_3^- \qquad\qquad - d_3^+ \qquad\qquad &= 320 \\
d_{21}^- \qquad\qquad + d_2^+ \; - d_{21}^+ &= 100 \\
x_1, x_2, d_1^-, d_2^-, d_3^-, d_{21}^-, d_1^+, d_2^+, d_3^+, d_{21}^+ &\geqslant 0
\end{aligned}
$$

The problem involves four priority factors and 10 variables—two choice and eight deviational variables. Let us follow the steps of the simplex method of goal programming outlined above.

## 1. SET UP THE INITIAL TABLE FOR THE PROBLEM

Table 5.7 provides the initial tableau for the problem. All the negative deviational variables are in the initial solution base, as we assume that the initial solution is at the origin. The initial table indicates that the second and the fourth goals are completely attained but the first and third goals are not achieved. At the origin we are not providing any working hours for full-time and part-time salesmen. Therefore, the second and fourth goals, which are concerned with the overtime of salesmen, must be completely attained in the initial solution.

## 2. DETERMINE THE NEW ENTERING VARIABLE

Through an examination of the $Z_j$–$C_j$ values at the $P_1$ level, it is apparent that the $x_1$ column is the optimum column, since $x_1$ decreases the unattainment of the first goal at the fastest rate (5).

Table 5.7

| $C_j$ | | | | | $P_1$ | $2P_3$ | $P_3$ | | $P_4$ | $3P_4$ | $P_2$ | |
|---|---|---|---|---|---|---|---|---|---|---|---|---|
| | $V$ | $C$ | $x_1$ | $x_2$ | $d_1^-$ | $d_2^-$ | $d_3^-$ | $d_{21}^-$ | $d_1^+$ | $d_2^+$ | $d_3^+$ | $d_{21}^+$ |
| $P_1$ | $d_1^-$ | 5500 | 5 | 2 | 1 | | | | -1 | | | |
| $2P_3$ | $d_2^-$ | 800 | ① | | | 1 | | | | -1 | | |
| $P_3$ | $d_3^-$ | 320 | | 1 | | | 1 | | | | -1 | |
| | $d_{21}^-$ | 100 | | | | | | 1 | | 1 | | -1 |
| $Z_j - C_j$ | $P_4$ | 0 | | | | | | | -1 | -3 | | |
| | $P_3$ | 1920 | 2 | 1 | | | | | -2 | -1 | | |
| | $P_2$ | 0 | | | | | | | | | | -1 |
| | $P_1$ | 5500 | 5 | 2 | | | | | -1 | | | |

## 3. DETERMINE THE EXITING VARIABLE FROM THE SOLUTION BASE

After we determine the optimum column, the key row should be identified by finding the row with the minimum value when constants are divided by the coefficients in the optimum column. In Table 5.7, the $d_2^-$ row is clearly the key row. Therefore, in the second-stage table, $d_2^-$ will be replaced by $x_1$ in the solution base.

## 4. DETERMINE THE NEW BASIC FEASIBLE SOLUTION

The revised tableau is presented in Table 5.8. It is simple to derive new values for the key row ($d_2^-$) since the old values are divided by the intersectional element of 1. It is even easier to derive the new values in $d_3^-$ and $d_{21}^-$ rows as the old values remain the same because of zero intersectional elements. The only row where some calculations are required is $d_1^-$ because there exists a coefficient in the optimum column $x_1$.

Table 5.8

| $C_j$ | | | | $P_1$ | $2P_3$ | $P_3$ | | | $P_4$ | $3P_4$ | $P_2$ |
|---|---|---|---|---|---|---|---|---|---|---|---|
| | $V$ | $C$ | $x_1$ | $x_2$ | $d_1^-$ | $d_2^-$ | $d_3^-$ | $d_{21}^-$ | $d_1^+$ | $d_2^+$ | $d_3^+$ | $d_{21}^+$ |
| $P_1$ | $d_1^-$ | 1500 | | 2 | 1 | −5 | | | −1 | 5 | | |
| | $x_1$ | 800 | 1 | | | 1 | | | −1 | | | |
| $P_3$ | $d_3^-$ | 320 | | 1 | | | 1 | | | −1 | | |
| | $d_{21}^-$ | 100 | | | | | | 1 | | (1) | | −1 |
| | $P_4$ | 0 | | | | | | | −1 | −3 | | |
| $Z_j - C_j$ | $P_3$ | 320 | | 1 | | −2 | | | | −1 | | |
| | $P_2$ | 0 | | | | | | | | | | −1 |
| | $P_1$ | 1500 | | 2 | | −5 | | | −1 | 5 | | |

## 5. DETERMINE WHETHER THE CURRENT SOLUTION IS OPTIMAL

Through an analysis of goal attainment ($Z_j$ values) in the constant column in Table 5.8, it should be evident that the second-stage solution is not the optimal solution. The first goal is still far from completely attained. Therefore, the solution is not the optimal one. We must repeat steps two to five.

Table 5.9 presents the remaining tableaus of the simplex solution to the problem. By now the reader should be able to follow the modified simplex procedure presented in the table without any difficulty. It is interesting to note that initially the solution is directed toward the achievement of the most important goal while ignoring the rest of the goals. For example, in this problem, the solution utilizes full-time salesmen hours because it is the fastest way to fulfill the goal for record sales. When the solution has utilized all the normal working hours of the full-time salesmen and the sales goal has not yet been fully attained, the solution then employs overtime of full-time salesmen rather than introducing the regular working hours of the part-time salesmen. In fact the simplex procedure utilizes 300 hours of overtime from the full-time salesmen in order to attain the sales goal completely. Once the first goal is attained as desired, the solution base is adjusted according to the second goal while maintaining the achievement of the first goal. In order to reduce the 200 hours of overtime for full-time salesmen in excess of the desired 100 hours of overtime, the regular working hours of the part-time salesmen are fully utilized.

Table 5.9

| $C_j$ | | | $P_1$ | $2P_3$ | $P_3$ | | | $P_4$ | $3P_4$ | $P_2$ | | |
|---|---|---|---|---|---|---|---|---|---|---|---|---|
| | $V$ | $C$ | $x_1$ | $x_2$ | $d_1^-$ | $d_2^-$ | $d_3^-$ | $d_{21}^-$ | $d_1^+$ | $d_2^+$ | $d_3^+$ | $d_{21}^+$ |
| $P_1$ | $d_1^-$ | 1000 | | 2 | 1 | -5 | | | -5 | -1 | | (5) |
| | $x_1$ | 900 | 1 | | | 1 | | | 1 | | | -1 |
| $P_3$ | $d_3^-$ | 320 | 1 | | | | 1 | | | | -1 | |
| $P_4$ | $d_2^+$ | 100 | | | | | | | 1 | 1 | | -1 |
| | $P_4$ | 100 | | | | | | | 1 | | -3 | -1 |
| $Z_j - C_j$ | $P_3$ | 320 | 1 | | | -2 | | | | | -1 | |
| | $P_2$ | 0 | | | | | | | | | | -1 |
| | $P_1$ | 1000 | | 2 | | -5 | | | -5 | -1 | | 5 |
| $P_2$ | $d_{21}^+$ | 200 | | 2/5 | 1/5 | -1 | | | -1 | -1/5 | | 1 |
| | $x_1$ | 1100 | 1 | 2/5 | 1/5 | | | | | -1/5 | | |
| $P_3$ | $d_3^-$ | 320 | (1) | | | | 1 | | | | -1 | |
| $P_4$ | $d_2^+$ | 300 | | 2/5 | 1/5 | -1 | | | | -1/5 | 1 | |
| | $P_4$ | 300 | | 2/5 | 1/5 | -1 | | | | -1/5 | -3 | |
| $Z_j - C_j$ | $P_3$ | 320 | 1 | | | -2 | | | | | -1 | |
| | $P_2$ | 200 | | 2/5 | 1/5 | -1 | | | -1 | -1/5 | | |
| | $P_1$ | 0 | -1 | | | | | | | | | |
| $P_2$ | $d_{21}^+$ | 72 | | | 1/5 | -1 | -2/5 | | -1 | -1/5 | (2/5) | 1 |
| | $x_1$ | 972 | 1 | | 1/5 | | -2/5 | | | -1/5 | 2/5 | |
| | $x_2$ | 320 | | 1 | | | 1 | | | | -1 | |
| $P_4$ | $d_2^+$ | 172 | | | 1/5 | -1 | -2/5 | | | -1/5 | 1 | 2/5 |
| | $P_4$ | 172 | | | 1/5 | -1 | -2/5 | | | -1/5 | -13/5 | |
| $Z_j - C_j$ | $P_3$ | 0 | | | | -2 | -1 | | | | | |
| | $P_2$ | 72 | | | 1/5 | -1 | -2/5 | | -1 | -1/5 | 2/5 | |
| | $P_1$ | 0 | -1 | | | | | | | | | |
| $3P_4$ | $d_3^+$ | 180 | | | 1/2 | -5/2 | -1 | | -5/2 | -1/2 | 1 | 5/2 |
| | $x_1$ | 900 | 1 | | | 1 | | | 1 | | | -1 |
| | $x_2$ | 500 | | 1 | 1/2 | -5/2 | | | -5/2 | -1/2 | | -5/2 |
| $P_4$ | $d_2^+$ | 100 | | | | | | | 1 | 1 | | -1 |
| | $P_4$ | 640 | | | 3/2 | -15/2 | -3 | | -13/2 | -3/2 | | 13/2 |
| $Z_j - C_j$ | $P_3$ | 0 | | | | -2 | -1 | | | | | |
| | $P_2$ | 0 | | | | | | | | | | -1 |
| | $P_1$ | 0 | -1 | | | | | | | | | |

When this adjustment does not completely eliminate the overtime in excess of 100 hours as desired in the second goal for the full-time salesmen, overtime of part-time salesmen is introduced to achieve this solution. The optimum solution indicates that the first three goals are completely attained but the fourth goal (minimization of overtime hours for salesmen) could not be achieved. The solution is, therefore, $x_1 = 900$, $x_2 = 500$, $d_2^+ = 100$, and $d_3^+ = 180$.

## Example 3

Let us now examine a more complex problem that involves three choice variables, 16 deviational variables, and five preemptive priority factors. The problem formulated in Example 4 of Chapter 3 will be presented. The problem involves a decision by the president of American Computer Hardwares, Inc. The model formulated was:

$$(5.6) \quad \text{Min } Z = P_1 d_1^- + 20P_2 d_2^- + 18P_2 d_3^- + 21P_2 d_4^- + P_3 d_{11}^+ + 20P_4 d_5^- + 18P_4 d_6^- + 21P_4 d_7^- + P_5 d_1^+$$

$$
\begin{array}{llr}
\text{s.t.} & 5x_1 + 8x_2 + 12x_3 + d_1^- \quad\quad\quad - d_1^+ & = 170 \\
& x_1 \quad\quad\quad\quad\quad + d_2^- \quad - d_2^+ & = 5 \\
& \quad x_2 \quad\quad\quad\quad + d_3^- \quad - d_3^+ & = 5 \\
& \quad\quad x_3 \quad\quad + d_4^- \quad - d_4^+ & = 8 \\
& x_1 \quad\quad\quad\quad\quad + d_5^- \quad - d_5^+ & = 10 \\
& \quad x_2 \quad\quad\quad\quad + d_6^- \quad - d_6^+ & = 12 \\
& \quad\quad x_3 \quad\quad + d_7^- \quad - d_7^+ & = 10 \\
& \quad\quad\quad\quad\quad d_{11}^- + d_1^+ \quad - d_{11}^+ & = 20 \\
\end{array}
$$

$$x_1, x_2, x_3, d_1^-, d_2^-, d_3^-, d_4^-, d_5^-, d_6^-, d_7^-, d_{11}^-, d_1^+, d_2^+, d_3^+, d_4^+, d_5^+, d_6^+, d_7^+, d_{11}^+ \geqslant 0$$

The solution to the above problem is presented in Tables 5.10 to 5.18. It will be left as an exercise for the reader to verify the above solution by the simplex procedure.

# Table 5.10

| $C_j$ | | | $P_1$ | | | $P_1$ | $20P_2$ | $18P_2$ | $21P_2$ | $20P_4$ | $18P_4$ | $21P_4$ | | $P_5$ | | | | | | | $P_3$ |
|---|---|---|---|---|---|---|---|---|---|---|---|---|---|---|---|---|---|---|---|---|---|
| | $V$ | $C$ | $x_1$ | $x_2$ | $x_3$ | $d_1^-$ | $d_2^-$ | $d_3^-$ | $d_4^-$ | $d_5^-$ | $d_6^-$ | $d_7^-$ | $d_{11}^-$ | $d_1^+$ | $d_2^+$ | $d_3^+$ | $d_4^+$ | $d_5^+$ | $d_6^+$ | $d_7^+$ | $d_{11}^+$ |
| $P_1$ | $d_1^-$ | 170 | 5 | 8 | 12 | 1 | | | | | | | | $-1$ | | | | | | | |
| $20P_2$ | $d_2^-$ | 5 | 1 | 1 | | | 1 | | | | | | | | $-1$ | | | | | | |
| $18P_2$ | $d_3^-$ | 5 | | 1 | 1 | | | 1 | | | | | | | | $-1$ | | | | | |
| $21P_2$ | $d_4^-$ | 8 | 1 | | ① | | | | 1 | | | | | | | | $-1$ | | | | |
| $20P_4$ | $d_5^-$ | 10 | | 1 | | | | | | 1 | | | | | | | | $-1$ | | | |
| $18P_4$ | $d_6^-$ | 12 | | 1 | | | | | | | 1 | | | | | | | | $-1$ | | |
| $21P_4$ | $d_7^-$ | 10 | | | 1 | | | | | | | 1 | 1 | | | | | | | $-1$ | |
| | $d_{11}^-$ | 20 | | | | | | | | | | | 1 | 1 | | | | | | | $-1$ |
| $P_5$ | | 0 | | | | | | | | | | | | $-1$ | | | | | | | |
| $P_4$ | | 626 | 20 | 18 | 21 | | | | | | | | | | | | | $-20$ | $-18$ | $-21$ | |
| $Z_j - c_j$ $P_3$ | | 0 | | | | | | | | | | | | | | | | | | | $-1$ |
| $P_2$ | | 358 | 20 | 18 | 21 | | | | | | | | | | $-20$ | $-18$ | $-21$ | | | | |
| $P_1$ | | 170 | 5 | 8 | 12 | | | | | | | | | $-1$ | | | | | | | |

112

Table 5.11

| $C_j$ | | | | | | $P_1$ | $20P_2$ | $18P_2$ | $21P_2$ | $20P_4$ | $18P_4$ | $21P_4$ | | $P_5$ | | | | | | | $P_3$ |
|---|---|---|---|---|---|---|---|---|---|---|---|---|---|---|---|---|---|---|---|---|---|
| | $V$ | $C$ | $x_1$ | $x_2$ | $x_3$ | $d_1^-$ | $d_2^-$ | $d_3^-$ | $d_4^-$ | $d_5^-$ | $d_6^-$ | $d_7^-$ | $d_{11}^-$ | $d_1^+$ | $d_2^+$ | $d_3^+$ | $d_4^+$ | $d_5^+$ | $d_6^+$ | $d_7^+$ | $d_{11}^+$ |
| $P_1$ | $d_1^-$ | 74 | 5 | 8 | | 1 | | | −12 | | | | | −1 | | | 12 | | | | |
| $20P_2$ | $d_2^-$ | 5 | 1 | | | | 1 | | | | | | | | −1 | | | | | | |
| $18P_2$ | $d_3^-$ | 5 | 1 | 1 | | | | 1 | | | | | | | | −1 | | | | | |
| | $x_3$ | 8 | 1 | | 1 | | | | 1 | | | | | | | | −1 | | | | |
| $20P_4$ | $d_5^-$ | 10 | 1 | 1 | | | | | | 1 | | | | | | | | −1 | | | |
| $18P_4$ | $d_6^-$ | 12 | | 1 | | | | | | | 1 | | | | | | | | −1 | | |
| $21P_4$ | $d_7^-$ | 2 | | | | | | | −1 | | | 1 | | 1 | | | | | | −1 | |
| | $d_{11}^-$ | 20 | | | | | | | | | | | 1 | 1 | | | ①(1) | | | | −1 |
| $P_5$ | | 0 | | | | | | | | | | | | −1 | | | | | | | |
| $P_4$ | | 458 | 20 | 18 | | | | | −21 | | | | | | | | 21 | −20 | −18 | −21 | |
| $P_3$ | | 0 | | | | | | | | | | | | | | | | | | | −1 |
| $P_2$ | | 190 | 20 | 18 | | | | | −21 | | | | | −1 | −20 | −18 | | | | | |
| $P_1$ | | 74 | 5 | 8 | | | | | −12 | | | | | | | | 12 | | | | |

$z_j - c_j$

113

Table 5.12

| $C_j$ | | | | | | $P_1$ | $20P_2$ | $18P_2$ | $21P_2$ | $20P_4$ | $18P_4$ | $21P_4$ | | $P_5$ | | | | | | | | $P_3$ |
|---|---|---|---|---|---|---|---|---|---|---|---|---|---|---|---|---|---|---|---|---|---|---|
| | V | C | $x_1$ | $x_2$ | $x_3$ | $d_1^-$ | $d_2^-$ | $d_3^-$ | $d_4^-$ | $d_5^-$ | $d_6^-$ | $d_7^-$ | $d_{11}^-$ | $d_1^+$ | $d_2^+$ | $d_3^+$ | $d_4^+$ | $d_5^+$ | $d_6^+$ | $d_7^+$ | $d_{11}^+$ |
| $P_1$ | $d_1^-$ | 50 | 5 | 8 | | 1 | | | | | | −12 | | −1 | | | | | | ⑫ | |
| $20P_2$ | $d_2^-$ | 5 | 1 | 1 | | | 1 | | | | | | | | −1 | | | | | | |
| $18P_2$ | $d_3^-$ | 5 | | 1 | | | | 1 | | | | | | | | −1 | | | | | |
| | $x_3$ | 10 | 1 | | 1 | | | | | | | 1 | | | | | | | | | |
| $20P_4$ | $d_5^-$ | 10 | | 1 | | | | | | 1 | | | | | | | | −1 | | | |
| $18P_4$ | $d_6^-$ | 12 | | 1 | | | | | | | 1 | | | | | | | | −1 | | |
| | $d_4^+$ | 2 | | | | | | | −1 | | | 1 | | | | | 1 | | | −1 | |
| | $d_{11}^-$ | 20 | | | | | | | | | | | 1 | 1 | | | | | | | −1 |
| | $P_5$ | 0 | | | | | | | | | | | | −1 | | | | | | | |
| | $P_4$ | 416 | 20 | 18 | | | | | | | | −21 | | | | | | −20 | −18 | | −1 |
| | $P_3$ | 0 | | | | | | | | | | | | | | | | | | | |
| $z_j - c_j$ | $P_2$ | 190 | 20 | 18 | | | | | −21 | | | | | −1 | −20 | −18 | | | | | |
| | $P_1$ | 50 | 5 | 8 | | | | | | | | −12 | | | | | | | | 12 | |

114

## Table 5.13

| $C_j$ | | | | | | $P_1$ | $20P_2$ | $18P_2$ | $21P_2$ | $20P_4$ | $18P_4$ | $21P_4$ | | $P_5$ | | | | | | | $P_3$ |
|---|---|---|---|---|---|---|---|---|---|---|---|---|---|---|---|---|---|---|---|---|---|
| | $V$ | $C$ | $x_1$ | $x_2$ | $x_3$ | $d_1^-$ | $d_2^-$ | $d_3^-$ | $d_4^-$ | $d_5^-$ | $d_6^-$ | $d_7^-$ | $d_{11}^-$ | $d_1^+$ | $d_2^+$ | $d_3^+$ | $d_4^+$ | $d_5^+$ | $d_6^+$ | $d_7^+$ | $d_{11}^+$ |
| | $d_7^+$ | 25/6 | 5/12 | 2/3 | | 1/12 | | | | | | −1 | | −1/12 | | | | | | 1 | |
| $20P_2$ | $d_2^-$ | 5 | ① | | | | 1 | | | | | | | | −1 | | | | | | |
| $18P_2$ | $d_3^-$ | 5 | | 1 | | | | 1 | | | | | | | | −1 | | | | | |
| | $x_3$ | 85/6 | 5/12 | 2/3 | 1 | 1/12 | | | | | | | | −1/12 | | | | | | | |
| $20P_4$ | $d_5^-$ | 10 | 1 | 1 | | | | | | 1 | | | | | | | | −1 | | | |
| $18P_4$ | $d_6^-$ | 12 | | 1 | | | | | | | 1 | | | | | | | | −1 | | |
| | $d_4^+$ | 37/6 | 5/12 | 2/3 | | 1/12 | | | −1 | | | | | | | | 1 | | | | |
| | $d_{11}^-$ | 20 | | | | | | | | | | | 1 | 1 | | | | | | | −1 |
| $z_j - c_j$ | $P_5$ | 0 | | | | | | | | | | | | −1 | | | | | | | |
| | $P_4$ | 416 | 20 | 18 | | | | | | | | −21 | | | | | | −20 | −18 | | |
| | $P_3$ | 0 | | | | | | | | | | | | | | | | | | | −1 |
| | $P_2$ | 190 | 20 | 18 | | | | | −21 | | | | | | −20 | −18 | | | | | |
| | $P_1$ | 0 | | | | −1 | | | | | | | | | | | | | | | |

115

Table 5.14

| $C_i$ | $V$ | $C$ | $x_1$ | $x_2$ | $x_3$ | $P_1$ $d_1^-$ | $20P_2$ $d_2^-$ | $18P_2$ $d_3^-$ | $21P_2$ $d_4^-$ | $20P_4$ $d_5^-$ | $18P_4$ $d_6^-$ | $21P_4$ $d_7^-$ | $P_5$ $d_{11}^-$ | $P_5$ $d_1^+$ | $d_2^+$ | $d_3^+$ | $d_4^+$ | $d_5^+$ | $d_6^+$ | $P_3$ $d_7^+$ | $P_3$ $d_{11}^+$ |
|---|---|---|---|---|---|---|---|---|---|---|---|---|---|---|---|---|---|---|---|---|---|
| | $d_7^+$ | 25/12 | | (2/3) | | −1/12 | −5/12 | | | | | −1 | | −1/12 | 5/12 | | | | | 1 | |
| | $x_1$ | 5 | 1 | | | | | | | | | | | | | | | | | | |
| $18P_2$ | $d_3^-$ | 5 | | 1 | | | | 1 | | | | | | | −1 | −1 | | | | | |
| | $x_3$ | 145/12 | | 2/3 | 1 | 1/12 | −5/12 | | | | | | | −1/12 | 5/12 | | | | | | |
| $20P_4$ | $d_5^-$ | 5 | | | | | | | | 1 | | | | | 1 | | | −1 | | | |
| $18P_4$ | $d_6^-$ | 12 | | 1 | | | | | | | 1 | | | | | | | | −1 | | |
| | $d_4^+$ | 45/12 | | 2/3 | | 1/12 | −5/12 | | −1 | | | | | | 5/12 | | 1 | | | | |
| | $d_{11}^-$ | 20 | | | | | | | | | | | 1 | 1 | | | | | | | −1 |
| | $P_5$ | 0 | | | | | | | | | | | | −1 | | | | | | | |
| | $P_4$ | 316 | | 18 | | −20 | −20 | | −21 | | | −21 | | | 20 | −18 | | −20 | −18 | | −1 |
| $z_j - c_j$ | $P_3$ | 0 | | | | | | | | | | | | | | | | | | | |
| | $P_2$ | 90 | | 18 | | −20 | −20 | | −21 | | | | | | 20 | −18 | | | | | |
| | $P_1$ | 0 | | | | −1 | | | | | | | | | | | | | | | |

116

Table 5.15

| $C_j$ | | | | | | $P_1$ | $20P_2$ | $18P_2$ | $21P_2$ | $20P_4$ | $18P_4$ | $21P_4$ | | $P_5$ | | | | | | | $P_3$ |
|---|---|---|---|---|---|---|---|---|---|---|---|---|---|---|---|---|---|---|---|---|---|
| | $V$ | $C$ | $x_1$ | $x_2$ | $x_3$ | $d_1^-$ | $d_2^-$ | $d_3^-$ | $d_4^-$ | $d_5^-$ | $d_6^-$ | $d_7^-$ | $d_{11}^-$ | $d_1^+$ | $d_2^+$ | $d_3^+$ | $d_4^+$ | $d_5^+$ | $d_6^+$ | $d_7^+$ | $d_{11}^+$ |
| | $x_2$ | 25/8 | | 1 | | 1/8 | −5/8 | | | | | −3/2 | | −1/8 | 5/8 | | | | | 3/2 | |
| | $x_1$ | 5 | 1 | | | | 1 | | | | | | | | −1 | | | | | | |
| $18P_2$ | $d_3^-$ | 15/8 | | | | −1/8 | 5/8 | 1 | | | | (3/2) | | 1/8 | −5/8 | −1 | | | | −3/2 | |
| | $x_3$ | 10 | | | 1 | | | | | | | 1 | | | | | | | | −1 | |
| $20P_4$ | $d_5^-$ | 5 | | | | | −1 | | | 1 | | | | | 1 | | | −1 | | | |
| $18P_4$ | $d_6^-$ | 71/8 | | | | −1/8 | 5/8 | | | | 1 | 3/2 | | 1/8 | −5/8 | | | | −1 | −3/2 | |
| | $d_4^+$ | 2 | | | | | | | −1 | | | 1 | | 1/12 | | | 1 | | | −1 | |
| | $d_{11}^-$ | 20 | | | | | | | | | | | 1 | 1 | | | | | | | −1 |
| $z_j - c_j$ | $P_5$ | 0 | | | | | | | | | | | −1 | −1 | | | | | | | |
| | $P_4$ | 259¾ | | | | −9/4 | 45/4 | | | | | 6 | | 9/4 | 35/4 | | | −20 | −18 | −27 | |
| | $P_3$ | 0 | | | | | | | | | | | | | | | | | | | −1 |
| | $P_2$ | 135/4 | | | | −9/4 | 35/4 | | −21 | | | 27 | | 9/4 | −45/4 | −18 | | | | −27 | |
| | $P_1$ | 0 | | | | −1 | | | | | | | | | | | | | | | |

117

Table 5.16

| $C_j$ | | | | | | $P_1$ | $20P_2$ | $18P_2$ | $21P_2$ | $20P_4$ | $18P_4$ | $21P_4$ | | $P_5$ | | | | | | | $P_3$ |
|---|---|---|---|---|---|---|---|---|---|---|---|---|---|---|---|---|---|---|---|---|---|
| | $V$ | $C$ | $x_1$ | $x_2$ | $x_3$ | $d_1^-$ | $d_2^-$ | $d_3^-$ | $d_4^-$ | $d_5^-$ | $d_6^-$ | $d_7^-$ | $d_{11}^-$ | $d_1^+$ | $d_2^+$ | $d_3^+$ | $d_4^+$ | $d_5^+$ | $d_6^+$ | $d_7^+$ | $d_{11}^+$ |
| | $x_2$ | 5 | | 1 | | | | 1 | | | | | | | 5/8 | −1 | | | | | |
| | $x_1$ | 5 | 1 | | | | | | | | | | | | −1 | | | | | | |
| $21P_4$ | $d_7^-$ | 5/4 | | | | −1/12 | 5/12 | 2/3 | | | | 1 | | 1/12 | −5/12 | −2/3 | | | | −1 | |
| | $x_3$ | 35/4 | | | 1 | 1/12 | −5/12 | −2/3 | | | | | | −1/12 | | 2/3 | | | | 1 | |
| $20P_4$ | $d_5^-$ | 5 | | | | | | | | 1 | | | | | | | | −1 | | | |
| $18P_4$ | $d_6^-$ | 7 | | | | | | | | | 1 | | | | | | | | −1 | | |
| | $d_4^+$ | 3/4 | | | | | | | −1 | | | | | | −5/8 | 1 | 1 | | | | |
| | $d_{11}^-$ | 20 | | | | 1/12 | −5/12 | −2/3 | −1 | | | | 1 | | | (2/3) | 1 | | | | −1 |
| | $P_5$ | 0 | | | | | | | | | | | | −1 | | | | | | | |
| $z_j - c_j$ | $P_4$ | 252¼ | | | | −21/12 | −45/4 | −4 | | −20 | −18 | −21 | | 21/12 | | 4 | | −20 | −18 | −21 | −1 |
| | $P_3$ | 0 | | | | | | | | | | | | | | | | | | | |
| | $P_2$ | 0 | | | | | | | | | | | | | | | | | | | |
| | $P_1$ | 0 | | | | −1 | | | | | | | | | | | | | | | |

118

Table 5.17

| $C_j$ | | | | | | $P_1$ | $20P_2$ | $18P_2$ | $21P_2$ | $20P_4$ | $18P_4$ | $21P_4$ | | $P_5$ | | | | | | | $P_3$ |
|---|---|---|---|---|---|---|---|---|---|---|---|---|---|---|---|---|---|---|---|---|---|
| | $V$ | $C$ | $x_1$ | $x_2$ | $x_3$ | $d_1^-$ | $d_2^-$ | $d_3^-$ | $d_4^-$ | $d_5^-$ | $d_6^-$ | $d_7^-$ | $d_{11}^-$ | $d_1^+$ | $d_2^+$ | $d_3^+$ | $d_4^+$ | $d_5^+$ | $d_6^+$ | $d_7^+$ | $d_{11}^+$ |
| | $x_2$ | 49/8 | | 1 | | 1/8 | −5/8 | 3/2 | | | | | | 5/8 | | | 3/2 | | | | |
| | $x_1$ | 5 | 1 | | | | | | | | | | | | | | | | | | |
| $21P_4$ | $d_7^-$ | 2 | | | | | 1 | −2/3 | | | | 1 | | 1/12 | −5/12 | | 1 | | | −1 | |
| | $x_3$ | 8 | | | 1 | | | 3/2 | 3/2 | | | | | −1/12 | | | −2/3 | | | | |
| $20P_4$ | $d_5^-$ | 5 | | | | | −1 | | 1 | 1 | | | | | 1 | | −2/3 | −1 | | | |
| $18P_4$ | $d_6^-$ | 47/8 | | | | −1/8 | 5/8 | 3/2 | | | 1 | | | −5/8 | −5/8 | | −3/2 | | −1 | | |
| | $d_3^+$ | 9/8 | | | | 1/8 | −5/8 | −1 | | | | | | | | 1 | 3/2 | | | | |
| | $d_{11}^-$ | 20 | | | | | | −3/2 | | | | | 1 | (1) | | | 3/2 | | | | −1 |
| | $P_5$ | 0 | | | | | | | | | | | | −1 | | | | | | | |
| | $P_4$ | 189 | | | | 9/4 | −35/4 | 13 | | | | | | 7/4 | | | −6 | −20 | −18 | −21 | |
| $z_j - c_j$ | $P_3$ | 0 | | | | | | | | | | | | | | | | | | | −1 |
| | $P_2$ | 0 | | | | | −20 | −18 | −21 | | | | | | | | | | | | |
| | $P_1$ | 0 | | | | −1 | | | | | | | | | | | | | | | |

119

## Table 5.18

| $C_j$ | | | $P_1$ | $20P_2$ | $18P_2$ | $21P_2$ | $20P_4$ | $18P_4$ | $21P_4$ | | $P_5$ | | | | | | | $P_3$ |
|---|---|---|---|---|---|---|---|---|---|---|---|---|---|---|---|---|---|---|
| V | C | $x_1$ | $x_2$ | $x_3$ | $d_1^-$ | $d_2^-$ | $d_3^-$ | $d_4^-$ | $d_5^-$ | $d_6^-$ | $d_7^-$ | $d_{11}^-$ | $d_1^+$ | $d_2^+$ | $d_3^+$ | $d_4^+$ | $d_5^+$ | $d_6^+$ | $d_7^+$ | $d_{11}^+$ |
| $x_2$ | 49/8 | | 1 | | 1/8 | −5/8 | 3/2 | | | | | | | | 5/8 | | 3/2 | | | | |
| $x_1$ | 5 | 1 | | | | 1 | | | | | | | | | −1 | | | | | | |
| $d_7^-$ | 1/3 | | | | | | | −2/3 | | | 1 | −1/12 | | −5/12 | | 1 | | | −1 | 1/12 |
| $x_3$ | 39/4 | | | 1 | | | | 3/2 | | | | 1/12 | | | | −2/3 | | | 1 | −1/12 |
| $d_5^-$ | 5 | | | | | | | | 1 | | | | | 1 | | −1 | | | | |
| $d_6^-$ | 47/8 | | | | −1/8 | 5/8 | 3/2 | −3/2 | | 1 | | | | −5/8 | | −3/2 | | −1 | | |
| $d_3^+$ | 9/8 | | | | 1/8 | −5/8 | −3/2 | | | | | | | | 1 | 3/2 | | | | |
| $d_1^+$ | 20 | | | | | | | | | | | 1 | 1 | | | | | | | −1 |
| $z_j - c_j$ | | | | | | | | | | | | | | | | | | | | |
| $P_5$ | 20 | | | | | | | | | | | | 1 | | | | | | | −1 |
| $P_4$ | 154 | | | | −9/4 | −35/4 | 13 | | | | | | | | | −6 | −20 | −18 | −21 | 7/4 |
| $P_3$ | 0 | | | | | | | | | | | | | | | | | | | −1 |
| $P_2$ | 0 | | | | | | | | −20 | −18 | −21 | | | | | | | | | |
| $P_1$ | 0 | | | | −1 | | | | | | | | | | | | | | | |

120

## SOME COMPLICATIONS AND THEIR RESOLUTION

The basic approach of goal programming, which is an attempt to achieve a set of prescribed goals according to priorities by the minimization of deviational variables, enables us to avoid many complications encountered in regular linear programming.[4] For example, problems encountered in linear programming due to inequality in the wrong direction, equality, variables unconstrained in sign, etc., in the original model do not present any difficulty in goal programming. This is because we evaluate goals in terms of both underachievement and overachievement. However, there are, several complications that often emerge in goal programming problems. They are discussed as follows:

### 1. Nonpositive Constant

To explain the nonpositive constant (rhs) problem, let us consider the goal constraint shown below.

$$(5.7) \qquad -5x_1 - x_2 + d_1^- - d_1^+ = -25$$

In the initial table of simplex goal programming, it is assumed that the solution is at the origin. Therefore, the deviational variable $d_1^-$ will take on the value of $-25$. However, the simplex method requires condition of non-negative variables $(x_i, d_i^-, d_i^+ \geqslant 0)$; thus, $d_1^- = -25$ is not permissible. In order to facilitate the initial solution we can multiply both sides by $(-1)$. The goal constraint becomes:

$$(5.8) \qquad 5x_1 + x_2 + d_1^+ - d_1^- = 25$$

If the goal is to achieve exactly $-25$ from the original constraint, it can easily be achieved by minimizing both $d_1^-$ and $d_1^+$ at the same priority level. However, if the goal is to make the constraint produce $-25$ or greater, $d_1^-$ must be minimized in (5.7) but in the revised goal constraint (5.8) $d_1^+$ should be minimized to derive the same effect. Similarly to assume the value of $-25$ or less from the constraint, $d_1^+$ must be minimized in (5.7) and $d_1^-$ must be minimized in the revised equation (5.8).

## 2. Tie for Entering Variable

In any goal programming problem it can easily happen during the iteration that two or more columns have exactly the same positive $Z_j-C_j$ value at the highest unattained goal level. As explained previously in this chapter, when there is such a case, determination of the optimum column and consequently the entering basic variable is based on the $Z_j-C_j$ values at the lower priority levels. If the tie cannot be broken, selection between the contending variables may be made arbitrarily. The other variable will generally be introduced into the solution base in the subsequent iterations.

## 3. Tie for Exiting Variable

To determine the variable that will leave the solution base, constraints must be divided by the coefficients in the optimum column and we must then determine the row with the minimum positive quotient. If there are two or more rows with identical minimum positive values, this raises the problem of degeneracy. The resolution of degeneracy should be decided by determining which row has the variable with the higher priority factor. By selecting the variable with the higher priority factor as the exiting variable, the solution process can be shortened as the higher priority goals will be attained faster.

## 4. Unbounded Solution

It is possible that because of an unrealistic priority structure of the decision maker or lack of constraints, the problem may allow one or more variables to increase without any limit. In most real-world problems, however, this situation rarely occurs, since goals tend to be set higher than easily attainable levels within the existing decision environment. The unbounded solution, if it does occur, also provides some insight in analyzing the decision maker's goal structure. It is often the case that important constraints are omitted in the problem when an unbounded solution is obtained.

## 5. Alternate Optimal Solutions

It is possible that two or more points provide optimal solutions that attain exactly the same level of goals. Such an occasion never occurs as long as (1) there is only a single deviational variable (single goal) at each preemptive priority level and (2) differential weights are assigned among subgoals at the same priority level when there is more than a single goal at each priority level. To illustrate the point, let us examine the record-shop case we discussed as Example 2 in this chapter. Suppose the manager has altered his priority sturcture in such a way that the model is formulated as below:

(5.9)   $\text{Min } Z = P_1 d_1^- + P_2 d_{21}^+ + 2P_3 d_2^- + P_3 d_3^- + P_4 d_2^+ + P_4 d_3^+$

s.t.
$$
\begin{array}{rcl}
5x_1 + 2x_2 + d_1^- \quad\quad - d_1^+ \quad\quad\quad\quad & = & 5{,}500 \\
x_1 \quad\quad + d_2^- \quad\quad - d_2^+ \quad\quad & = & 800 \\
x_2 \quad + d_3^- \quad\quad - d_3^+ & = & 320 \\
d_{21}^- + d_2^+ - d_{21}^+ & = & 100 \\
x_1, x_2, d_1^-, d_2^-, d_3^-, d_{21}^-, d_1^+, d_2^+, d_3^+, d_{21}^+ & \geqslant & 0
\end{array}
$$

The objective function of the model indicates that the manager does not assign differential weights in the minimization of overtime hours between full-time and part-time salesmen. The graphical solution of the problem is presented in Figure 5.2. The first three goals can be completely attained at either point A or B or any point on the straight line connecting A and B, but the fourth goal cannot be achieved, since the sum of overtime cannot be eliminated completely. The alternate optimal solution would be avoided if the manager of the record shop assigned differential weights in the minimization of overtime according to opportunity costs involved.

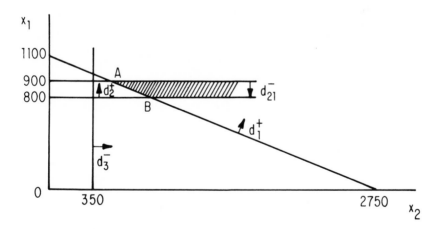

**Figure 5.2**

## CONCLUSION

In this chapter we have discussed a modified simplex procedure to solve goal programming problems. The simplex method is a powerful algebraic procedure, yet it is relatively easy to understand and apply. Although the generalized inverse technique has been introduced as a possible solution procedure for goal programming problems, it is not discussed in this chapter. It has been proved repeatedly that the generalized inverse technique is beyond the understanding of many students. Consequently, the technique often discourages students' interest in goal programming. However, those who are interested in exploring goal programming in greater depth should study the works by Ijiri.[5]

There are many aspects of goal programming solution that deserve our attention, such as the duality, postoptimal sensitivity analysis, parametric programming, solution under the condition of uncertainty, etc. These advanced topics of goal programming will be discussed in length in Chapter 7. A computer-based solution of goal programming problems based on the simplex procedure will be our topic of discussion in Chapter 6.

## PROBLEMS

1.    Solve the following problem by the simplex method of goal programming.

$$\text{Min } Z = P_1 d_1^- + P_2 d_3^+ + P_3 d_2^+$$
$$\begin{array}{llll}
\text{s.t.} & 5x_1 + 10x_2 + d_1^- & - d_1^+ & = 10{,}000 \\
& x_1 & + d_2^- & - d_2^+ & = 700 \\
& x_2 & + d_3^- & - d_3^+ = 600 \\
& x_1, x_2, d_1^-, d_2^-, d_3^-, d_1^+, d_2^+, d_3^+ \geq & 0
\end{array}$$

2.    Solve the following problem by the simplex method of goal programming.

$$\text{Min } Z = P_1(d_1^- + d_1^+) + P_2 d_2^- + P_3 d_3^- + 5P_4 d_3^+ + 3P_4 d_2^+$$
$$\begin{array}{llll}
\text{s.t.} & x_1 + x_2 + d_1^- & - d_1^+ & = 800 \\
& 5x_1 & + d_2^- & - d_2^+ & = 2{,}500 \\
& 3x_2 & + d_3^- & - d_3^+ & = 1{,}400 \\
& x_1, x_2, d_1^-, d_2^-, d_3^-, d_1^+, d_2^+, d_3^+ \geq & 0
\end{array}$$

3.    Solve the following problem by the simplex method of goal programming.

$$\text{Min } Z = P_1 d_2^- + P_2 d_3^- + P_3 d_1^- + P_4 d_{31}^+ + P_5 d_1^+$$

$$
\begin{array}{lr}
\text{s.t.} \quad 10x_1 + 12x_2 + d_1^- \qquad\qquad -d_1^+ \qquad\qquad = & 1{,}000 \\
x_1 \qquad + d_2^- \qquad\qquad -d_2^+ \qquad\qquad = & 40 \\
x_2 \qquad - d_3^- \qquad\qquad - d_3^+ \qquad\qquad = & 40 \\
d_{31}^- \qquad\qquad + d_3^+ - d_{31}^+ = & 5 \\
x_1, x_2, d_1^-, d_2^-, d_3^-, d_{31}^-, d_1^+, d_2^+, d_3^+, d_{31}^+ \geq & 0
\end{array}
$$

4.  Solve Example 3 in Chapter 3 by the simplex method of goal programming.

5.  Solve Problem 1 at the end of Chapter 3 by the simplex method of goal programming. In order for the production manager to achieve all of his goals, (1) how should the goal structure be changed; (2) if the manager does not wish to change his goal structure, how should normal working hours for the two teams be changed?

6.  Solve Problem 3 at the end of Chapter 3 by the simplex method of goal programming.

## REFERENCES

Charnes, A., and Cooper, W. W. *Management Models and Industrial Applications of Linear Programming.* 2 vols. New York: John Wiley & Sons, Inc., 1961.

Dantzig, George B. *Linear Programming and Extensions.* Princeton, N. J.: Princeton University Press, 1963.

Garvin, W. W. *Introduction to Linear Programming.* New York: McGraw-Hill, Inc., 1962.

Hadley, G. *Linear Programming.* Reading, Mass.: Addison-Wesley, Inc., 1962.

Hillier, F., and Lieberman, G. J. *Introduction to Operations Research.* San Francisco: Holden-Day, Inc., 1967.

Ijiri, Y. *Management Goals and Accounting for Control.* Chicago: Rand-McNally, 1965.

Teichroew, D. *An Introduction to Management Science: Deterministic Models.* New York: John Wiley & Sons, Inc., 1964.

Wagner, H. *Principles of Operations Research with Applications to Managerial Decisions.* Englewood Cliffs, N. J.: Prentice-Hall, Inc., 1969.

# Chapter 6

## Computer-Based Solution
## of
## Goal Programming

For any management science technique to be a truly valuable tool for decision analysis, it must accommodate itself to a computer-based solution. The complexity of real-world problems usually compels the use of computers. Many simple hypothetical problems being discussed and taught in classrooms exist only in textbooks. This by no means suggests that simple examples are of no value. Actually, they provide the foundation for understanding complex concepts of various management science techniques. Nevertheless, in order to apply a technique to practical problems, which is indeed the very purpose of management science training, computer-based analysis is usually required. We have developed many powerful and mathematically sophisticated techniques—nonlinear programming, dynamic programming, game theory, etc.—that have found a disappointingly limited scope for practical applications to real-world problems. Modeling with such techniques is extremely hard for complex problems, and consequently a computer analysis is of little value.

In order for goal programming to be a useful management science technique for decision analysis, a computer-based solution is an essential requirement. Thus far, however, there has not been an efficient computer program for goal programming that has been widely circulated.[1] As a matter of fact, this may well be one of the reasons for the limited application of goal programming in spite of its advantages. In an eagerness to apply goal programming, some researchers have used the conventional linear programming package by converting the preemptive priority factors to numerical values. Their studies may have produced some interesting results and their desire to apply goal programming to practical decision problems may have been satisfied. However, the use of linear programming by estimating the numerical values of multiple conflicting goals brings us right back to where we started. Conceptually, goal programming is based on ordinal solution on the basis of the priority of goals (i.e., $P_j > > > P_{j+1}$). Consequently, employing the linear programming package for a goal programming problem is not only defeating the very purpose of the goal programming approach but also violates the basic rule of goal programming. Furthermore, since the estimation of numerical values for multidimensional goals can only be achieved through a fabrication and distortion of the existing information, the linear programming solution will have only very limited value, if even that, to the decision maker.[2] Quite possibly and justifiably, the solution may not be accepted for implementation since it does not reflect the true decision environment.

Suppose we assign certain numerical values to represent the preemptive priority factors, such as 1,000 utiles for $P_1$, 500 utiles for $P_2$, and so on. In the process what we have implied is that the first goal is exactly twice as important as the second goal. In reality, this may not be the case. In fact, quite possibly, the decision maker does not even know how much more important the first goal is than the second goal. The only thing he is sure of may be that in his judgment the first goal is more important than the second goal. Then, how can we interpret the model solution? Some may insist that as long as the computer program allows the solution according to the importance rank of goals it should be sufficient. That may be so. But quite frequently, complex decision problems have all types of technological coefficients. Consequently, it is often possible that low-order goals become more important than a higher-order goal or the higher-order goals become less important than the lower-order goals in the solution process. The extreme cases are related to the sensitivity analysis and dual solution of goal programming problems based on the numerical objective function. This topic will be thoroughly examined in Chapter 7.

This chapter presents a computer-based solution procedure of goal programming. It presents a detailed description of a computer program of the simplex algorithm for goal programming. More specifically, it discusses the data input for the computer solution, the input process, the process for calculating the results, and finally the procedure for printout of the results. This computer program has been tested through many practical applications, and it has been proved efficient as long as the hardware has sufficient memory capacity for the problem.

## SETTING UP THE GOAL PROGRAM FOR THE COMPUTER

For illustrative purposes, let us consider the textile-manufacturing problem discussed as model (4.13) in Chapter 4.

$$
\begin{aligned}
(6.1) \quad \text{Min } Z = \; & P_1 d_1^- + P_2 d_4^+ + 5P_3 d_2^- + 3Pd_3^- + P_4 d_1^+ \\
\text{s.t.} \quad & x_1 + x_2 + d_1^- && - d_1^+ = 80 \\
& x_1 && + d_2^- && = 70 \\
& x_2 && + d_3^- && = 45 \\
& x_1 + x_2 && + d_4^- - d_4^+ = 90
\end{aligned}
$$

The computer program is designed in such a way that it will take care of the deviational and slack variables automatically. However, we have to specify the regular choice variables and their technological coefficients, the direction of constraints (equality or inequality), and the objective function. Thus, for the above-stated problem, the information listed below is all that is required to formulate the problem.

| Col # | 1 | 2 | | |
|---|---|---|---|---|
| Row # | $x_1$ | $x_2$ | Sign | rhs |
| 1 | 1 | 1 | B | 80 |
| 2 | 1 | | L | 70 |
| 3 | | 1 | L | 45 |
| 4 | 1 | 1 | B | 90 |

The symbols in the sign column will be explained below under the heading "The Sign Card."

The computer solution setup of a goal programming requires five basic parts: (1) problem card, (2) the sign card, (3) the objective function, (4) the substitution rates, and (5) the right-hand side.

## 1. The Problem Card

The problem card describes the parameters of the problem under consideration. The problem card setup is:

Card 1:       the problem card
Col. 1-4:     punch **PROB**
Col. 5-7:     the number of rows
Col. 8-10:    the number of columns
Col. 11-13:   the number of priority factors

The rows, of course, refer to the number of constraint and goal equations. The columns refer to the number of real variables used in the problem (not including deviational and slack variables). The number of priority factors represents the number of actual priority levels only. Artificial priorities are automatically created by the program in order to create the first basis.

Example: The above-stated problem is used to explain the problem card setup. There are four rows, two real variables, and four priority factors in our problem.

Col    1 2 3 4 5 6 7 8 9 10 11 12 13

P R O B 0 0 4 0 0 2   0   0   4

## 2. The Sign Card

The sign card describes the direction of constraints. There are four possibilities:

a. "E" for "exactly equal." No deviation in either direction is possible.

b. "G" for "greater than." This sign allows only the positive deviation from the right-hand side.

c. "L" for "less than." This sign allows only the negative deviation from the right-hand side.

d. "B" for "both directions are possible." This sign allows the minimization of either or both the negative and positive deviations from the goal or constraints.

One or both deviational variables of a constraint must appear in the objective function. If neither deviational variable appears in the objective function, it is possible that both deviational variables may end up in the basis and the constraint $d_i^- \cdot d_i^+ = 0$ will not be met. If both deviational variables appear in the objective function, they may be assigned different priority factors.

Example: For our problem, the sign card will be:

| Col. | 1 | 2 | 3 | 4 |
|------|---|---|---|---|
|      | B | L | L | B |

## 3. The Objective Function Cards

The objective function cards specify the priority factors, their locations, and the type of deviational variable (either positive or negative). The data cards for the objective function are prefaced by a name card with "OBJ" punched in the first three columns.

| Col. | 1 | 2 | 3 | 4 | 5 |
|------|---|---|---|---|---|
|      | 0 | B | J |   |   |

The data cards of the objective function define each element in the objective function in the following manner:

Col. 1-3:  Either the word "POS" for positive or "NEG" for negative should be punched. This specifies whether the positive or negative deviational variable is to be minimized at the stated priority level.

Col. 4-7:  Blank

Col. 8-9:  The row (ith) in which the deviational variable mentioned above appears.

Col. 10-12:  Blank

Col. 13-14:  The priority level at which the deviational variable is to be minimized. The lower the subscript, the higher the priority.

Col. 15-25:  The coefficient of the priority factor (differential weights). There must always be a value here. The program will not assume one. The decimal number should be punched, which enables one both to avoid having to right-justify the variable and to put in weights that are less than one.

Example: For our example, which had the objective function Min $Z = P_1 d_1^- + P_2 d_4^+ + 5P_3 d_2^- + 3P_3 d_3^- + P_4 d_1^+$, the objective function data cards will be:

| Deviation | Row deviation Appeared | Priority | Weights |
|---|---|---|---|
| Col   1 2 3 4 5 6 7 8 9 10 11 | 12 13 14 15 16 17 | 18 19 20 | 21 22 23 24 |
| NEG | 0 1 | 0 1 | 1 . 0 |
| POS | 0 4 | 0 2 | 1 . 0 |
| NEG | 0 2 | 0 3 | 5 . 0 |
| NEG | 0 3 | 0 3 | 3 . 0 |
| POS | 0 1 | 0 4 | 1 . 0 |

## 4. The Data Section Cards (Technological Coefficients)

The data section cards specify the technological coefficients of the choice variables. The data section cards must be prefaced by a card with the word "DATA" punched in the first four columns. This should be followed by cards with the following information:

Col. 1-7:    Blank
Col. 8-9:    The row (ith) in which the coefficient is located.
Col. 10-12:  Blank
Col. 13-14:  The column (jth) in which the coefficient is located.
Col. 15-25:  The value of the coefficient to be placed in the above indexed location.

Example: For our example, the technological coefficients were as follows:

| Col. # | 1 | 2 |
|---|---|---|
| Row # | $x_1$ | $x_2$ |
| 1 | 1 | 1 |
| 2 | 1 | |
| 3 | | 1 |
| 4 | 1 | 1 |

The above should be punched in the following manner:

| | | Matrix<br>Row | | | Matrix<br>Col | | | | | | | | | Value In<br>Indicated<br>Position | | | | | | |
|---|---|---|---|---|---|---|---|---|---|---|---|---|---|---|---|---|---|---|---|---|

Col | 1 2 3 4 5 6 7 8 9 10 11 12 13 14 15 16 17 18 19 20 21 22 23 24 25

|     Row     |     Col     |   Value   |
|-------------|-------------|-----------|
| 0 1         | 0 1         | 1 . 0     |
| 0 1         | 0 2         | 1 . 0     |
| 0 2         | 0 1         | 1 . 0     |
| 0 3         | 0 1         | 1 . 0     |
| 0 4         | 0 1         | 1 . 0     |
| 0 4         | 0 2         | 1 . 0     |

## 5. The Right-Hand-Side Cards

The right-hand side is the last item to be read into the computer. The right-hand-side cards should be proceeded by a card labled "RGHT" punched in the first four columns. The label card will be followed by the right-hand-side cards in the following manner:

Col. 1-10:   right-hand-side value for the first row
Col. 11-20:   right-hand-side value for the second row
Col. 20-30:   right-hand-side value for the third row

.    .
.    .
.    .

Col. 71-80:   right-hand-side value for the eighth row.

If there are more than eight rows, simply go to the next card.

Example: For our example, the right-hand-side cards will be:

| Col | 10 | 20 | 30 | 40 |
|-----|------|------|------|------|
|     | 80.0 | 70.0 | 45.0 | 90.0 |

There is one more aspect of the computer solution that deserves our attention. If there exists goal decomposition, we may have two or more separate goals concerning a certain desired aspect of the decision problem. For example, in the textile manufacturing problem, we could reformulate the model as follows:

$$(6.2) \quad \text{Min } Z = P_1 d_1^- + P_2 d_{11}^+ + 5P_3 d_2^- + 3P_3 d_3^- + P_4 d_1^+$$

$$\text{s.t.} \quad x_1 + x_2 + d_1^- \qquad\qquad - d_1^+ \qquad = 80$$

$$x_1 \qquad\quad + d_2^- \qquad\qquad\qquad = 70$$

$$x_2 \qquad\quad + d_3^- \qquad\qquad\qquad = 45$$

$$d_{11}^- + d_1^+ - d_{11}^+ = 10$$

The fourth equation in the model represents a specified goal concerning the overtime operation of the manufacturing plant. Since this equation does not contain any choice variables, it is impossible to prepare data selection cards. In such a case, we can treat $d_1^+$ as if it were a choice variable. Therefore, the first and the fourth equations of the model can be expressed as $x_1 + x_2 + d_1^- - x_3$ $= 80$ and $X_3 + d_{11}^- - d_{11}^+ = 10$. This simple modification of the model enables one to set up the goal programming computer deck.

## EXAMPLES OF COMPUTER DECK SETUP

The complete computer deck of goal programming should be arranged as explained above, in addition to necessary system cards specified by each computing facility. The order of cards for the computer deck setup can be simply demonstrated by Figure 6.1. The problem input deck represents the computer cards from **PROB** card to the right-hand-side data card(s). The numbers of sign cards, objective data cards, data selection cards, and right-hand-side data cards vary according to the complexity and characteristics of the problem under analysis.

## ANALYSIS OF THE COMPUTER OUTPUT

The computer solution of goal program provides the following output: complete printout of input data (the right-hand side, the substitution rates, and the objective function), the final simplex solution table (including $Z_j - C_j$ matrix and evaluation of objective function), slack analysis, variable analysis, and the analysis of the objective. In order to assist the potential user of this program, let us analyze each item of the output.

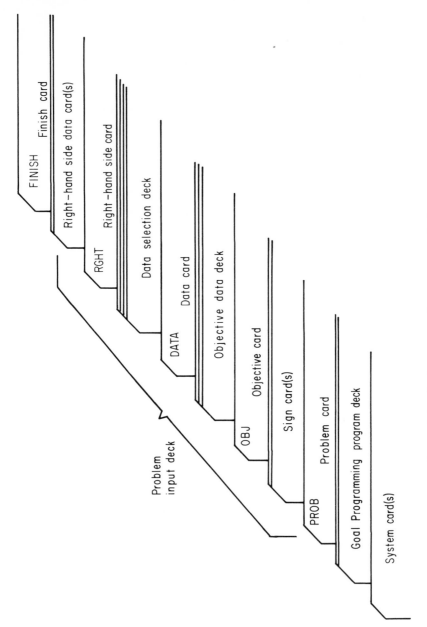

**Figure 6.1.  Goal programming computer deck setup.**

## 1. Input Data

The computer prints out the complete input data so that the user of the program can easily recheck the data he has fed into the computer.

For our example, the computer printout of input data appears as follows:

|  | THE RIGHT HAND SIDE-INPUT | PAGE 01 |
|---|---|---|
| 1 | 80.00000 | |
| 2 | 70.00000 | |
| 3 | 45.00000 | |
| 4 | 90.00000 | |

THE SUBSTITUTION RATES-INPUT                    PAGE 02

Row 1

| 1.000 | 0.0 | 0.0 | 0.0 | -1.000 | 0.0 | 1.000 | 1.000 |
|---|---|---|---|---|---|---|---|

ROW 2

| 0.0 | 1.000 | 0.0 | 0.0 | 0.0 | 0.0 | 1.000 | 0.0 |
|---|---|---|---|---|---|---|---|

ROW 3

| 0.0 | 0.0 | 1.000 | 0.0 | 0.0 | 0.0 | 0.0 | 1.000 |
|---|---|---|---|---|---|---|---|

ROW 4

| 0.0 | 0.0 | 0.0 | 1.000 | 0.0 | -1.000 | 1.000 | 1.000 |
|---|---|---|---|---|---|---|---|

THE OBJECTIVE FUNCTION-INPUT                    PAGE 03

PRIORITY 4

| 0.0 | 0.0 | 0.0 | 0.0 | 1.000 | 0.0 | 0.0 | 0.0 |
|---|---|---|---|---|---|---|---|

PRIORITY 3

| 0.0 | 5.000 | 3.000 | 0.0 | 0.0 | 0.0 | 0.0 | 0.0 |
|---|---|---|---|---|---|---|---|

PRIORITY 2

| 0.0 | 0.0 | 0.0 | 1.000 | 0.0 | 0.0 | 0.0 | 0.0 |
|---|---|---|---|---|---|---|---|

PRIORITY 1

| 1.000 | 0.0 | 0.0 | 0.0 | 0.0 | 0.0 | 0.0 | 0.0 |
|---|---|---|---|---|---|---|---|

```
          SUMMARY OF INPUT INFORMATION              PAGE 04

          NUMBER OF ROWS..................4

          NUMBER OF VARIABLES.............8

          NUMBER OF PRIORITIES............4

          ADDED PRIORITIES................0

          ITERATIONS......................3
```

Everything described above should be self-explanatory. It should be pointed out that in the substitution rates and the objective function matrices the first four columns represent negative deviational variables, followed by the two positive deviational variable columns and the two choice variables at the end.

## 2. The Final Simplex Solution

The computer program prints out the final simplex solution of the problem in the form of simplex matrix notation. This table is extremely useful for the user if he desires to perform postoptimal sensitivity analysis or parametric goal programming.

For our example, the final simplex solution printout appears as follows:

```
                  THE SIMPLEX SOLUTION                    PAGE 05

THE RIGHT HAND SIDE
     8.      20.

     7.      70.

     3.      25.

     5.      10.

THE SUBSTITUTION RATES
     1.000   -1.000    0.0    1.000    0.0   -1.000    0.0    1.000

     0.0     1.000    0.0    0.0      0.0    0.0      1.000   0.0

    -1.000   1.000    1.000  -1.000   0.0    1.000    0.0     0.0

     0.0     0.0      0.0    1.000    1.000  -1.000    0.0     0.0
```

THE ZJ-CJ MATRIX

| | | | | | | | |
|---|---|---|---|---|---|---|---|
| 0.0 | 0.0 | 0.0 | 1.000 | 0.0 | -1.000 | 0.0 | 0.0 |
| -3.000 | -2.000 | 0.0 | -3.000 | 0.0 | 3.000 | 0.0 | 0.0 |
| 0.0 | 0.0 | 0.0 | 0.0 | 0.0 | -1.000 | 0.0 | 0.0 |
| -1.000 | 0.0 | 0.0 | 0.0 | 0.0 | -1.000 | 0.0 | 0.0 |

AN EVALUATION OF THE OBJECTIVE FUNCTION

| | |
|---|---|
| 4 | 10.00 |
| 3 | 75.00 |
| 2 | 0.0 |
| 1 | 0.0 |

The above printout can be interpreted this way:

A. The right-hand side. The numbers on the left-hand side are variable numbers for the basic variables (i.e. variable $8 = x_2$ variable $7 = x_1$, variable $3 = d_3^-$, and variable $5 = d_1^+$). The values on the right-hand side represent constants of the basic variables.

B. The substitution rates. The substitution rates matrix is again based on the column arrangement of $d_i^-$, $d_i^+$, $x_j$, in that order.

C. The ZJ-CJ matrix. This simplex criterion should be self-explanatory.

D. An evaluation of the objective function. This evaluation simply presents the $Z_j$ value of goals. In other words, the values present the underattained portion of goals. It should be apparent in our example that the first two goals are completely attained, while the third and fourth goals are only partially attained.

## 3. The Slack Analysis

The slack analysis presents the values of the right-hand side and also values of the negative and positive variables for each equation. The slack analysis is an excellent vehicle to analyze the details of goal attainments when the problem is complex. Also, this analysis often proves to be very helpful in identifying errors when the model does not represent the exact decision environment.

For our example, the slack analysis appears as follows:

SLACK ANALYSIS                    PAGE 06

| ROW | AVAILABLE | POS-SLK | NEG-SLK |
|-----|-----------|---------|---------|
| 1 | 80.00000 | 10.0000 | 0.0 |
| 2 | 70.00000 | 0.0 | 0.0 |
| 3 | 45.00000 | 0.0 | 25.00000 |
| 4 | 90.00000 | 0.0 | 0.0 |

## 4. Variable Analysis

The variable analysis presents the constants of only the basic choice variables. When the problem under consideration is a very complex one, the variable analysis is especially helpful because it presents only the constants of the basic choice variables, as compared to the final simplex solution table.

The variable analysis of our example appears as follows:

VARIABLE ANALYSIS                    PAGE 07

| VARIABLE | AMOUNT |
|----------|--------|
| 1 | 70.00000 |
| 2 | 20.00000 |

It should be pointed out here that the variable numbers are rearranged. These numbers are variable numbers of the choice variables (i.e., $x_j$).

## 5. Analysis of the Objective

An analysis of the objective presents the $Z_j$ value for the goals. These values represent the underattained portion of goals. If the model requires assignment of artificial priority to set up the initial table, the artificial priority will also be printed out.

For our example, the print-out of the analysis of the objective appears as follows:

ANALYSIS OF THE OBJECTIVE        PAGE 08

| PRIORITY | UNDERACHIEVEMENT |
|----------|------------------|
| 4 | 10.00000 |
| 3 | 75.00000 |
| 2 | 0.0 |
| 1 | 0.0 |

The appendix to Chapter 6 presents the complete goal programming computer deck. This program is capable of solving problems with: the number of variables (including deviational variables) $\leq 125$, the number of rows $\leq 60$, and the number of priority levels $\leq 10$. If a problem is bigger than the above specified size, it can easily be accommodated by expanding appropriate dimensions.

## APPENDIX TO CHAPTER 6

## A FORTRAN PROGRAM FOR GOAL PROGRAMMING PROBLEMS

```
      DIMENSION KEPT(60)
      DIMENSION   RHS1(60)
      DIMENSION VALY(60,10)
      DIMENSION Y(60)
      DIMENSION PRDT(60)
      DIMENSION AMT(60)
      DIMENSION   ZVAL(10)
      DIMENSION C(60,125)
      DIMENSION   DOD(60)
      DIMENSION   DUD(125)
      DIMENSION VALX(10,125)
      DIMENSION X(125)
      DIMENSION RVLX(10,125)
      DIMENSION D(60,125)
C     GOAL PROGRAMMING
      CALL START(N,M,L,C,VALX,VALY,PRDT,RHS1,KPCK,KEPT,TEST)
      DO 21 J=1,M
21    X(J)=J
      DO 20 I=1,N
20    Y(I)=I
15    FORMAT(13,F12.2)
12    FORMAT(10F8.3)
```

```
13    FORMAT(8F9.0)
      DO 25 K=1,L
      DO 25 I=1,N
      VALY (I,K)=VALX(K,I)
25    CONTINUE
      ITAB=0
C     BRING IN NEW VARIABLES
      ITER=0
C     CALCULATE NET CONTRIBUTION OF EACH VARIABLE (RVLX(K,J))
31    L1=0
32    K3=L-L1
33    IF(K3-1) 800,40,40
40    DO 60 K=1,K3
      DO 60 J=1,M
      SUMP=0.
      DO 50 I=1,N
      P=VALY(I,K)*C(I,J)
      SUMP=SUMP+P
50    CONTINUE
      RVLX(K,J)=SUMP-VALX(K,J)
60    CONTINUE
      ITER=ITER+1
C     BRING IN X(K2)
      ZMAX=0.
      DO 90 J=1,M
      IF(K3-L) 92,70,70
92    K4=K3+1
```

```
      DO 91 K=K4,L
      IF(RVLX(K,J)) 90,91,91
91    CONTINUE
70    IF(RVLX(K3,J)-ZMAX) 90,90,80
80    ZMAX=RVLX(K3,J)
      K2=J
90    CONTINUE
95    IF(ZMAX) 790,790,100
C     WHICH VARIABLE IS REMOVED FROM THE BASIS
C     CALCULATE LIMITING AMT FOR EACH BASIS VARIABLE
100   DO 150 I=1,N
      IF(PRDT(I)) 110,120,120
110   WRITE(6,13) PRDT(I)
      GO TO 830
120   IF(C(I,K2)) 130,130,140
130   AMT(I)=-1.
      GO TO 150
140   AMT(I)=PRDT(I)/C(I,K2)
150   CONTINUE
C     SELECT SMALLEST POSITIVE LIMITING AMT
      I=1
160   IF(AMT(I)) 170,210,210
170   I=I+1
      IF(I-N) 160,160,180
180   WRITE(6,13) AMT(N)
      GO TO 830
210   ZMIN=AMT(I)
```

```
      K1=I
  220 I=I+1
      IF(I-N) 230,230,300
  230 IF(AMT(I)) 220,240,240
  240 IF(ZMIN-AMT(I)) 220,220,210
C     REMOVE Y(K1)
  300 Y(K1)=X(K2)
      DO 310 K=1,L
      VALY(K1,K)=VALX(K,K2)
  310 CONTINUE
C     CALCULATE NEW RIGHT-HAND SIDES
      DO 400 I=1,N
      PRDT(I)=PRDT(I)-ZMIN*C(I,K2)
  400 CONTINUE
      PRDT(K1)=ZMIN
C     CALCULATE NEW SUBSTITUTION RATES
      DO 500 J=1,M
      DO 500 I=1,N
      D(I,J)=C(I,J)-C(K1,J)*(C(I,K2)/C(K1,K2))
  500 CONTINUE
      DO 510 J=1,M
      D(K1,J)=C(K1,J)/C(K1,K2)
  510 CONTINUE
      DO 520 J=1,M
      DO 520 I=1,N
      C(I,J)=D(I,J)
  520 CONTINUE
```

```
C          WRITE ALL TABLES OR JUST OPTIMAL TABLE
           IF(ITAB) 40,40,600
C          WRITE EACH TABLE
  600      DO 610 I=1,N
           WRITE(6,13) Y(I),PRDT(I)
  610      CONTINUE
           DO 620 I=1,N
           WRITE(6,12) (C(I,J),J=1,M)
  620      CONTINUE
           GO TO 40
C          MOVE TO NEXT LOWER PRIORITY LEVEL
  790      L1=L1+1
           GO TO 32
C          WRITE FINAL RESULTS
  800      WRITE(6,1014)ITER
           WRITE(6,1015)
 1015      FORMAT(1H1)
 1014      FORMAT(10X,'ITERATIONS.............',I5)
           WRITE(6,5000)
 5000      FORMAT(55X,'THE SIMPLEX SOLUTION',25X,'PAGE 05')
           WRITE(6,5001)
 5001      FORMAT(' THE RIGHT HAND SIDE')
  801      DO 810 I=1,N
           WRITE(6,13) Y(I), PRDT(I)
  810      CONTINUE
           WRITE(6,5002)
 5002      FORMAT(' THE SUBSTITUTION RATES')
```

```
811   DO 812 I=1,N
      WRITE(6,12)(C(I,J),J=1,M)
812   CONTINUE
      WRITE(6,5003)
5003  FORMAT(' THE ZJ-CJ MATRIX'  )
813   DO 814 K=1,L
      WRITE(6,12) (RVLX(K,J) , J=1,M)
814   CONTINUE
C     EVALUATE OBJECTIVE FUNCTION
      DO 820 K=1,L
      ZVAL(K)=0.
      DO 820 I=1,N
      ZVAL(K)=ZVAL(K)+PRDT(I)*VALY(I,K)
820   CONTINUE
      WRITE(6,5004)
5004  FORMAT(' AN EVALUATION OF THE OBJECTIVE FUNCTION')
      DO 821 K=1,L
      KK=L-K
      IF(TEST.EQ.1.0)GO TO 89
      KK=KK+1
89    WRITE(6,15) KK,ZVAL(K)
821   CONTINUE
      CALL FINISH(RHS1,PRDT,VALY,L,KPCK,Y,N,KEPT,TEST)
830   STOP
      END
      SUBROUTINE START(NROWS,NVAR,NPRT,C,VALX,VALY,RHS  ,RHS1,KPCK,KEPT,T
     %EST)
```

```
C         THE START SUBROUTINE IS DESIGNED TO TAKE INFORMATION IN A SPEC-
C         IFIED FORMAT AND TRANSFORM IT INTO A SERIES OF USABLE MATRICES.
C.........................................................................
          REAL NEG
          REAL L
          NV=125
          NR=60
        1 FORMAT(A4,3I3)
          DATA POS,NEG/'POS ','NEG '/
          DATA DATA/'DATA'/
          DATA OBJ/'OBJ '/
          DATA PROB/'PROB'/
          DATA B /'B'/
          DATA E,G,L/'E','G','L'/
          DIMENSION RHS(60)
          DATA  RGHT/'RGHT'/
          DIMENSION VALY(60,10)
          DIMENSION C(60,125),VALX(10,125)
          DIMENSION EQUALS(60),RVLX(10,125)
          DIMENSION KEPT(60)
          DIMENSION RHS1(60)
          TEST=0.0
C
C
C
C         READ THE PROBLEM CARD FOR THE NUMBER OF ROWS,VARIABLES, AND
```

```
C.....................................................................
   10 READ(5,1)ANAME,NROWS,NVAR,NPRT
      LISP=NPRT+1
      IF(NVAR.LE.0) GO TO 1020
      IF(NPRT.LE.0) GO TO 1020
      IF(NROWS.LE.0) GO TO 1020
      IF(ANAME.NE.PROB) GO TO 901
C
C
C     READ THE SIGN CARD.
C     IT WILL CONTAIN ONE OF THE FOLLOWING LETTERS FOR EACH ROW
C     FOR EQUALS                               E
C     FOR LESS THAN OR EQUAL TO                L
C     FOR GREATER THAN OR EQUAL TO             G
C     FOR BOTH DEVIATIONS                      B
C.....................................................................
      READ(5,11)(EQUALS(I),I=1,NROWS)
   11 FORMAT(80A1)
C.....................................................................
C
C
      NART=0
C     COUNT THE NUMBER OF POSITIVE SLACK VARIABLES
C.....................................................................
      NFLDS=0
      DO 12 I=1,NROWS
      IF(EQUALS(I).EQ.B)NFLDS=NFLDS+1
   12 IF(EQUALS(I).EQ.G)NFLDS=NFLDS+1
```

```
      C
      C
      C
      C.........................................
      C                TEST FOR SIZE
      C.........................................
            NSIZE=NFLDS+NROWS+NVAR
            IF(NROWS.GT.NR) GO TO 911
            IF(NSIZE.GT.NV) GO TO 911
      C
      C
      C.........................................
      C             CLEAR ALL MATRICES
      C.........................................
            KDUD=NPRT+1
            DO 16 J=1,NSIZE
            DO 16 I=1,NROWS
            KEPT(I)=0
            IF(IGT.KDUD) GO TO 17
            K=I
            RVLX(K,J)=0.0
            VALX(K,J)=0.0
         17 (IF(I.EQ.J) C(I,J)=1.0
            VALY(I,K)=0.0
            IF(I.NE.J) C(I,J)=0.0
         16 CONTINUE
            KPCK=0
            K=KDUD
      C
```

```
      :
      :
      :
C
C        ADJUST THE SLACK VARIABLES AND OBJECTIVE FUNCTION TO MEET THE
C        REQUIREMENTS OF THE SIGN
C.........:.......................................................................
         DO 13 I=1,NROWS
         IF(EQUALS(I).EQ.E) GO TO 14
         IF(EQUALS(I).EQ.G) GO TO 15
         IF(EQUALS(I).EQ.L) GO TO 13
         IF(EQUALS(I).EQ.B)GO TO 18
         GO TO 910
   14    J=I
         VALX(K,J)=1.0
         NART=NART+1
         TEST=1.0
         GO TO 13
   15    KPCK=KPCK+1
         J=NROWS+KPCK
         C(I,J)=-1.0
         KEPT(I)=J
         J=I
         VALX(K,J)=1.
         NART=NART+1
         TEST=1.0
         GO TO 13
   18    KPCK=KPCK+1
         J=KPCK+NROWS
```

```
          C(I,J)=-1.0
          KEPT(I)=J
     13   CONTINUE
     C
     C
     C         READ THE OBJECTIVE FUNCTION
     C
     19   READ(5,21)ANAME
          I=0
          IF(ANAME.NE.OBJ) GO TO 920
          IF(ANAME.EQ.OBJ) GO TO 20
     20   READ(5,21)ANAME,I,M,TEMP
          IF(ANAME.EQ.DATA) GO TO 30
          IF(M.LE.0) GO TO 1022
          K=LISP-M
     21   FORMAT(A4,2I5,F16.0)
          IF(J.LE.0) GO TO 1022
          IF(K.GT.NPRT) GO TO 1024
          IF(ANAME.EQ.NEG) GO TO 26
          IF(ANAME.EQ.POS) GO TO 25
          GO TO 27
     26   J=I
          VALX(K,J)=TEMP
          GO TO 20
     25   J=KEPT(I)
          IF (KEPT(I).EQ.0) GO TO 1026
          VALX(K,J)=TEMP
```

```
      GO TO 20
   27 IF(TEMP)926,20,926
C
C
C        READ THE DATA MATRIX IN
C.....................................
   30 READ(5,21)ANAME,I,J,TEMP
      IF(ANAME.EQ.RGHT) GO TO 40
      IF(I.LE.0) GO TO 1090
      IF(J.EQ.0) GO TO 1090
      J=KPCK+NROWS+J
      C(I,J)=TEMP
      GO TO 30
C
C
C        READ THE RIGHT HAND SIDE
C.....................................
   40 READ(5,44)(RHS(I),I=1,NROWS)
   44 FORMAT(8F10.0)
C
C
C        WRITE THE ABOVE RESULTS
C.....................................
      WRITE(6,5015)
 5015 FORMAT(55X,'THE RIGHT HAND SIDE-INPUT',33X,'PAGE 01')
      DO 41 I=1,NROWS
      IF(RHS(I))941,42,43
```

```
   42 RHS(I)=.00001
   43 RHS1(I)=RHS(I)
      WRITE(6,1111)I,RHS(I)
 1111 FORMAT(10X,I3,2X,F15.5 )
   41 CONTINUE
      WRITE(6,620)
  620 FORMAT(1H1)
      WRITE(6,5016)
 5016 FORMAT(55X,'THE SUBSTITUTION RATES-INPUT',18X,'PAGE 02')
      DO 1112 I=1,NROWS
      WRITE(6,2519) I
 2519 FORMAT(1X,'ROW',I5)
 1112 WRITE(6,1113)(C(I,J),J=1,NSIZE)
 1113 FORMAT(10F8.3)
      WRITE(6,620)
      WRITE(6,5017)
 5017 FORMAT(55X,'THE OBJECTIVE FUNCTION-INPUT',19X,'PAGE 03')
      DO 1114 K=1,NPRT
      M=LISP-K
      WRITE(6,2150) M
 2150 FORMAT(' PRIORITY',I5)
 1114 WRITE(6,1113)(VALX(K,J),J=1,NSIZE)
      WRITE(6,620)
      WRITE(6,5018)
 5018 FORMAT(55X,'SUMMARY OF INPUT INFORMATION ',19X,'PAGE',' 04')
      NVAR=NSIZE
      WRITE(6,2017) NROWS,NVAR,NPRT,NART
```

```
2017 FORMAT(10X,'NUMBER OF ROWS.........',I5,/,10X,'NUMBER OF VARIABLES
     *....',I5,/,10X,'NUMBER OF PRIORITIES...',I5,/,10X,'ADDED PRIOR
     2ITIES........',I5)
        IF(NART.GT.0)    NPRT=NPRT+1
        RETURN
 910 WRITE(6,914)
9140FORMAT('PROGRAM CONTAINS AN ERROR EITHER IN THE NUMBER OF ROWS PUN
     1CHED OR IN THE SIGN CARD.THE VALUE IS SOMETHING OTHER THAN "E", "G"
     2,OR"L"')
        GO TO 999
1090 WRITE (6,1091)
1091 FORMAT('   IMPROPER DATA COLUMN OR ROW DEFINITION')
        GO TO 999
 920 WRITE(6,921)
9210FORMAT('   AN OBJECTIVE CARD WITH THE VALUE',F16.3,'     I
     1S FOUND BUT INSTRUCTIONS AS TO WHICH DEVIATION HAS BEEN NEGLECTED.
     2EXAMINE YOUR DATA ')
        GO TO 999
1020 WRITE (6,1021)
1021 FORMAT('   NUMBER OF ROWS, VARIABLES, OR PRIORITIES CANNOT BE EQUA
     1L TO ZERO UNDER ANY CIRCUMSTANCES')
        GO TO 999
1022 WRITE (6,1023)
1023 FORMAT('   COLUMN VALUE OR PRIORITY VALUE IS EQUAL TO OR LESS THAN
     1ZERO     ')
        GO TO 999
 911 WRITE(6,912)
```

```
9120FORMAT('   THE NUMBER OF VARIABLES NEEDED TO COMPUTE THIS PROGRAM
     1IS TOO GREAT UNDER PRESENT DIMENSIONS.  SEE YOUR PROGRAMMER FOR AL
     2TERING THIS RESTRICTION TO MEET YOUR NEEDS')
          GO TO 999
1026 WRITE(6,1027)
1027 FORMAT('  ATTEMPT IS MADE TO MINIMIZE NON EXISTANT POSITIVE   DEVIA
     1TION')
          GO TO 999
1024 WRITE(6,1025)
1025 FORMAT('  OBJECTIVE FUNCTION PRIORITY EXCEEDS STATED NUMBER OF PRI
     1ORITIES')
          GO TO 999
 901 WRITE(6,902)
 902 FORMAT(' PROBLEM CARD MISSING OR MISPUNCHED')
          GO TO 999
 926 WRITE(6,927)
 927 FORMAT('  A CARD IN THE OBJECTIVE SECTION DEFINED SOME VALUE FOR T
     1HE OBJECTIVE FUNCTION BUT FAILED TO DEFINE WHETHER THIS WAS TO AP
     2PLY TO THE POSITIVE OR NEGATIVE DEVIATION')
 941 WRITE(6,942)
 942 FORMAT('  NEGATIVE VALUES ARE NOT ALLOWED ON THE RIGHT HAND SIDE.
     1 CORRECT PROBLEM BY MULTIPLYING ENTIRE CONSTRAINT THROUGH BY MINU
     2S ONE.')
          GO TO 999
 999 STOP
     END
     SUBROUTINE FINISH(RHS1,RHS,VALY,NPRT,KPCK,Y,NROWS,KEPT,TEST)
```

```
      REAL NEGSLK
      DIMENSION VALY(60,10)
      DIMENSION ZVAL(10)
      DIMENSION RHS(60)
      DIMENSION KEPT(60)
      DIMENSION Y(60),RHS1(60)
C     RHS1 IS THE RESERVED VECTOR OF RHS VALUES FROM THE BEGINNING.
C     THE ENDING RHS VALUES ARE SUBTRACTED FROM THE BEGINNING ONES
C     AND THE RESULT IS PLACED INTO THE APPROPRIATE SLACK COLUMN.
C     THE REMAINDER OF THE VALUES ARE PRINTED ON PAGE TWO OF THE RE-
C     SULTS.
C
C     SLACK ANALYSIS
C
      WRITE(6,21)
   21 FORMAT(1H1,120X,'PAGE  06'//,50X,'SLACK ANALYSIS')
    1 FORMAT(////)
      WRITE(6,1)
      WRITE(6,8)
    8 FORMAT(10X,'ROW',6X,'AVAILABLE',12X,'POS-SLK',12X'NEG-SLK')
      WRITE(6,1)
      DO 19 I=1,NROWS
      NEGSLK=0.0
      POSSLK=0.0
      DO 11 J=1,NROWS
      M=Y(J)
```

```
      IF(I-M) 9,10,9
    9 IF(M-KEPT(I))) 11,12,11
   11 CONTINUE
      GO TO 13
   10 NEGSLK=RHS(J)
      GO TO 13
   12 POSSLK=RHS(J)
   13 WRITE(6,14)I,RHS1(I),POSSLK,NEGSLK
   14 FORMAT(10X,I3,3F20.5)
   19 CONTINUE
   43 FORMAT(10X,I3,3X,F15.5)
C
C     VARIABLE AMOUNTS
C
      WRITE(6,44)
   44 FORMAT(1H1,120X,'PAGE 07'//,50X,'VARIABLE ANALYSIS')
      WRITE(6,45)
   45 FORMAT(////,7X,'VARIABLE        AMOUNT',//)
      DO 41 I=1,NROWS
      NCHCK=Y(I)-KPCK-NROWS
      IF(NCHCK)41,41,42
   42 WRITE(6,43)NCHCK,RHS(I)
   41 CONTINUE
      WRITE(6,72)
   72 FORMAT(1H1)
      WRITE(6,50)
   50 FORMAT(//,55X,'ANALYSIS OF THE OBJECTIVE',23X,'PAGE 8',////,50X,'P
```

```
      %RIORITY',10X,'UNDER-ACHIEVEMENT',/)
         DO 52 K=1,NPRT
         ZVAL(K)=0.0
         DO 51 I=1,NROWS
   51    ZVAL(K)=ZVAL(K)    +VALY(I,K)*RHS(I)
         LISP=NPRT+1
         KK=LISP-K
         IF(TEST.EQ.0.0)   GO TO 52
         KK=NPRT-K
         IF(KK.GT.0) GO TO 52
         WRITE(6,78) ZVAL(K)
   78    FORMAT(/,45X,'ARTIFICIAL',5X,F20.5)
         GO TO 77
   52    WRITE(6,53) KK,ZVAL(K)
   53    FORMAT(1H0,52X,I2,5X,F20.5)
   77    CONTINUE
         STOP
         END
```

## PROBLEMS

1.  Solve Problem 1 at the end of Chapter 5 by using the computer program.

2.  Solve Problem 2 at the end of Chapter 5 by using the computer program.

3.  Solve Problem 3 at the end of Chapter 5 by using the computer program.

4.  Solve Example 6 in Chapter 3 by using the computer program.

5.  A state-supported university faces the problem of admission planning for the coming academic year. The primary problem results from multiple conflicting goals of the institution concerning the desired student body. The university administration has had a long-standing policy that requires that at least 80 percent of the entering new students meet the state residence requirement (i.e., be in-state students).

    Another important criterion for granting admission is the applicant's college board examination scores and his rank in his high school graduating class. Past college work and records are evaluated to determine the eligibility of transferring and readmitted students.

    The university estimates that the number of eligible applicants for the coming year are:

    |                    | In-State | Out-of-State |
    | ------------------ | -------- | ------------ |
    | Freshmen men       | 2400     | 1500         |
    | Freshment women    | 1000     | 500          |
    | Transfer men       | 400      | 300          |
    | Transfer women     | 225      | 75           |
    | Readmitted men     | 250      | 50           |
    | Readmitted women   | 50       | 20           |

    The university has an enormous investment in its physical facilities. The institution also supports a large number of faculty, administrative, and staff members to operate the total university system. The volume of these two activities is largely determined by the number of students attending the university. A wide fluctuation in either direction from the projected enrollment needed to support these facilities and personnel will result in either overcrowding of the facilities and overworking the personnel, or underutilization of the facilities and laying-off of personnel. With these considerations in mind, the university has projected that between 3,000 and 4,500 new students must be admitted if the correct level of operation is to be maintained.

The university has made an effort to develop long-range projections concerning the classes of students to be admitted during each academic year. These projections apply primarily to the admission of males against females and the admission of freshmen against transfers. These ratios determine the long-term growth patterns of the university. During the academic year under consideration, the university has decided that it must not admit more than 1,000 new women students. The university receives many more applications for admission from women than it can possibly accept.

In addition to the male-female ratio, the university must also consider the ratio between freshmen and transfers in the entering group of new students. While the university receives a large number of applications from students desiring to transfer from other institutions, the number of transfers must be limited. For the coming year, the university believes that it can admit no more than 600 transfer students.

The university maintains a large number of on-campus residence halls for undergraduate students. These facilities are self-supporting, with the residents paying the cost of retiring the bonded debt plus the normal operating expenses. With no outside funds coming into the system to cover operating deficits, the system must not lose money. Thus it is most important to achieve the desired level of occupancy.

In order to achieve the optimum level for the coming year, the university must admit 3,500 new students (1,100 females and 2,400 males). The actual numbers of students can exceed these estimates by 3 percent or be 1.5 percent lower without causing serious overcrowding or revenue loss. The various classes of students tend to choose to live in the residence halls in different proportions. Generally, the university requires all freshmen and sophomore men to reside on campus and all women under the age of 21 to live on campus. With this in mind, the following percentages of students choose to live in the university residence halls.

|            | Men   | Women |
|------------|-------|-------|
| Freshmen   | 100%  | 100%  |
| Transfers  | 60    | 90    |
| Readmitted | 30    | 60    |

The director of admissions lists the following goals for admission planning in the rank of importance:

1. At least 80 percent of the total number of new students must meet the state residence requirement.
2. Avoid lowering the university admission standards.
3. Avoid residence hall occupancy less than 98.5 percent of capacity.

4.   Avoid underutilization of the physical facilities.
5.   Avoid overadmission of women.
6.   Avoid overutilization of physical facilities.
7.   Avoid occupancy of the residence halls beyond 103 percent of capacity.
8.   Avoid overutilization of admission of transfer students.

A. Soive the above goal programming problem by using the computer program. Determine the number of students in each classification listed below and discuss the goal attainment.

| | *In-State* | *Out-of-State* |
|---|---|---|
| Freshmen men | | |
| Freshmen women | | |
| Transfer men | | |
| Transfer women | | |
| Readmitted men | | |
| Readmitted women | | |

B. The director of admissions has decided to introduce an additional goal as follows:

Male-Female Ratio:

| Male students | 70 percent |
|---|---|
| Female students | 30 percent |

Academic Standing:

| Freshmen students | 80 percent |
|---|---|
| Transfer students | 15 |
| Readmitted students | 5 |

The above goal should be attained while meeting the 80 percent state residence requirement. If the director places this new goal as the third priority goal, how would it change the solution? Solve this problem by using the computer program.

# Chapter 7

## Advanced Topics in Goal Programming

Thus far the basic concepts and solution methods of goal programming have been presented. Goal programming is indeed a powerful and flexible technique that can be applied to a variety of decision problems. It should, however, be pointed out that goal programming is by no means a panacea for contemporary decision problems that involve multiple conflicting goals. The fact is that goal programming is applicable only under certain specified assumptions and conditions. Most goal programming applications have thus far been limited to well-defined deterministic problems. Furthermore, the primary analysis has been limited to the identification of an optimal solution that maximizes goal attainment to the extent possible within specified constraints. The primary reason for the narrow scope of goal programming analysis is that the technique is still in its early development. In order to develop goal programming as a powerful technique for modern decision analysis, many refinements and further research are necessary. In this sense, goal programming is still an academic virgin territory; many areas need future research before it can become a widely accepted and applied technique like linear programming. This chapter explores some of the potential research areas in goal programming in order to suggest directions for future studies.

## POSTOPTIMAL SENSITIVITY ANALYSIS

Deriving the optimal solution has been the primary solution procedure of goal programming. However, an analysis of the effects of parameter changes after determining the optimal solution is also a very important part of any solution process. This procedure is broadly defined as the postoptimal sensitivity analysis. Because there usually exists some degree of uncertainty in real-world problems concerning the model parameters—i.e., priority factor $(C_j)$, technological coefficients of choice variables $(a_{ij})$, and the available resource or goal constraints $(b_i)$—sensitivity analysis should be an important part of the goal programming solution. If the optimal solution is relatively sensitive to changes in certain parameters, special efforts should be directed to forecasting the future values of these parameters. By the same token, if the optimal solution has very little sensitivity to changes in certain parameters, it might be a waste of effort and resources to try to estimate the values of parameters more accurately. Although sensitivity analysis should be an important part of goal programming analysis, there has been a total lack of research in this area.

By way of introducing some ideas concerning sensitivity analysis in goal programming, let us review the textile manufacturing problem presented in Chapters 4 and 5. In the example, the firm has a normal production capacity of 80 hours a week. With that capacity the firm produces $x_1$ quantity of upholstery and $x_2$ quantity of dress materials. The production rate for both types of material is 1,000 yards per hour. The president of the firm has established sales goals of 70,000 and 45,000 yeards weekly for upholstery and dress materials respectively. The goal programming problem was formulated as:

$$(7.1) \quad \text{Min } Z = P_1 d_1^- + P_2 d_{11}^+ + 5P_3 d_2^- + 3P_3 d_3^- + P_4 d_1^+$$

$$\begin{aligned}
\text{s.t.} \quad x_1 + x_2 + d_1^- \qquad\quad - d_1^+ \qquad\qquad &= 80 \\
x_1 \qquad\quad + d_2^- \qquad\qquad\qquad &= 70 \\
x_2 \qquad + d_3^- \qquad\qquad\qquad &= 45 \\
d_{11}^- + d_1^+ - d_{11}^+ &= 10 \\
x_1, x_2, d_1^-, d_2^-, d_3^-, d_{11}^-, d_1^+, d_{11}^+ &\geqslant 0
\end{aligned}$$

The final simplex solution to the above problem is presented in Table 7.1. The solution indicates that the first two goals are completely attained while the third and fourth goals are only partially attained. Let us examine some simple but pertinent aspects of the solution that may provide more insight about the solution we have derived. One question might be: What would be required in the

Table 7.1

| $C_j$ | | | | | $P_1$ | $5P_3$ | $3P_3$ | | | $P_4$ | $P_2$ |
|---|---|---|---|---|---|---|---|---|---|---|---|
| | $V$ | $C$ | | $x_1$ | $x_2$ | $d_1^-$ | $d_2^-$ | $d_3^-$ | $d_{11}^-$ | $d_1^+$ | $d_{11}^+$ |
| | $x_2$ | 20 | | | 1 | 1 | $-1$ | | 1 | | $-1$ |
| | $x_1$ | 70 | | 1 | | | 1 | | | | |
| $3P_3$ | $d_3^-$ | 25 | | | | $-1$ | 1 | 1 | $-1$ | | 1 |
| $P_4$ | $d_1^+$ | 10 | | | | | | | 1 | 1 | $-1$ |
| | $P_4$ | 10 | | | | | | | 1 | | $-1$ |
| $C_j$ | $P_3$ | 75 | | | | $-3$ | $-2$ | | $-3$ | | 3 |
| $Z_j - C_j$ | $P_2$ | 0 | | | | | | | | | $-1$ |
| | $P_1$ | 0 | | | | $-1$ | | | | | |

present decision system to attain the third goal completely? This question can be easily answered from a brief analysis of Table 7.1. The only positive value at the $P_3$ level in the simplex criterion is in the $d_{11}^+$ column. This indicates that the only way to attain the third goal completely (i.e., minimize the $Z_j$ value of 75 at $P_3$ to zero) is by employing more overtime production. The coefficients of the $d_{11}^+$ column in the main body of the table suggest that the substitution rate between $d_3^-$ and $d_{11}^+$ is one. Consequently, 25 more hours of overtime would be required to attain the third goal completely.

There is one problem associated with the attainment of the third goal. There is a negative coefficient at the $P_2$ level of $d_{11}^+$ column in the simplex criterion $(Z_j - C_j)$. This indicates that the only way to attain the third goal is at the expense of the second goal. With the given priority structure of goals, therefore, it is simply impossible to attain the third goal without violating some priorities. This type of analysis provides information as to how and in what range the priority structure must be altered in order to attain the third goal.

We can investigate the fourth goal in the same manner. The only positive coefficient at the $P_4$ level in the simplex criterion is in $d_{11}^-$ column. This indicates that attainment of the fourth goal would require replacement of overtime operation ($d_1^+$) by underutilization of the allowed overtime ($d_{11}^-$). However, there exists a direct conflict between the third and fourth goals. The only way to attain the fourth goal is at the expense of the third goal. Unless the priority structure of goals is changed, there is no way the fourth goal can be attained.

The dual solution procedure of goal programming has not been explored at the writing of this text. Consequently, the usual sensitivity analysis employed in linear programming cannot be applied to goal programming. Indeed, duality is a very important future research area of goal programming. With this limitation in mind, let us explore some other aspects of sensitivity analysis applicable to goal programming problems.

## 1. Change in Priority Factor ($C_j$)

Unlike linear programming, an analysis of effects of changes in $C_j$ is not a simple task for goal programming problems. For linear programming problems, the unidimensionality of the objective function allows a relatively easy sensitivity analysis of changes in $C_j$. In goal programming, however, changes of $C_j$ are actually changes in the preemptive priority factors and thereby changes in goal dimensions. Undoubtedly, the conventional sensitivity analysis used in linear programming is perfectly applicable as long as changes in $C_j$ are limited to the differential weights at the same preemptive priority level. A change here does not involve a change of dimensions but rather a change within a dimension.

Another difficulty involved in changes with $C_j$ is that it usually results in a rearrangement of the entire priority structure of goals. For example, in the above problem, we cannot simply change $P_1$ assigned to $d_1^-$ (underachievement of production capacity) to $P_2$. The priority factor $P_2$ has already been assigned to $d_{11}^+$ (minimization of overtime operation of the plant in excess of 10 hours). Since these two goals are not commensurable and also may not be of the same importance, the same priority factor should not be assigned. It is, therefore, necessary either to rearrange the entire priority structure of goals or simply to switch those two priorities if it is so desired by the decision maker.

Let us first examine the effects of changes in $C_j$ for nonbasic variables (variables not in the solution base of the optimal solution). Suppose the decision maker would like to know the effects of the following changes in priorities:

$$\text{from: } \begin{array}{l} P_1 d_1^- \\ P_2 d_{11}^+ \end{array} \qquad \text{to: } \begin{array}{l} P_1 d_{11}^+ \\ P_2 d_1^- \end{array}$$

Since variables $d_1^-$ and $d_{11}^+$ are both nonbasic variables, changes in their priority factors would have no effect on the values of the basic variables. The only immediate changes will involve the coefficients of the $d_1^-$ and $d_{11}^+$ columns in the simplex criterion (there will be a $-1$ at the $P_2$ level in the $d_1^-$ column and a $-1$ at the $P_1$ level in the $d_{11}^+$ column).

If changes in $C_j$ involve combinations of basic variables, or basic and nonbasic variables, there would most likely be some change in the optimal solution. For example, let us assume that $P_3$ is assigned to $d_1^+$, $5P_4$ to $d_2^-$, and $3P_4$ to $d_3^-$ (switch of $P_3$ and $P_4$). This change involves basic variables $d_3^-$ and $d_1^+$ and a nonbasic variable $d_2^-$. This change results in a $Z_j$ of $10P_3$ and $75P_4$, as shown in Table 7.2. Furthermore, since there exists a positive $Z_j - C_j$ coefficient at the $P_3$ level in the $d_{11}^-$ column with no negative $Z_j - C_j$ coefficients at higher priority levels, the solution is no longer the optimal solution. An additional iteration will provide the optimal solution of $x_1 = 70$, $x_2 = 10$, $d_3^- = 35$, and $d_{11}^- = 10$. All of the goals would be completely attained except the fourth goal.

## 2. Change in Resources or Goal Levels ($b_i$)

Changes in the right-hand side of the constraints may have some effect on the optimal solution. Since the right-hand side is the constant value of the basic variable of each equation, changes in $b_i$ will result in changes in the constant of the basic variables of the optimal solution. It is possible that new constants of the basic variables could be negative after the right-hand side of one equation is changed. Thus, the solution may become infeasible. The basic problem, therefore, is to determine the feasibility of the new solution. If the solution is infeasible, some adjustments must be made to identify the new optimal solution. We will discuss this point in the latter part of this section.

Table 7.2

| $C_j$ | | | | | $P_1$ | $5P_4$ | $3P_4$ | | $P_3$ | $P_2$ |
|---|---|---|---|---|---|---|---|---|---|---|
| | $V$ | $C$ | $x_1$ | $x_2$ | $d_1^-$ | $d_2^-$ | $d_3^-$ | $d_{11}^-$ | $d_1^+$ | $d_{11}^+$ |
| | $x_2$ | 20 | | 1 | 1 | −1 | | 1 | | −1 |
| | $x_1$ | 70 | 1 | | | 1 | | | | |
| $3P_4$ | $d_3^-$ | 25 | | | −1 | 1 | 1 | −1 | | 1 |
| $P_3$ | $d_1^+$ | 10 | | | | | | ①  | 1 | −1 |
| $Z_j - C_j$ | $P_4$ | 75 | | | −3 | −2 | | −3 | | 3 |
| | $P_3$ | 10 | | | | | | 1 | | −1 |
| | $P_2$ | 0 | | | | | | | | −1 |
| | $P_1$ | 0 | −1 | | | | | | | |

If the new solution is feasible—that is, if all new constants of the basic variables are non-negative—it will be the optimal solution. The only additional calculations required will be in the analysis of goal attainment. Suppose the right-hand side of an equation ($b_i$) has been changed to a new value ($b_i'$). Let us denote $d_i^-*$ as the negative deviational variable coefficient in the equation where the right-hand side has been changed and $a_{k,i}^*$ as the coefficient of $d_i^-*$ in the final simplex solution for various equations. Then, $a_{k,i}^*(b_i' - b_i)$ should be added to the right-hand side of the previous final constants of the basic variables (for k = 1, 2, ..., m), in order to obtain the new constants.[1]

To illustrate the procedure, let us review the problem described in (7.1). Suppose the right-hand side in the first constraint, $x_1 + x_2 + d_1^- - d_1^+ = 80$, is increased by 10: $x_1 + x_2 + d_1^- - d_1^+ = 90$. From the final simplex table presented in Table 7.1, we can derive the new constants of the basic variables as follows:

(7.2)
$$x_2 + d_1^- - d_2^- + d_{11}^- + d_{11}^+ = 20 + 1(90 - 80) = 30$$
$$x_1 + d_2^- = 70$$
$$- d_1^- + d_2^- + d_3^- - d_{11}^- + d_{11}^+ = 25 - 1(90 - 80) = 15$$
$$d_{11}^- + d_1^+ - d_{11}^+ = 10$$

Note that the second and the fourth constraints do not contain $d_1^-$ and thus their constraints remain as they were. The new optimal solution will be $x_1 = 70$, $x_2 = 30$, $d_3^- = 15$, and $d_1^+ = 10$. This solution indicates that the increased normal production capacity results in an increase in production for $x_2$ by 10. Consequently, underachievement of sales goal for $x_2$ decreases from 25 to 15. However, overtime operation of the plant remains at 10 hours, as the plant is in operation for 100 hours, which is still 10 hours over the increased normal operation capacity of 90 hours.

Next, let us analyze the effects of decrease in $b_i$ on the optimal solution. Suppose the normal production capacity has decreased from 80 to 70 hours. We can calculate new constants of the basic variables in the following manner:

(7.3)
$$x_2 + d_1^- - d_2^- + d_{11}^- - d_{11}^+ = 20 + 1(70 - 80) = 10$$
$$x_1 + d_2^- = 70$$
$$-d_1^- + d_2^- + d_3^- - d_{11}^- + d_{11}^+ = 25 - 1(70 - 80) = 35$$
$$d_{11}^- + d_1^+ - d_{11}^+ = 10$$

With the reduced normal production capacity, production of dress material $(x_2)$ has been decreased to 10. Accordingly, underachievement of sales goal for the dress material $(d_3^-)$ has been increased to 35. Again, there were no changes in production of upholstery material $(x_1)$ and in the overtime operation of the plant $(d_1^+)$. The optimal solution is now $x_1 = 70$, $x_2 = 10$, $d_3^- = 35$, and $d_1^+ = 10$. There are slight changes in goal attainments. The first two goals are completely attained. However, the third goal is achieved to a lesser degree $(P_3 = 105)$, and attainment of the fourth goal remains at the same level.

The above two illustrations provided positive new constants and consequently the new solutions were feasible and optimal. Let us examine a case where the revised solution is not feasible. Suppose the normal production capacity has decreased to 50 hours. The new constants of the final solution will be changed to:

$$(7.4) \qquad \begin{aligned} x_2 + d_1^- - d_2^- + d_{11}^- - d_{11}^+ &= 20 + 1(50 - 80) = -10 \\ x_1 + d_2^- &= 70 \\ -d_1^- + d_2^- + d_3^- - d_{11}^- + d_{11}^+ &= 25 - 1(50 - 80) = 55 \\ d_{11}^- + d_1^+ - d_{11}^+ &= 10 \end{aligned}$$

It should be apparent that the above solution is infeasible, as the constant of the first basic variable is now negative. In such a case, the solution must be adjusted. With the new production constraint, it is impossible to produce 70 units of the upholstery material $(x_1 = 70)$. It is also evident that achievement of sales goal for the dress material $(d_3^-)$ of 55 units is an impossiblity. Consequently, constants of the first three equations must be adjusted. The negative constant for the first equation indicates that a revised solution should be derived by entering a variable that has a negative coefficient. Furthermore, the entering variable should have positive coefficients in the second and third equations so as to adjust the constants. It is clear that $d_2^-$ should enter to replace $x_2$ in the basic solution. The procedure is shown in Tables 7.3 and 7.4. By entering $d_2^-$ as the basic variable, the new optimal solution can be derived. The new optimal solution is both feasible and optimal.

In complex problems with a large number of constraints and variables, adjustment of the infeasible solution would be a very difficult task. In fact, it may require more effort and time to make the necessary adjustments to avoid the infeasibility than starting a new solution from the beginning with the changed right-hand side. This is especially true if a computer program is readily available for model solution.

Table 7.3

| $C_j$ | | | $P_1$ | | 5$P_3$ | 3$P_3$ | | | $P_4$ | $P_2$ |
|---|---|---|---|---|---|---|---|---|---|---|
| | $V$ | $C$ | $x_1$ | $x_2$ | $d_1^-$ | $d_2^-$ | $d_3^-$ | $d_{11}^-$ | $d_1^+$ | $d_{11}^+$ |
| | $x_2$ | $-10$ | 1 | 1 | $-1$ | | | 1 | | $-1$ |
| | $x_1$ | 70 | 1 | | | 1 | | | | |
| 3$P_3$ | $d_3^-$ | 55 | | | $-1$ | 1 | 1 | $-1$ | | 1 |
| $P_4$ | $d_1^+$ | 10 | | | | | | 1 | 1 | $-1$ |
| | $P_4$ | 10 | | | | | | 1 | | $-1$ |
| $Z_j - C_j$ | $P_3$ | 165 | | | $-3$ | $-2$ | | $-3$ | | 3 |
| | $P_2$ | 0 | | | | | | | | $-1$ |
| | $P_1$ | 0 | | | $-1$ | | | | | |

Table 7.4

| $C_j$ | | | $P_1$ | | 5$P_3$ | 3$P_3$ | | | $P_4$ | $P_2$ |
|---|---|---|---|---|---|---|---|---|---|---|
| | $V$ | $C$ | $x_1$ | $x_2$ | $d_1^-$ | $d_2^-$ | $d_3^-$ | $d_{11}^-$ | $d_1^+$ | $d_{11}^+$ |
| 5$P_3$ | $d_2^-$ | 10 | | $-1$ | $-1$ | 1 | | $-1$ | | 1 |
| | $x_1$ | 60 | 1 | 1 | 1 | | | 1 | | $-1$ |
| 3$P_3$ | $d_3^-$ | 45 | 1 | | | | 1 | | | |
| $P_4$ | $d_1^+$ | 10 | | | | | | 1 | 1 | $-1$ |
| | $P_4$ | 10 | | | | | | 1 | | $-1$ |
| $Z_j - C_j$ | $P_3$ | 185 | | $-2$ | $-5$ | | | $-5$ | | 5 |
| | $P_2$ | 0 | | | | | | | | $-1$ |
| | $P_1$ | 0 | | | $-1$ | | | | | |

## 3. Change in Technological Coefficients ($a_{ij}$)

Changes in technological coefficients are not only frequent in reality but may also have profound effects on the problem solution. The procedure for investigating the effect of changes in $a_{ij}$ is very similar to that of changes in $b_i$. For example, if $a_{ij}$ is changed to $a'_{ij}$ for variable $x_j$, $a^*_{k,i}(a'_{ij}-a_{ij})$ should be added to the current coefficient of $x_j$ in the final solution tableau (where $a^*_{k,i}$ is the coefficient of the negative deviational variable $d_i^{-*}$ in the equation where a technological coefficient has been changed). If $x_j$ is a nonbasic variable ($x_j = 0$), the change in its coefficient will have no effect on the constants of the basic variables. In other words, the same feasible solution should be obtained with the new coefficients. The only question that requires our attention is whether the solution is still optimal or not. This question can be easily answered by analyzing the simplex criterion.

To illustrate the procedure explained above, let us first examine a change in $a_{ij}$ when $x_j$ is a nonbasic variable. Going back to Table 7.4, let us assume that the first equation has been changed from the original $x_1 + x_2 + d_1^- - d_1^+ = 80$ to $x_1 + 0.5x_2 + d_1^- - d_1^+ = 80$. Since $d_1^-$ has coefficients in the first two equations of the final simplex solution, there will be a change in the coefficients of $x_2$ in these equations. The coefficient of $x_2$ in the first equation will be changed from $-1$ to $-1 + (-1)(0.5-1) = -0.5$. The new solution is presented in Table 7.5.

Table 7.5

| $C_j$ | | | | | $P_1$ | $5P_3$ | $3P_3$ | | | $P_4$ | $P_2$ |
|---|---|---|---|---|---|---|---|---|---|---|---|
| | $V$ | $C$ | | $x_1$ | $x_2$ | $d_1^-$ | $d_2^-$ | $d_3^-$ | $d_{11}^-$ | $d_1^+$ | $d_{11}^+$ |
| $5P_3$ | $d_2^-$ | 10 | | | $-1/2$ | $-1$ | 1 | | $-1$ | | 1 |
| | $x_1$ | 60 | | 1 | $1/2$ | 1 | | | 1 | | $-1$ |
| $3P_3$ | $d_3^-$ | 45 | | | ① | | | 1 | | | |
| | $d_1^+$ | 10 | | | | | | | 1 | 1 | $-1$ |
| $Z_j - C_j$ | $P_4$ | 10 | | | | | | | 1 | | |
| | $P_3$ | 185 | | | $1/2$ | $-5$ | | | $-5$ | | 5 |
| | $P_2$ | 0 | | | | | | | | | $-1$ |
| | $P_1$ | 0 | | | | $-1$ | | | | | |

This solution is feasible, as all constants of the basic variables are non-negative. However, it is no longer optimal. An additional iteration provides the optimal solution as shown in Table 7.6.

A brief analysis of the optimal solution ($x_1 = 37.5$, $x_2 = 45$, $d_2 = 33.5$, $d_1^+ = 10$) indicates that the solution is indeed in line with the change in the technological coefficient of $x_2$. The change of $a_{12}$ from 1 to 0.5 suggests that the relative efficiency of production for the dress material ($x_2$) has doubled. Consequently, the production of dress material becomes more desirable than before. Therefore, the product mix was adjusted and goal attainments were improved. It should be pointed out that a change in the technological coefficient may also result in change in the priority factor. For example, in our example the priority factor assigned to achievement of sales goal for the dress material should actually be increased from $3P_3$ to $6P_3$, if the differential weight is to be determined by the combination of profit per unit and production efficiency. This change has no effect on the optimal solution presented in Table 7.6. However, the optimal solution tends to be more sensitive when $C_j$ changes simultaneously with $a_{ij}$.

Analysis of changes in technological coefficients is more complex when $x_j$ is a basic variable (and nondegenerate). It should be sufficiently clear that, since $x_j$ is positive, a change in the coefficient of $x_j$ will have some effect on constants of other basic variables. Therefore, the revised solution must be checked for its feasibility as well as its optimality. First, we should calculate the new

Table 7.6

| $C_j$ | | | | | $P_1$ | $5P_3$ | $3P_3$ | | $P_4$ | $P_2$ |
|---|---|---|---|---|---|---|---|---|---|---|
| | $V$ | $C$ | $x_1$ | $x_2$ | $d_1^-$ | $d_2^-$ | $d_3^-$ | $d_{11}^-$ | $d_1^+$ | $d_{11}^+$ |
| $5P_3$ | $d_2^-$ | 32.5 | | | −1 | 1 | 1/2 | −1 | | 1 |
| | $x_1$ | 37.5 | 1 | | | 1 | −1/2 | 1 | | −1 |
| | $x_2$ | 45 | | 1 | | | 1 | | | |
| $P_4$ | $d_1^+$ | 10 | | | | | | 1 | 1 | −1 |
| $Z_j - C_j$ | $P_4$ | 10 | | | | | | 1 | | −1 |
| | $P_3$ | 162.5 | | | −5 | | −1/2 | −5 | | 5 |
| | $P_2$ | 0 | | | | | | | | −1 |
| | $P_1$ | 0 | | | −1 | | | | | |

coefficients of $x_j$ for each equation in the final simplex solution by the above-described method. Since $x_j$ is a basic variable, the previous simplex table when $x_j$ just entered the solution should have shown a zero coefficient for $x_j$ in all equations except the one containing $x_j$ as its basic variable. This condition must be restored algebraically so as to test the feasibility and optimality of the revised solution.

To illustrate the complex procedure, let us go back to Table 7.1. Suppose the first equation is changed from the original $x_1 + x_2 + d_1^- - d_1^+ = 80$ to $x_1 + 1/2x_2 + d_1^- - d_1^+ = 80$. This change results in a modified solution as shown in Table 7.7. The new solution is feasible since all constants of the basic variables are non-negative. However, this solution must be revised so that the coefficient of $x_2$ can be eliminated from every equation except the first equation where $x_2$ is the basic variable. The revised solution, shown in Table 7.8, is feasible, and it appears to be perfectly in line with the improved production efficiency for the material. However, the solution is not optimal. An additional iteration yields the optimal solution as shown in Table 7.9.

If the revised solution was infeasible, necessary adjustments could be made algebraically for simple goal programming problems as described previously. In complex problems, however, it appears that it would be simpler to solve the problem from the beginning with the new technological coefficients. This is especially true if the priority factors also change simultaneously with the technological coefficients.

Table 7.7

| $C_j$ | | | $P_1$ | $5P_3$ | $3P_3$ | | | | $P_4$ | $P_2$ |
|---|---|---|---|---|---|---|---|---|---|---|
| | $V$ | $C$ | $x_1$ | $x_2$ | $d_1^-$ | $d_2^-$ | $d_3^-$ | $d_{11}^-$ | $d_1^+$ | $d_{11}^+$ |
| | $x_2$ | 20 | (1/2) | 1 | −1 | | | | 1 | −1 |
| | $x_1$ | 70 | 1 | | | 1 | | | | |
| $3P_3$ | $d_3^-$ | 25 | | | 1/2 | −1 | 1 | 1 | −1 | 1 |
| $P_4$ | $d_1^+$ | 10 | | | | | | 1 | 1 | −1 |
| $Z_j - C_j$ | $P_4$ | 10 | | | | | | | 1 | −1 |
| | $P_3$ | 75 | | | 1/2 | −3 | −2 | | −3 | 3 |
| | $P_2$ | 0 | | | | | | | | −1 |
| | $P_1$ | 0 | | | −1 | | | | | |

Table 7.8

| $C_j$ | | | | | $P_1$ | $5P_3$ | $3P_3$ | | $P_4$ | $P_2$ |
|---|---|---|---|---|---|---|---|---|---|---|
| | $V$ | $C$ | $x_1$ | $x_2$ | $d_1^-$ | $d_2^-$ | $d_3^-$ | $d_{11}^-$ | $d_1^+$ | $d_{11}^+$ |
| | $x_2$ | 40 | | 1 | 2 | $-2$ | | 2 | | $-2$ |
| | $x_1$ | 70 | 1 | | | 1 | | | | |
| $3P_3$ | $d_3^-$ | 5 | | | $-2$ | ②  | 1 | | $-2$ | 2 |
| $P_4$ | $d_1^+$ | 10 | | | | | | 1 | 1 | $-1$ |
| | $P_4$ | 10 | | | | | | | 1 | $-1$ |
| $Z_j - C_j$ | $P_3$ | 15 | | | | $-6$ | 1 | | $-6$ | 6 |
| | $P_2$ | 0 | | | | | | | | $-1$ |
| | $P_1$ | 0 | | | $-1$ | | | | | |

Table 7.9

| $C_j$ | | | | | $P_1$ | $5P_3$ | $3P_3$ | | $P_4$ | $P_2$ |
|---|---|---|---|---|---|---|---|---|---|---|
| | $V$ | $C$ | $x_1$ | $x_2$ | $d_1^-$ | $d_2^-$ | $d_3^-$ | $d_{11}^-$ | $d_1^+$ | $d_{11}^+$ |
| | $x_2$ | 45 | | 1 | | | 1 | | | |
| | $x_1$ | 67.5 | 1 | | | 1 | $-1/2$ | 1 | | $-1$ |
| $3P_3$ | $d_3^-$ | 2.5 | | | | $-1$ | 1 | $1/2$ | $-1$ | 1 |
| $P_4$ | $d_1^+$ | 10 | | | | | | 1 | 1 | $-1$ |
| | $P_4$ | 10 | | | | | | | 1 | $-1$ |
| $Z_j - C_j$ | $P_3$ | 7.5 | | | | $-3$ | $-2$ | $-3/2$ | $-3$ | 3 |
| | $P_2$ | 0 | | | | | | | | $-1$ |
| | $P_1$ | 0 | | | $-1$ | | | | | |

## 4. Addition of a New Constraint or Goal

If the decision maker decides to add a new constraint or a goal after the optimal solution is reached, it may change the optimality of the solution and also the degree of goal attainments. Identification of the new optimal solution becomes more complicated if the addition of a new constraint or goal requires the formulation of an entirely different objective function. The procedure to make the necessary adjustment is, first, to include the new constraint or goal equation in the final simplex solution by entering the negative deviational variable as a basic variable. Next, we must eliminate any variables in the new equation that appear as basic variables in the simplex table. If the new solution is feasible, we should check its optimality. If the solution is infeasible because of a negative constant associated with the new negative deviational variable, which just entered the solution, again necessary adjustments may be made according to the previously described procedure.

To illustrate, let us add the new goal constraint $x_2 + d_4^- - d_4^+ = 30$. Suppose priority factor $P_2$ is assigned to $d_4^-$ and all the previous priority factors are reduced by one level except $P_1$. The new objective function would read: Minimize $Z = P_1 d_1^- + P_2 d_4^- + P_3 d_{11}^+ + 5P_4 d_2^- + 3P_4 d_3^- + P_5 d_1^+$. The new simplex table, including the new goal constraint, is shown in Table 7.10. By algebraically

### Table 7.10

| $C_j$ | | | | | $P_1$ | $5P_4$ | $3P_4$ | $P_2$ | | $P_5$ | | $P_3$ |
|---|---|---|---|---|---|---|---|---|---|---|---|---|
| | $V$ | $C$ | $x_1$ | $x_2$ | $d_1^-$ | $d_2^-$ | $d_3^-$ | $d_4^-$ | $d_{11}^-$ | $d_1^+$ | $d_4^+$ | $d_{11}^+$ |
| | $x_2$ | 20 | | ①  | 1 | −1 | | | 1 | | | −1 |
| | $x_1$ | 70 | 1 | | | 1 | | | | | | |
| $3P_4$ | $d_3^-$ | 25 | | | | −1 | 1 | 1 | −1 | | | 1 |
| $P_5$ | $d_1^+$ | 10 | | | | | | | 1 | 1 | | −1 |
| $P_2$ | $d_4^-$ | 30 | | 1 | | | | 1 | | | −1 | |
| | $P_5$ | 10 | | | | | | | | 1 | | −1 |
| $Z_j - C_j$ | $P_4$ | 75 | | | −3 | −2 | | | −3 | | | 3 |
| | $P_3$ | 0 | | | | | | | | | | −1 |
| | $P_2$ | 30 | | 1 | | | | | | | −1 | |
| | $P_1$ | 0 | | | −1 | | | | | | | |

Table 7.11

| $C_j$ | | | | | $P_1$ | $5P_4$ | $3P_4$ | $P_2$ | | $P_5$ | | | $P_3$ |
|---|---|---|---|---|---|---|---|---|---|---|---|---|---|
| | $V$ | $C$ | $x_1$ | $x_2$ | $d_1^-$ | $d_2^-$ | $d_3^-$ | $d_4^-$ | $d_{11}^-$ | $d_1^+$ | $d_4^+$ | $d_{11}^+$ |
| | $x_2$ | 20 | | 1 | 1 | −1 | | | 1 | | | −1 |
| | $x_1$ | 70 | 1 | | | 1 | | | | | | |
| $3P_4$ | $d_3^-$ | 25 | | | −1 | 1 | 1 | | −1 | | | 1 |
| $P_5$ | $d_1^+$ | 10 | | | | | | | 1 | 1 | | −1 |
| $P_2$ | $d_4^-$ | 10 | | | −1 | (1) | | 1 | −1 | | −1 | 1 |
| | $P_5$ | 10 | | | | | | | 1 | | | −1 |
| | $P_4$ | 75 | | | −3 | −2 | | | −3 | | | 3 |
| $Z_j - C_j$ | $P_3$ | 0 | | | | | | | | | | −1 |
| | $P_2$ | 10 | | | −1 | 1 | | | −1 | | −1 | 1 |
| | $P_1$ | 0 | | | −1 | | | | | | | |

Table 7.12

| $C_j$ | | | | | $P_1$ | $5P_4$ | $3P_4$ | $P_2$ | | $P_5$ | | | $P_3$ |
|---|---|---|---|---|---|---|---|---|---|---|---|---|---|
| | $V$ | $C$ | $x_1$ | $x_2$ | $d_1^-$ | $d_2^-$ | $d_3^-$ | $d_4^-$ | $d_{11}^-$ | $d_1^+$ | $d_4^+$ | $d_{11}^+$ |
| | $x_2$ | 30 | | 1 | | | | 1 | | −1 | | |
| | $x_1$ | 60 | 1 | | | 1 | | −1 | 1 | 1 | | −1 |
| $3P_4$ | $d_3^-$ | 15 | | | | | 1 | −1 | | 1 | | |
| $P_5$ | $d_1^+$ | 10 | | | | | | | 1 | 1 | | −1 |
| $5P_4$ | $d_2^-$ | 10 | | | −1 | 1 | | 1 | −1 | | −1 | 1 |
| | $P_5$ | 10 | | | | | | 1 | | | | −1 |
| | $P_4$ | 95 | | | −5 | | | 2 | −5 | | −2 | 5 |
| $Z_j - C_j$ | $P_3$ | 0 | | | | | | | | | | −1 |
| | $P_2$ | 0 | | | | | | −1 | | | | |
| | $P_1$ | 0 | | | −1 | | | | | | | |

eliminating any other basic variable for the new equation we derive Table 7.11. The new solution is feasible but not optimal. An additional iteration yields the optimal solution as presented in Table 7.12.

## 5. Addition of a New Variable

It may be possible that the decision maker discovers, after the optimal solution has been identified, that some of the pertinent variables are not included in the model. Addition of a new variable may also involve changes in the objective function, technological coefficients, and constants of the basic variables. We have already discussed procedures to analyze the effects of changes in $C_j$, $a_{ij}$, and $b_i$. These procedures can be applied to the addition of a new variable to the model.


## PARAMETRIC GOAL PROGRAMMING

Another important area of future research is parametric goal programming. Sensitivity analysis is limited to determination of the effects of a single change in parameters in the optimal solution. Parametric goal programming, on the other hand, is a systematic analysis of changes in a goal programming solution associated with the magnitude of simultaneous changes in the model parameters. Parametric goal programming is a useful tool in determining trade-offs among decision variables and goals. Our particular interests are the effects of systematic changes in $C_j$ (priority structure of goals) and $b_i$ (right-hand side of constraints and goals).

As long as changes in $C_j$ are limited to goals at the same priority level, the conventional parametric linear programming is readily applicable. If changes in $C_j$ involve changes in the preemptive priority factor, it is considerably more difficult to determine their effects. Fortunately, however, management goals are not subject to random or irrational changes as contrary to changes of $C_j$ in linear programming problems. Stated differently, if a decision maker regards a goal as most important in the model solution, he would not be interested in the systematic analysis of changes in the optimal solution as that goal descends all the way to the least important level. If he wishes to analyze trade-offs between this goal and the second or third goal, this can be easily accommodated by sensitivity analysis.

A systematic change of the right-hand side $(b_i)$ may result in different optimal solutions, and this information may be valuable for decision analysis. Adopting the same procedure utilized in sensitivity analysis, one can determine how much change in $b_i$ is required before some constants of the basic variables become negative and thereby result in an infeasible solution. Up to that point, a change in $b_i$ would still yield a feasible solution. Thus, the range of optimality for the solution can easily be determined. It should be reiterated here, however, that in complex goal programming problems it is extremely tedious work to determine the effects of systematic changes in $b_i$. This becomes even more so if the objective function changes according to changing $b_i$ values. It is the opinion of the author that parametric goal programming can easily be facilitated by a computer program. By attaching a parametric analysis subroutine to the computer program of goal programming, it can be offered as an option in addition to the regular optimal solution. Perhaps this can be accomplished through future research.

## GOAL PROGRAMMING UNDER UNCERTAINTY

One of the frequent problems in the practical application of goal programming is the difficulty of determining proper values of model parameters, especially the right-hand side of constraints ($b_i$) and technological coefficients ($a_{ij}$). If the values of certain parameters are based on random events that are difficult to predict, the problem becomes goal programming under the condition of uncertainty. In real-world problems, stochastic conditions are indeed quite frequent. In this regard, Elmaghraby points out the limitation of deterministic linear programming:

> A characteristic of linear programming problems is the deterministic nature of the parameters involved. For example, the allocation of capital resources is always to a fixed and known demand, the vitamin content of various diets (in diet problems) is known deterministically with no possible variation, etc. Clearly, such an ideal state of knowledge is scarcely, if ever, realized. In many situations, production is to varying demand, while almost all of the parameters (representing, cost yield, effectiveness, etc.) exhibit stochastic variations.[2]

For many deterministic models, such as linear programming and goal programming, one of the universal problems in practical applications is the validity of the model parameters used in the solution. Often the true values of the model parameters become known only after the model result has been implemented.

## 1. Linear Programming Under Uncertainity

Numerous approaches have been attempted to introduce the effects of uncertainty into linear programming models. Some of these methods permit retention of the simplex method as a solution algorithm, while others require a more rigorous quadratic programming procedure for their solution. Since a major part of the work on goal programming to date has utilized the simplex algorithm for solution, the emphasis in this chapter will be on determining any parallels between linear programming methods under uncertainty and goal programming subject to uncertain conditions that permit the retention of the simplex algorithm.

The general linear programming problem is:

(7.5)    $\text{Max } Z = \sum_{j=1}^{n} c_j x_j$

s.t.

$\sum_{j=1}^{n} a_{ij} x_j \leq b_i \ (i = 1, 2, \ldots, m)$

$x_j \geq 0, \ (j = 1, 2, \ldots, n)$

or in matrix form,

(7.6)    $\text{Max } Z = cx$

s.t.     $Ax \leq b$

$x \geq 0$

Under this notation, A represents the technological coefficient matrix, x is the structural row vector, c is the contribution vector associated with each x, and b is the resource availablity respectively.

Madansky has outlined three general and commonly used methods of dealing with the effects of uncertainty in linear programming[3]:

a. replacing the random elements by their expected values;
b. replacing the random elements by pessimistic estimates of their values;
c. recasting the problem into a two-stage problem, where in the second stage, one can compensate for inaccuracies in the first-stage activities.

These three approaches have been called the expected value solution, the "fat" solution, and the "slack" solution. In one-stage solutions, we may approach the problem on a "wait-and-see" basis or, alternatively, we may use a "here-and-now" approach. With the former method we observe the behavior of the random elements and then incorporate the results of the observation into a nonstochastic linear program. Models of this nature have been discussed by Freund and by Tintner.[4] In the "here-and-now" approach, the x-vector is solved in order to determine a certain probability of occurrence. This probability must equal or exceed a certain feasibility constraint of probability, P. The formulation of this ·problem can be made as follows:

(7.7)    Max $Z$ = expected value of cx = $E(cx)$
         s.t.    Prob $(Ax \geqslant b) \geqslant P$

This is the general form of chance-constrained programming. Chance-constrained programming has been developed extensively by Charnes and Cooper, and a variation of this method has been suggested by Evers.[5] In the Evers model, the A-matrix could be partitioned into both a stochastic component and a deterministic component. The new objective function contained not only the contributions resulting from deterministic factors but also contributions resulting from any infeasibility and shortages.

Referring back to Madansky, we are given an alternative in the event that the solution vector to a chance-constrained problem provides too low a confidence estimate. In this case, the "fat" solution can be used. With the "fat" solution, a pessimistic A-matrix and b-vector are chosen so that we can be assured of a high probability of success. This might be characterized as a "brute force" method and would not seem to be very efficient in terms of resource allocation.

The so-called "slack" solution seems to be somewhat similar to goal programming in its approach. In fact, goal programming may even be a special case of the slack method under conditions of fairly certain expected values in the A-matrix and b-vector. But with the slack method, A and b are subject to random variation. The only apparent difference between this approach and goal programming is the random variation in A and b and the fact that the choice variables appear in the objective function. Methods for solving two-stage "here-and-now" problems such as this have been discussed by Dantzig and by Elmaghraby.[6]

Some of these methods of linear programming under uncertainty assume that the random elements can be described by continuous normal distributions. These models are very complex and require specialized solution algorithms. Other methods attempt to approximate stochastic variations in a piecewise sense by assuming discrete probability distributions. These methods permit approximations to stochastic behavior in linear programming and sometimes permit the use of the simplex algorithm.

## 2. Some Approaches to Goal Programming Under Uncertainty

Some very limited work has been done to date toward offering approaches to goal programming under uncertainty. Charnes, Cooper, Neihaus, and Sholtz have jointly developed a manpower planning model which considers the effects of Markov processes from period to period.[7] Contini has suggested a form of chance-constrained goal programming using normally distributed random variables associated with the deviational variables.[8] The programming method requires the use of quadratic programming techniques.

In an effort to render a better understanding of the effects of uncertainty on goal programming, we will attempt to limit the range of possible approaches. Two approaches will be offered that permit the retention of the simplex algorithm, and that will perhaps provide a starting point for future research in this area.

The first of these examples will resemble to some extent the piecewise linear approximation of Elmaghraby, using discrete distributions for the A-matrix. The second approach will attempt to offer some added sophistication by including variance-covariance effects. This form of approximation has been discussed by Dantzig with respect to linear programming.[9]

### A.  THE PIECEWISE APPROXIMATION APPROACH[10]

Let us assume that a production manager must allocate available production time between his two teams of workers, team 1 and team 2, in order to meet as closely as possible a production goal of 15 units per day. The two teams produce the same product, but team 1 is more experienced at the operation than team 2. Therefore, their processing rates are different. The production manager knows from past performance records that these two processing rates do not remain constant but can vary depending on a number of known and unknown factors. From his records, the production manager has been able to determine some discrete probability distributions of processing rates for the two teams. Assume these distributions appear as follows:

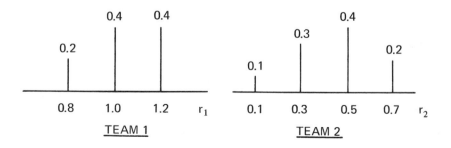

The available production time for each team is eight hours per day under normal operations. The manager assigns the highest priority to the achievement of the production goal of 15 units per day. Second, the manager would like to avoid any underutilization of regular operation hours for either team. If overtime is required by either team in order to meet the production goal for the day, the manager would rather give the overtime to the more efficient team 1. In fact, the average processing rate for team 2 works out to be about half that of team 1. Thus avoiding overtime is twice as important as for team 1. The manager assigns the third goal to the avoidance of overtime. Any overproduction will be acceptable within a reasonable limit, so the lowest priority will be placed on minimizing that.

If we denote $x_1$ and $x_2$ as the number of hours allocated to team 1 and team 2 respectively, the goal programming problem may be formulated as:

$$
\begin{aligned}
(7.8) \quad \text{Min } Z = {}& P_1 d_1^- + P_2 d_2^- + P_2 d_3^- + 2P_3 d_3^+ + P_3 d_2^+ + P_4 d_1^+ \\
\text{s.t.} \quad & s_1 x_1 + s_2 x_2 + d_1^- \quad\quad - d_1^+ \quad\quad\quad = 15 \\
& x_1 \quad\quad\quad + d_2^- \quad\quad - d_2^+ \quad\quad = 8 \\
& x_2 \quad\quad\quad + d_3^- \quad\quad\quad - d_3^+ = 8 \\
& x_i, d_i^-, d_i^+ \geqslant 0
\end{aligned}
$$

where $s_1$ and $s_2$ represent possible ranges of the processing rates for stochastic variables $r_1$ and $r_2$. The $d_1^-$ represents expected value of the underachievement of the production goal, given all possible combinations of processing rates and production times. Similarly, $d_1^+$ represents the expected value of overproduction. The determinant of all of the values of the other variables is the stochastic negative deviation, $d_1^-$. When written in terms of stochastic variables, the objective function, $Z$, can take on an expected value. Naturally, we would like to minimize the expected value of the objective function. The first partial derivative of $Z$ with respect to $d_1^-$ should allow us to determine the marginal "cost" of $d_1^-$ over various ranges of $s_1$ and $s_2$. The objective function can thus be restated as:

$$
\begin{aligned}
(7.9) \quad \text{Min } Z = {}& P_1(15 - s_1 x_1 - s_2 x_2 + d_1^+) + P_2 d_2^- + P_2 d_3^- \\
& + 2P_3 d_3^+ + P_3 d_2^+ + P_4(s_1 x_1 + s_2 x_2 + d_1^- - 15)
\end{aligned}
$$

Following the methods of Elmaghraby, the expected value of $s_1 x_1$ is:

$$(7.10) \quad E(s_1 x_1) = x_1 \sum_{0}^{s_1} r_1 p(r_1) + x_1 s_1 \sum_{s_1}^{\infty} p(r_1)$$

The expected value of $s_2 x_2$ is written in a similar fashion. By a chain rule of partial differentiation:

$$(7.11) \qquad \frac{\partial Z}{\partial(d_1^-)} = \frac{\partial Z}{\partial s_1} \cdot \frac{ds_1}{d(d_1^-)} + \frac{\partial Z}{\partial s_2} \cdot \frac{ds_2}{d(d_1^-)}$$

In order to maintain consistency throughout the new goal program while eliminating nonlinearities, the values chosen for each partitioned s will be the means of their corresponding ranges. Thus $s_{11}$, $s_{12}$, and $s_{13}$ will be the means of the three ranges for $s_1$. Accordingly, $s_{21}$, $s_{22}$, $s_{23}$, and $s_{24}$ will be the means of the four ranges of $s_2$. The range combinations, means, and weights for $P_1$ and $P_4$ can be tabulated as below:

| Range of $s_1$ | Mean | Range of $s_2$ | Mean | Weights |
|---|---|---|---|---|
| $s_{11} < .8$ | .4 | $s_{21} < .1$ | .05 | 2.0 |
| $.8 < s_{12} < 1.0$ | .9 | $s_{21} < .1$ | .05 | 1.8 |
| $1.0 < s_{13} < 1.2$ | 1.1 | $s_{21} < .1$ | .05 | 1.4 |
| $s_{11} < .8$ | .4 | $.1 < s_{22} < .3$ | .2 | 1.9 |
| $.8 < s_{12} < 1.0$ | .9 | $.1 < s_{22} < .3$ | .2 | 1.7 |
| $1.0 < s_{13} < 1.2$ | 1.1 | $.1 < s_{22} < .3$ | .2 | 1.3 |
| $s_{11} < .8$ | .4 | $.3 < s_{23} < .5$ | .4 | 1.6 |
| $.8 < s_{12} < 1.0$ | .9 | $.3 < s_{23} < .5$ | .4 | 1.4 |
| $1.0 < s_{13} < 1.2$ | 1.1 | $.3 < s_{23} < .5$ | .4 | 1.0 |
| $s_{11} < .8$ | .4 | $.5 < s_{24} < .7$ | .6 | 1.2 |
| $.8 < s_{12} < 1.0$ | .9 | $.5 < s_{24} < .7$ | .6 | 1.0 |
| $1.0 < s_{13} < 1.2$ | 1.1 | $.5 < s_{24} < .7$ | .6 | .6 |

These values provide the parameters for the first twelve rows of a new goal programming problem. For example, the first row would be written as follows:

$$(7.12) \qquad .4x_1 + .05x_2 + d_1^{1-} - d_1^{1+} = 15$$

The last two rows of the model are the same eight-hour constraints that appeared in the original model.

The computed results of this program specify that meeting our production goal of 15 units will require team 1 to work 36.5 hours and team 2 to work eight hours. This, of course, is an impossible situation. But the results do indicate that this condition was brought about by the influence of the first row in the model. This row had very low combined processing rates coupled with a high weight on the $P_1$ level. In order to satisfy this goal first, the $x_1$ value chosen had to be very high.

By contrast, when a nonstochastic goal program of this problem is solved using the expected values of $s_1$ and $s_2$ ($s_1 = 1.04$ and $s_2 = .44$), the results are: $x_1 = 11.05$ hours and $x_2 = 8$ hours. This solution appears more "reasonable" to us, of course. In considering possible causes of the difficulty, one factor stood out. The weighting factors derived for each of the partitioned production goals were actually based on the cumulative probability that $s_1$ and $s_2$ would be *greater* than the range upper bound specified. One possible alternative that might solve the problem is to base the weights on the cumulative probability of $s_1$ and $s_2$ being *less* than the range limit specified. Under this new arrangement, the weights would appear in this order: 0, .2, .6, .1, .3, .7, .4, .6, 1.0, .8, 1.0, 1.4. Under these weights, it is likely that the production goal would best be met by satisfying the last row. The solution then would be about $x_1 = 9$, $x_2 = 8$. Another alternative would be actually to partition the x's according to the s with which they were combined. This method would seem more closely to approach that of Elmaghraby. The values $x_{11}$, $x_{12}$, and $x_{13}$ in the solution would then be summed to form $x_1$, and the same procedure would apply to $x_2$.

Although this partitioning and weighting method we have just described may appear feasible with small numbers of variables, it would probably become too cumbersome to work with if the size of the problem expanded.

## B. THE VARIANCE-COVARIANCE APPROACH

We will now examine an alternative approach to goal programming under uncertainty. Let $x_i$ be the amount of scarce resource (i.e., man-hours) devoted to the production of some product. We are able to determine, from past records, the mean value of production per man-hour, $E_i$, for each of three production operations for the same product. We also can compute from these records the standard deviation and variance from these means. The productivity of each of the three operations is a stochastic variable. In addition, it is assumed that these productivities are correlated to some extent. We have only enough man-hours available in the day to produce most efficiently on fewer than three operations.

That is, if operations 1 and 2 are producing near peak capacity, operation 3 will probably not be. We will assume the correlation coefficients between the processing rates of the three operations to be as follows: $r_{12} = 0.8$, $r_{13} = 0.4$, and $r_{23} = 0.6$. Let us also assume that the processing rates are all normally distributed. Operations 2 and 3 have lower processing rates but also lower standard deviations. Operation 3 can be very efficient at times, but it is not too predictable and has a higher standard deviation. If we employed all of our available man-hours on operation 3, we might be able to meet our production quota with time to spare, but we might also fall considerably short. Employing all of our man-hours on operations 1 and 2 would give a much more predictable yield, but it might run into overtime. We would like to allocate our scarce man-hours among all three of these operations in the day so that we will be reasonably sure of meeting our production schedule with a minimum of risk.

We know from investment portfolio theory that a boundary exists in which the trade-off between desired return and risk is the most efficient possible.[11] This efficient boundary generally appears as a parabola, with risk increasing rapidly as expected return increases. The risk here is generally measured in terms of the variance. This is very similar to our production problem in which just the right mix of man-hours allocated to each operation is required to satisfy our feelings about risk. We may schedule for a certain expected production, but the associated risk may not allow us sufficient confidence in meeting our schedule. Under these circumstances, a rescheduling of production may provide us with a more satisfactory risk measurement (variance).

The formula for the variance of production is:

$$(7.13) \quad s_p^2 = x_1^2 s_1^2 + x_2^2 s_2^2 + x_3^2 s_3^2 + 2x_1 x_2 r_{12} s_1 s_2 + 2x_1 x_3 r_{13} s_1 s_3 + 2x_2 x_3 r_{23} s_2 s_3$$

The expected value of production is:

$$(7.14) \quad E_p = E_1 x_1 + E_2 x_2 + E_3 x_3$$

These two equations combine most efficiently along the parabolic boundary we have just discussed. If the first partial differential of $s_p^2$ is taken with respect to $E_p$, we will have a linear equation for the slope of the parabola at any point. A chain rule for partial differentiation must be employed and the result will be:

$$(7.15) \quad \frac{\partial \left(s_p^2\right)}{\partial E_p} = \left(\frac{2s_1^2}{E_1} + \frac{2r_{12}s_1s_2}{E_2} + \frac{2r_{13}s_1s_3}{E_3}\right) x_1$$

$$+ \left(\frac{2s_2^2}{E_2} + \frac{2r_{12}s_1s_2}{E_1} + \frac{2r_{23}s_2s_3}{E_3}\right) x_2 + \left(\frac{2s_3^2}{E_3} + \frac{2r_{13}s_1s_3}{E_1} + \frac{2r_{13}s_1s_3}{E_2}\right) x_3$$

Suppose the following data is available along with the correlation coefficients already mentioned:

$$E_1 = 1 \text{ unit/man-hour,} \qquad s_1 = .1$$
$$E_2 = 2 \qquad\qquad\qquad\qquad s_2 = .2$$
$$E_3 = 5 \qquad\qquad\qquad\qquad s_3 = 2.0$$

When the above data are substituted into the equations we have developed, a model can be set forth having two goals and one constraint equation. Assume that we have 32 man-hours available in the day. The goal program can be formulated as:

$$(7.16) \quad \text{Min } Z = P_1 d_1^- + P_2 d_3^- + P_3 d_2^+ + P_4 d_3^+$$
$$\text{s.t.} \quad x_1 + 2x_2 + 5x_3 + d_1^- \qquad\quad - d_1^+ \qquad\qquad = 80$$
$$.0004x_1 - .024x_2 + 1.2x_3 + d_2^- - d_2^+ \qquad = 0$$
$$x_1 + \quad x_2 + \quad x_3 \qquad + d_3^- - d_3^+ = 32$$

The above model can be interpreted as follows: We are seeking to obtain the most efficient trade-off between risk and expected value of production. We have set a production goal of 80 units in expected value as the most important priority goal. The second goal is to avoid the underutilization of available manpower. The third goal is to minimize the risk associated with the production schedule for 80 units. The last goal is to minimize the overtime. The first goal will tend to "push" us out toward the efficient trade-off boundary. The second goal will tend to "pull" us down toward the efficient boundary in terms of minimum risk for a given production schedule. We are constrained by the scarce resource of 32 man-hours, but we could schedule overtime if necessary. The first approximation solution to the program will provide us with enough information to compute our total standard deviation. This, in turn, used with our expected production will allow us to compute a z-value to determine the confidence interval for production being greater than a specified level. If this probability is adequate and the specified level is adequate, we need go no further. We will simply schedule production on the basis of the first program run. If this information is not satisfactory, then we will have to state a new goal for production and make a second simulation, and so on.

In this section we have explored some of the factors that affect goal programming under the condition of uncertainty. Many of the approaches to linear programming under uncertainty were considered in an effort to determine what parallels, if any, one could draw with respect to goal programming. The scope of possibilities was admittedly narrowed in order to bring into sharper focus the ability of the simplex technique to be retained for goal programming under uncertainty. At this point, it is believed that more questions have been raised than conclusions drawn. It is hoped that this study will have suggested some possible avenues for further study within this narrowed scope.

## OTHER RESEARCH AREAS

In this chapter we have explored three areas where future research is needed in goal programming. Obviously, postoptimal sensitivity analysis, parametric goal programming, and goal programming under uncertainty do not exhaust future research areas. Indeed there are numerous aspects of goal programming that require improvements and refinements. Furthermore, applications of goal programming should also be explored in various decision problems.

A very important area of future research needs is integer goal programming. In many practical problems, the decision variables are required to be integer values. For example, in many assignment problems the decision variables are people, machines, vehicles, etc. This necessitates integer solution. Perhaps some approaches being used in integer linear programming could be applied to goal programming. Or, perhaps a distinct integer goal programming may be developed. Further research in this area may include such problems as zero-one goal programming and the branch-and-bound algorithm. It appears certain that goal programming will find wide applications in integer decision problems.

Another potential area of future research is network optimization through goal programming. Network models involving PERT, CPM, and GERT may be optimized by goal programming. Undoubtedly, a great deal of further research is required in order to apply goal programming to network models, especially for stochastic networks models, i.e., GERT.

The most immediate and important research area appears to be the duality theory of goal programming. It seems that the formulation of dual goal programming may open up many new areas of future development. In some goal programming studies we have seen duality. However, these studies used the conventional linear programming package as the solution vehicle. In the process, the authors simply assigned numerical weights in place of the preemptive priority factors. In other words, in order to use a linear programming package and also to analyze the duality, they went back to what they tried to avoid. This process defeated the very purpose of applying goal programming. We are not going to discuss here again the limitations of linear programming for decision problems that involve an elusive and abstract objective function composed of multiple objective criteria. If the only way to analyze the duality of goal programming is by going back to numerical linear programming, the results of the solution will most likely not justify the trouble. For complex decision problems involving abstract multiple goals, the only way to use the numerical approach is through an arbitrary fabrication and distortion of the decision environment.[12] Therefore, the model solution may have very little, if any, value to the decision maker. As it stands now, there is a great need of future research in the duality theory of goal programming.

## REFERENCES

Charnes, A. and Cooper, W. W. "Chance Constrained Programming." *Management Science,* vol. 6, no. 2 (October 1959), pp. 73-79.

_____.*Management Models and Industrial Applications of Linear Programming.* New York: John Wiley & Sons, Inc., 1961.

_____, Niehaus, R. J., and Scholtz, D. "An Extended Goal Programming Model for Manpower Planning." Management Science Research Report No. 156. Pittsburg: Carnegie-Mellon University, Graduate School of Industrial Administration, December 1968.

Contini, Bruno. "A Stochastic Approach to Goal Programming." *Operations Research,* vol. 16, no. 3 (May-June 1968), pp. 576-86.

Dantzig, G. "Linear Programming Under Uncertainty." *Management Science,* vol. 1 (April 1955), pp. 196-207.

_____."Recent Advances in Linear Programming." *Management Science,* vol. 2 (July 1956), pp. 131-44.

_____. *Linear Programming and Extensions.* Princeton, N. J.: Princeton University Press, 1963.

Elmaghraby, S. E. A. "Programming Under Uncertainty." Ph.D. dissertation, Cornell University, 1958, pp. 115-26.

_____. "An Approach to Linear Programming Under Uncertainty." *Operations Research,* vol. 7, no. 2 (March-April 1959), pp. 208-16.

_____."Allocation Under Uncertainty When the Demand Has Continuous D.F. " *Management Science,* vol. 6, no. 8 (April 1960), pp. 208-24.

Evers, W. H. "A New Model for Stochastic Linear Programming." *Management Science,* vol. 8, no. 9 (May 1967), pp. 680-693.

Freund, R. J. "The Introduction of Risk into a Programming Model." *Econometrica,* vol. 24 (July 1956), pp. 253-63.

Hillier, F. S., and Lieberman, G. J. *Introduction to Operations Research.* San Francisco: Holden-Day, Inc., 1967.

Ijiri, Yuji, *Management Goals and Accounting for Control.* Chicago: Rand-McNally, 1965.

Madansky, A. "Methods of Solution of Linear Programs Under Uncertainty." *Operations Research,* vol. 10, no. 4 (July-August 1962), pp. 463-64.

Markowitz, Harry M. *Portfolio Selection: Efficient Diversification of Investments.* Cowles Foundation Monograph 16. New York: John Wiley & Sons, Inc., 1959.

Tintner, G. "A Note on Stochastic Linear Programming." *Econometrica,* vol. 28 (April 1960), pp. 490-95.

# PART 3

## APPLICATIONS OF GOAL PROGRAMMING

# Chapter 8

# Goal Programming for Production Planning

## INTRODUCTION

For any individual firm, it is extremely important to achieve the most efficient utilization of available resources while meeting the restrictions imposed by the environment as well as by organizational policies concerning employment, inventories, production, and the use of outside capacity. Thus, the economic significance of aggregate production planning for the firm and the nation as a whole has been well recognized.

The most difficult problem encountered in aggregate production planning is the fluctuation in demand over time. Demand fluctuations can be absorbed by adopting one of a combination of the following strategies:

1. Adjusting the size of work force through hiring or laying off employees.
2. Allowing overtime or idle time while maintaining a constant work force.

*Note:* This chapter is based on the following studies concerning goal programming application to aggregate production planning: V. Jaaskelainen, "A Goal Programming Model of Aggregate Production Planning," *Swedish Journal of Economics,* vol. 71, no. 2 (1969), pp. 14-29; S. M. Lee and V. Jaaskelainen, "Goal Programming: Management's Math Model," *Industrial Engineering,* February 1971, pp. 30-35.

3. Altering the level of subcontracting while keeping the production rate constant.
4. Adjusting the inventory level to absorb fluctuations in demand.

In actual work settings, however, aggregate production planning is further complicated by other factors. For example, variability of material costs according to the size and volatile change in inventory levels of component parts as well as finished products should be analyzed. Another problem may be the seasonality factor associated with the employee's willingness to work overtime, i.e., employees usually do not wish to work overtime during vacation seasons such as in early summer or in December. The most serious problem, however, is associated with the degree of accuracy in sales forecasts. The impact of errors in forecasting should be carefully evaluated in production planning. As presented in Figure 8.1, the aggregate production planning strategy is indeed a dynamic process that relates demand and shipment of goods.

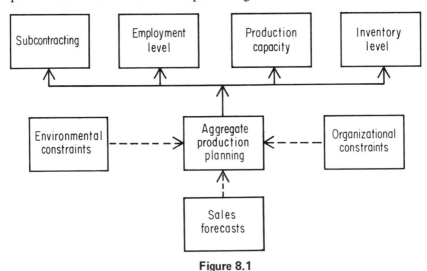

**Figure 8.1**

Management is faced with the problem of selecting the strategy that will result in the attainment of the firm's objectives. This decision problem has been the target of extensive research, and several methods for finding the optimal strategy have since been introduced. However, none of these suggested methods has found any widespread use in industry. One reason seems to be that industry is not yet ready to accept the use of formal mathematical models for aggregate production planning. But the primary reason seems to be that the proposed models are gross oversimplifications of reality, and moreover, they do not provide room to reflect management's preferences or policies in the solution.

Therefore, an effective application of such methods may be possible only at the expense of changing organizational policies. This chapter reviews limitations of existing methods and presents a goal programming model that eliminates most of these critical limitations.

## LIMITATIONS OF EXISTING METHODS

There have been three basic methods suggested as possible solutions to aggregate production planning. They are: the transportation method of linear programming, the simplex method of linear programming, and the linear decision rule model.[1]

The model utilizing the transportation method of linear programming has been criticized for its ignorance of costs associated with change in the production rate and the penalty of back ordering or lost sales. Also, it has been argued that the model does not provide a way to include a bound on the inventory level[2]. Moreover, the model implies production capacity of a single dimension. This means that all products of a multiproduct firm must first be reduced to units utilizing the single dimension of productive capacity before any analysis can be accomplished. It may be unrealistic to assume that such transformation is possible in practice.

The models using the simplex method of linear programming have been criticized because they assume a deterministic demand.[3] Also, it has been said that these models are unrealistic since they do not take subcontracting into account although it is widely used in actual operations.[4]

The linear decision rule model suffers from the same limitation as the model utilizing the transportation method, namely the necessity of unidimensional production capacity. This model also lacks constraints on work force, overtime, and inventories.[5]

Both the models utilizing the simplex method and the linear decision rule are subject to criticism for their inability to handle third-power and higher-order cost relationships. Another important weakness of these models is that they may lead to frequent changes in the work force. In these models, costs of changing the work force are assumed to be continuous although in reality they may be discontinuous.[6]

In general, it seems that little criticism has been directed at the models utilizing the simplex method of linear programming. Empirical studies in production planning have shown that a deterministic model can provide quite favorable results even under stochastic conditions.[7] Furthermore, subcontracting possibilities can easily be considered and second-degree cost curves can be approximated with piecewise linear curves.

Then why has the model not been widely accepted? A clue may be found in the experience gained by a large manufacturing firm when they applied the linear decision rule model to production planning. When the analysts returned to the firm a number of years after the implementation of the model, they found that the scheduling of production recommended by the model was not being observed. The reason was found to be that the frequent changes in employment level recommended in the model were not followed. Frequent change in employment level was conflicting with management policy, and consequently the suggested schedules of the model were ignored.

The difficulty with the general linear programming model is not so much in its inability to represent the complexities of reality. Rather, the difficulty lies in the fact that its application requires cost information that is often very hard to obtain. For example, costs of hiring and laying off work force are not easy to determine if low employee morale and public image of the organization are to be considered. It may be equally difficult to determine the correct costs of carrying inventory. The determination of the costs of warehouse operations, depreciation, taxes, etc., may not be a problem. On the other hand, it may be extremely difficult to estimate the costs for lost opportunity due to tying up capital in inventory. Finally, it is difficult to determine the actual costs of stockouts.

Since no satisfactory scientific methods are available to determine costs objectively, the procedure usually followed is to ask management to provide its "best estimate" of costs. Yet responsible managers who can provide concrete estimates of these costs are hard to find. In this situation, we can use neither the models cited above nor the heuristic and computer search methods that many management scientists have turned to because of the poor acceptance of proposed analytical methods.

If we assume, however, that management is capable of providing the analyst an ordinal measure for various costs considered in aggregate production planning, then it is possible to use an analytical framework to solve the problem without having to resort to the heuristic method. This study suggests that goal programming can provide an improved model to solve the problem of aggregate production planning.

## AGGREGATE PRODUCTION PLANNING AS A GOAL PROGRAMMING PROBLEM

Let us assume that, according to management, the costs of shortages are estimated to be higher than costs of changing the employment level, and the latter costs are higher than inventory costs. Based upon this assumption, we can set three separate goals: the levels of production, employment, and inventories. It is evident that aggregate production planning can be easily treated as a goal programming problem.

If we assume that the shortage of a given product is considered more critical than shortages of some other products, we can assign a larger weight to the deviational variable associated with this product. In other words, we can assign differential weights to each variable within the same priority level, provided that these variables are commensurable.

The same reasoning can be applied throughout the variables. If it were considered more important to avoid underemployment in some capacity groups than in others, we could give different weights to variables in the respective capacity groups in the same order level. Similarly, differential weights could be assigned to deviations from goals in the lowest-order group representing excess inventories in physical units. Now, in order to minimize the capital employed in production, we try to minimize the sum of each of the deviational variables in the objective function multiplied by the weights assigned to these variables.

Considering the complicated nature of aggregate production planning, especially when different time periods are considered, it seems wise to present a simple production problem first followed by a complex multi-time-period production planning problem.

## Example 1

Continental Electronics, Inc. produces two products: record players and tape recorders. The production of both products is done in two separate machine centers within the plant. Each record player requires two hours in machine center 1 and one hour in machine center 2. Each tape recorder, on the other hand, requires one hour in machine center 1 and three hours in machine center 2. In addition, each product requires some in-process inventory. The per-unit in-process inventory required is $50 for the record player and $30 for the tape recorder.

The firm has normal monthly operation hours of 120 for machine center 1 and 150 for machine center 2. The cost accounting department estimates that the average costs of per-hour operation are $80 and $20 for machine centers 1 and 2 respectively. The estimated profit per unit is $100 for the record player and $75 for the tape recorder. According to the marketing department, the forecasted sales for the record player and the tape recorder are 50 and 80 respectively for the coming month.

The president of the firm has established the following goals for production in the next month, in ordinal ranking of importance:

1. Limit the amount tied up in in-process inventory for the month to $4,600.
2. Achieve the sales goal of 50 record players for the month.
3. Avoid any underutilization of regular operation hours of both machine centers. (The president desires to assign weights according to the cost of idle time in each machine center.)
4. Limit the overtime operation of machine center 1 to 20 hours.
5. Achieve the sales goal of 80 tape recorders for the month.
6. Limit the sum of overtime operation for both machine centers (the president wants to assign weights according to the overtime cost of each machine center).

Let us formulate the above production planning problem as a goal programming model.

## 1. NORMAL OPERATION HOURS OF MACHINE CENTERS

The normal monthly operation capacity for machine center 1 is 120 hours, and for machine center 2 it is 150 hours.

$$(8.1) \qquad 2x_1 + x_2 + d_1^- - d_1^+ = 120$$
$$x_1 + 3x_2 + d_2^- - d_2^+ = 150$$

where $x_1$ = number of record players produced in the month
$x_2$ = number of tape recorders produced in the month.

## 2. IN-PROCESS INVENTORY

The first goal of the president is not to tie up more than $4,600 in in-process inventory for the month.

$$(8.2) \qquad 50x_1 + 30x_2 + d_3^- - d_3^+ = 4,600$$

## 3. SALES GOALS

The president desires to achieve the sales goals of 50 record players and 80 tape recorders.

$$(8.3) \qquad x_1 + d_4^- - d_4^+ = 50$$
$$x_2 + d_5^- - d_5^+ = 80$$

## 4. LIMIT ON OVERTIME OPERATION OF MACHINE CENTER 1

The president has decided that the maximum allowed overtime operation should be the 20 hours in machine center 1.

$$(8.4) \qquad d_1^+ + d_{11}^- - d_{11}^+ = 20$$

Now the complete production planning model can be formulated as below:

$$(8.5) \quad \text{Min } Z = P_1 d_3^+ + P_2 d_4^- + 4P_3 d_1^- + P_3 d_2^- + P_4 d_{11}^+ + P_5 d_5^- + 4P_6 d_1^+$$
$$+ P_6 d_2^+$$

$$
\begin{array}{lrcr}
\text{s.t.} & 2x_1 + \quad x_2 + d_1^- - d_1^+ & = & 120 \\
& x_1 + 3x_2 + d_2^- - d_2^+ & = & 150 \\
& 50x_1 + 30x_2 + d_3^- - d_3 & = & 4{,}600 \\
& x_1 \qquad\quad + d_4^- - d_4^+ & = & 50 \\
& x_2 \qquad\quad + d_5^- - d_5^+ & = & 80 \\
& d_{11}^- + d_1^+ - d_{11}^+ & = & 20
\end{array}
$$

$$x_1, x_2, d_1^-, d_2^-, d_3^-, d_4^-, d_5^-, d_{11}^-, d_1^+, d_2^+, d_3^+, d_4^+, d_5^+, d_{11}^+ \geq 0$$

In the above model, the differential weights of $P_3$ and $P_6$ levels are assigned according to the opportunity cost based on the costs of per-hour operation for the two machine centers. The solution of the problem yields the following results:

## 1. VARIABLES

$$x_1 = 50, x_2 = 40, d_1^+ = 20, d_2^+ = 20, d_3^- = 900, d_5^- = 40,$$
$$d_1^- = d_2^- = d_3^+ = d_4^- = d_4^+ = d_5^+ = d_{11}^+ = 0$$

The solution indicates that the firm produces 50 record players and 40 tape recorders with 20 hours of overtime operation in both machine centers. The total in-process inventory was $3,700, $900 short of the $4,600 limit as set by the president. The firm is not able to achieve the sales goal of 80 tape recorders. The total profit of the firm in the month is $8,000.

## 2. GOAL ATTAINMENT

With the above solution, the firm is able to achieve the following goals:

In-process inventory limit: Achieved
Sales goal for record players: Achieved
Avoiding underutilization of operation capacity: Achieved

Overtime operation limit in center 1: Achieved

Sales goal for tape recorders: Not achieved

Overtime operation limit of both centers: Not achieved

The above-stated goal attainments indicate that the firm is able to achieve the most important production-related goals during the month.

## Example 2

For a more complex aggregate production planning problems, let us first define variables, constants, and priority factors in order to design a general goal programming model. Once the general model is constructed, the application of the model will be explained through a numerical example.[8]

## A. VARIABLES

$q_i(t)$   = units of product i assembled in period t, (i=1,2,...,m, t=1,2,...,n)

$d_i^+(t)$   = closing inventory of product i at the end of period t, i.e. the positive deviation of initial inventory and final assembly from the respective demand, (t=1,2,...,n−1)

$d_i^-(t)$   = shortage of product i in period t, i.e. the number of units of product i with which the initial inventory and assembly of period t fall short of the demand in period t

$d_{i1}^+(n)$   = lost sales of product i in period n

$d_{i1}^-(n)$   = closing inventory of product i at the end of period n

$r_{ij}(t)$   = units of component j of product i produced in period t $(j=1,2,...,l_j)$

$s_{ij}(t)$   = units of component j of product i subcontracted in period t

$z_{ij}^+(t)$   = closing inventory of component j of product i at the end of period t (t=1,2,...,n−1), i.e. the positive deviation of cumulative production of component j from the amounts needed in the final assembly

$x^{k+}(t)$   = number of overtime operation hours in load center k in period t (k=1,2,...,p)

$x^{k-}(t)$   = number of unutilized hours of regular production capacity at load center k in period t, i.e. the negative deviation of the production plan from normal capacity

## B. CONSTANTS

$I_i(0)$   = initial inventory of product i

$I_i(n)$   = required closing inventory of product i at the end of period n

$H_{ij}(0)$    = initial inventory of component j of product i

$H_{ij}(n)$    = required closing inventory of component j of product i at the end of period n

$C^k(t)$    = regular time of load center k in period i

$B^k(t)$    = upper limit of overtime of load center k in period $t(B^k(t)$ is normally a given fraction of $C^k(t)$, but it need not be the same fraction in every period)

$G_{ij}(t)$    = upper limit of subcontracting possibilities of component j of product i in period t

$E(D_i(t))$ = demand forecast of product i in period t.

$a_{ij}^k$    = number of hours required in load center k to produce one unit of component j of product i

$b_i$    = gross margin of product i

$c_i$    = standard cost of product i

$c_{ij}$    = standard cost of component j of product i


## C. AGGREGATE PRODUCTION GOALS AND THEIR PRIORITIES

In addition to those variables and constants we need to define the aggregate production goals and their priorities. There can be a wide variety of goal structure for production goals based on the particular condition of the firm. Therefore, there is no one universal goal structure for aggregate production planning problems. To simplify the example, however, let us assume that the following goal structure represents the management's preemptive priority factors associated with goals in aggregate production:

$P_1$ = the preemptive priority factor assigned to the satisfaction of demand by production from final assemblies. The negative deviations of final assemblies from demands ($d_i^-(t)$ in periods t=1,2, . . . ,n−1 and $d_{i1}^+(t)$ in period n) should be minimized.

$P_2$ = the second highest priority factor assigned to the minimization of underutilization of normal production capacity at each load center ($t^{k-}(t)$).

$P_3$ = The third factor assigned to the minimization of overtime operation hours at each load center, $x^{k+}(t)$.

$P_4$ = the last priority factor assigned to the desired level of closing inventories for final assemblies ($d_i^+(t)$) and for components ($z_{ij}^+(t)$) in periods t=1,2, . . . ,n−1.

## D. THE MODEL

The general model for aggregate production planning can now be formulated. The objective is the minimization of deviations from certain goals with assigned preemptive priority factors.

### 1. Objective function

$$(8.6) \quad \text{Min } Z = P_1 \sum_{t=1}^{n-1} \sum_{i=1}^{m} b_i d_i^- (t) + P_1 \sum_{i=1}^{m} b_i d_{i1}^+ (n) + P_2 \sum_{t=1}^{n} \sum_{k=1}^{p} x^{k-}(t) +$$

$$P_3 \sum_{t=1}^{n} \sum_{k=1}^{p} x^{k+}(t) + P_4 \sum_{t=1}^{n-1} \sum_{i=1}^{m} c_i d_i^+(t) +$$

$$P_4 \sum_{t=1}^{n-1} \sum_{i=1}^{m} \sum_{j=1}^{l_i} c_{ij} z_{ij}^+ (t)$$

The objective function indicates that the most important goal of management is to achieve the maximum sales. Hence, the highest priority factor $P_1$ is assigned to the negative deviation from demands. The achievement of the maximum sales is not equally important for all products. Therefore, we assign differential weights according to the gross margin per unit of each product, $b_i$.

Secondly, management desires to avoid underutilization of normal production capacity of load centers. In other words, the second goal is to keep employment as close to the level set in long-range plans as possible. Therefore, $P_2$ is assigned to the negative deviations from normal capacities, $x^k(t)$.

The overtime operation is also to be minimized, but only after achieving full employment. The third priority $P_3$ is assigned to the part-time deviations from normal capacities, $x^{k+}(t)$. In a multiproduct firm with multidimensional production capacity, it is extremely difficult to maintain equal employment level at all load centers. In this model, a higher priority is assigned to underutilization than overtime in order to ensure full employment at load centers even if it means overtime operation may be required in some other load centers.

Since we assigned $P_2$ to $x^{k-}(t)$ variables, this will prevent the overexpansion of subcontracting. The structure of the model is such that subcontracting choices will be made in a way that will decrease the employment at load centers as little as possible.

Finally, the last goal is to minimize the capital employed in production. This goal can be achieved by limiting inventories to the lowest possible level. We assign $P_3$ to variables representing the closing inventories, $d_i^+(t)$ and $z_{ij}^+(t)$. Since some components may cost more than others, differential weights are assigned according to the standard costs, $c_i$ and $c_{ij}$.

## 2. Relationship between demand, final assemblies, and inventory of completed products

(8.7)   $I_i(t-1) + q_i(t) - d_i^+(t) + d_i^-(t) = E(D_i(t)), (t=1),$
$d_i^+(t-1) + q_i(t) - d_i^+(t) + d_i^-(t) = E(D_i(t)), (t=2,3,\ldots,n-1),$
$d_i^+(t-1) + q_i(t) \qquad + d_i^-(t) = E(D_i(t)) + I_i(t), (t=n)$
$(i=1,2,\ldots,m).$
$d_i^-(n) + d_{il}^-(n) - d_i^+(n) = I_i(n). (i=1,2,\ldots,m).$

The first equation of constraint (8.7) indicates that the initial inventory, $I_i(0)$, plus the units assembled in the first period, $Q_i(1)$, plus the shortage in the first period, $d_i^-(D)$, minus the closing inventory of the first period, $d_i^+(1)$, must equal the demand in the first period, $E(D_i(1))$. Other equations in (8.7) are interpreted in an analogous manner. For example, the second equation states that the initial inventory of the second period (which is the closing inventory of the first period), $d_i^+(1)$, plus the units assembled in the second period, $Q_i(2)$ plus the shortage in the second period, $d_i^-(2)$, minus the closing inventory of the second period, $d_i^+(2)$, must equal the demand for the second period, $E(D_i(2))$. In other words, it is assumed in the model that the sales lost in one period cannot be recovered in the next period.

The last period restriction does not allow any positive deviation from the sum of the demand in the period, $E(D_i(n))$, and the required closing inventory, $I_i(n)$. The variable $d_i^+(n)$ is omitted to ascertain that underutilization of normal capacity of load centers is not avoided simply by excessive production for inventory. The structure of the constraint implies that if the base period is the first month in a year, then there may be a large quantity of inventories for some components at the end of some months but the production for the entire year will be limited to the expected demand and the required closing inventory.

The last constraint in (8.7) divides the shortages in the final period, $d_i^-(n)$, into two components. If the shortage $d_i^-(n)$ is greater than the required closing inventory. $I_i(n)$, then $d_{i1}^+(n)$ will represent the number of units which are short in the final assemblies for the demand of the period. If the shortage is smaller than the required closing inventory, $d_{i1}^+(n)$ will indicate the number of units which the production program contributes to the closing inventory. In the present model, the highest priority is assigned to the minimization of only $d_{i1}^+(n)$. This implies that there will be a solution even when there is no closing inventory.

## 3. Production and subcontracting of components

$$(8.8) \quad H_{ij}(t-1) + r_{ij}(t) + s_{ij}(t) - z_{ij}^+(t) = q_i(t), (t=1)$$
$$z_{ij}^+(t-1) + r_{ij}(t) + s_{ij}(t) - z_{ij}^+(t) = q_i(t), (t=2,3,\ldots,n-1)$$
$$z_{ij}^+(t-1) + r_{ij}(t) + s_{ij}(t) \qquad = q_i(t) + H_{ij}(t), (t=n)$$
$$(j=1,2,\ldots,1_i, i=1,2,\ldots,m)$$

The first constraint in (8.8) indicates that the initial inventory of component $j$ of product $i$, $H_{ij}(0)$, plus the production, $r_{ij}(1)$, and the subcontracting, $s_{ij}^+(1)$, in the first period minus the closing inventory, $z_{ij}^+(1)$, must equal the number of units assembled in the first period, $Q_i(1)$. It should be apparent, for the sake of simplicity, that only one unit of a component is required for each unit assembled. In reality, several units of a component may be required. This can be easily accommodated by multiplying $Q_i(1)$ by the required number of units. Since it is impossible to have final assemblies without the necessary components, negative deviations do not have any meaning and therefore they are omitted.

The remaining equations in (8.8) should be self-explanatory. The variable $z_{ij}^+$ is omitted from the equation for the last period to prevent the excessive production for inventory in order simply to avoid unemployment.

## 4. Capacity constraints

$$(8.9) \quad \sum_{i=1}^{m} \sum_{j=1}^{1_i} a_{ij}^k r_{ij}(t) + x^{k-}(t) - x^{k+}(t) = C^k(t),$$
$$(k=1,2,\ldots,p; t=1,2,\ldots,n)$$

The above constraint stipulates that the capacity used at the kth load center, $\sum\sum a_{ij}^k r_{ij}(t)$, plus unutilized normal capacity, $x^{k-}(t)$, minus overtime operation, $x^{k+}(t)$, must equal the available normal capacity, $C^k(t)$.

## 5. Restrictions to overtime operation

(8.10)   $x^{k+}(t) \leqslant B^k(t), (k = 1,2, \ldots, p, t=1,2, \ldots, n)$

The above constraint simply specifies that the actual overtime operation at load center k in period t, $x^{k+}(t)$, must be less than or equal to the allowed upper limit, $B^k(t)$.

## 6. Subcontracting of components

(8.11)     $s_{ij}(t) \leqslant G_{ij}(t), (i=1,2, \ldots, m, j=1,2, \ldots, l_i, t=1,2, \ldots, n)$

The actual subcontracting of component j, $s_{ij}(t)$, must be less than or equal to the specified upper limit, $G_{ij}(t)$. In our model, it is assumed that every component has a subcontracting possibility. In reality, however, some components are never subcontracted, some are always subcontracted, and yet others have both possibilities. The model is capable of reflecting management policies concerning subcontracting. It is also possible that the same component can be purchased from several subcontractors within a given period. In order to facilitate the several simultaneous subcontractings, the left-hand side of the constraint (8.8) should include a variable $s_{ij}(t)$ for each subcontractor and additional restrictions for each new variable should also be introduced in (8.11). These adjustments are needed because a single subcontractor may not be able to supply the required quantity. Sometimes, several subcontractors are used simultaneously to reduce the risk of possible shortage due to unpredictable factors, such as strike, delay in delivery, etc.

## 7. Non-negative constraint

Finally, the non-negative constraint should be defined:

(8.12)    $Q_i(t), d_i^+(t), d_i^-(t), d_{i1}^+(n), d_{i1}^-(n), r_{ij}(t), s_{ij}(t), z_{ij}^+(t), x^{k+}(t),$
$x^{k-}(t) \geqslant 0$
$(i=1,2, \ldots, m; j=1,2, \ldots, l_i; k=1,2, \ldots, p; t=1,2, \ldots, n)$

## E. A NUMERICAL ILLUSTRATION

Now let us apply the model to a very simplified numerical illustration.[9] We assume that the firm under analysis is engaged in manufacturing only two products. There are only two periods of planning horizon in the example. We also simplified the model one step further by ignoring the constraints (8.8) that relate components to final assemblies because in this example we will consider only complete products. Also, the possibility of subcontracting will be excluded from the model. Therefore, we can drop variables $s_{ij}(t)$ and constraints (8.11).

We assume that there is a given initial inventory for both products and that a certain closing inventory level is set as a target that cannot be exceeded. However, the production plan may not achieve these targets if demand during the two periods under consideration is so high that it cannot be satisfied with the available production capacity. We assume that the constants of the example are given the following numerical values:

$b_1 = 1, c_1 = 1, E(D_1(1)) = 8, E(D_1(2)) = 11, b_2 = 1, c_2 = 2, E(D_2(1)) = 8,$
$E(D_2(2)) = 17, I_1(0) = 4, I_1(2) = 5, C^1(1) = 10, C^1(2) = 10, I_2(0) = 4,$
$I_2(2) = 3, C^2(1) = 10, C^2(2) = 10, B^1(1) = 2, B^1(2) = 2, a_1^1 = 1, a_1^2 = 1,$
$B^2(1) = 2, B^2(2) = 2, a_2^1 = 1/2, a_2^2 = 0$

Now, following the notation described previously we can formulate the complete model as follows:

(8.13) $\text{Min } Z = P_1 d_1^-(1) + P_1 d_{11}^+(2) + P_1 d_2^-(1) + P_1 d_{21}^+(2) + P_2 x^{1-}(1) +$
$\qquad P_2 x^{1-}(2) + P_2 x^{2-}(2) + P_3 x^{1+}(1) + P_3 x^{1+}(2) + P_3 x^{2+}(1)$
$\qquad + P_3 x^{2+}(2) + P_4 d_1^+(1)$

$\quad \text{s.t.} \quad Q_1(1) + d_1^-(1) - d_1^+(1) = E(D_1(1)) - I_1(0) = 4$
$\qquad\qquad Q_1(2) + d_1^-(2) = E(D_1(2)) + I_1(2) = 6$
$\qquad\qquad Q_2(1) + d_2^-(1) - d_2^+(1) = E(D_2(1)) - I_2(0) = 4$
$\qquad\qquad Q_2(2) + d_2^-(2) = E(D_2(2)) + I_2(2) = 14$
$\qquad\qquad d_1^-(2) + d_{11}^-(2) - d_{11}^+(2) = I_1(2) = 5$
$\qquad\qquad d_2^-(2) + d_{21}^-(2) - d_{21}^+(2) = I_2(2) = 3$
$\qquad\qquad Q_1(1) + 0.5Q_2(1) + x^1(1) - x^{1+}(1) = C^1(1) = 10$
$\qquad\qquad Q_1(1) + x^{2-}(1) - x^{2+}(1) = C^2(1) = 10$
$\qquad\qquad Q_1(2) + 0.5Q_2(2) + x^{1-}(2) - x^{1+}(2) = C^1(2) = 10$
$\qquad\qquad Q_1(2) + x^{2-}(2) - x^{2+}(2) = C^2(2) = 10$
$\qquad\qquad x^{1+}(1) + x^{11-}(1) = B^1(1) = 2$
$\qquad\qquad x^{2+}(1) + x^{21-}(1) = B^2(1) = 2$
$\qquad\qquad x^{1+}(2) + x^{11-}(2) = B^1(2) = 2$
$\qquad\qquad x^{2+}(2) + x^{21-}(2) = B^2(2) = 2$
$\qquad\qquad Q_1(1), Q_1(2), Q_2(2), Q_2(2), d_1^-(1), d_1^-(2), d_2^-(1), d_2^-(2),$
$\qquad\qquad d_{11}^-(2), d_{21}^-(2), d_1^+(1), d_2^+(1), d_{11}^+(2), d_{21}^+(2), x^{1-}(1), x^{1-}(2),$
$\qquad\qquad x^{2-}(1), x^{2-}(2), x^{11-}(2), x^{21-}(2), x^{1+}(1), x^{1+}(2), x^{2+}(1),$
$\qquad\qquad x^{2+}(2), x^{11+}(2), x^{21}(2) \geqslant 0$

The above problem is solved by the computer program presented in Chapter 6. The optimum solution presents the following results:

## 1. Variables

$Q_1(1) = 10, Q_1(2) = 3\frac{1}{2}, Q_2(1) = 4, Q_2(2) = 17$
$d_1^-(2) = 6\frac{1}{2}, d_2^-(2) = 3, d_1^+(1) = 6, d_2^+(1) = 0$
$d_{11}^+(2) = 1\frac{1}{2}, x^{2-}(2) = 6\frac{1}{2}, x^{21-}(1) = 2, x^{21-}(2) = 2$
$x^{1+}(1) = 2, x^{1+}(2) = 2$

## 2. Goal attainment

The optimum solution to the problem presents the following simplex criterion for priority factors: $P_1 = 3/2$, $P_2 = 6\ 1/2$, $P_3 = 4$, and $P_4 = 6$. This indicates that none of the goals has been completely attained. The forecasted demand is so high in relation to the existing production capacity that in the second period there is the shortage of 3/2 units of the first product. This in turn resulted in the underachievement of the closing inventory target. In fact, there is not a single unit of either product in the closing inventory. This result makes sense, however, because the production capacity is first directed toward the satisfaction of the current demand.

In analyzing the third goal, it is interesting to note that a part of the production capacity is unutilized although demand is not completely satisfied. In the second load center, 6½ units are unutilized in the second period. The first product requires the capacity of both load centers whereas the second product requires only the capacity of the first load center, which is already in full use. It would therefore seem perfectly logical to increase the production of the first product because its demand has not completely been satisfied and decrease the production of the second product. This would result in a more even utilization of the production capacity of load centers. However, this cannot be implemented because it will increase the total shortages, and in the present model the most important goal is to avoid shortages. The optimal solution is to avoid shortages and the underutilization of the normal production capacity of the second load center by scheduling overtime operation for the first load center whenever this is possible.

## CONCLUSION

In this chapter the goal programming approach is applied to production planning. It is evident that because goal programming is such a flexible technique it is superior to other techniques introduced thus far for aggregate production planning.

In the second example, a complex aggregate production planning is treated as a goal programming model. It is, of course, possible to develop the model even further in many respects. For example, in Example 2 there were upper limits of inventories only in the final planning period. It would be easy to include such restrictions for the individual components as well as the overall inventories of the final products.

It appears that aggregate production planning is an area where goal programming can be applied very effectively. The primary reason is, of course, that there are only limited human factors involved in decision analysis. Therefore, the future outcome can be forecast with a greater accuracy. Furthermore, the production process, inventories, and capital requirements are relatively easy to quantify.

## REFERENCES

Bowman, E. H. "Production Scheduling by the Transportation Method of Linear Programming." *Operations Research,* vol. 4, no. 1 (1956), pp. 100-03.

Buffa, Elwood S. *Production-Inventory Systems: Planning and Control.* Homewood, Ill.: Richard D. Irwin, 1968.

Dzielinski, B., Baker, C. and Manne, A. "Simulation Tests of Lot-Size Programming." *Management Science,* vol. 9, no. 3 (1963), pp. 229-53.

Hanssman, F., and Hess, S. W. "A Linear Programming Approach to Production and Employment Scheduling." *Management Technology,* 1960, pp. 46-51.

Holt, C., Modigliani, F., and Simon, H. A. "A Linear Decision Rule for Production and Employment Scheduling." *Management Science,* vol. 2, no. 1 (1955), pp. 1-30.

Jaaskelainen, V. "A Goal Programming Model of Aggregate Production Planning." *Swedish Journal of Economics,* vol. 71, no. 2 (1969), pp. 14-29.

Lee, S. M., and Jaaskelainen, V. "Goal Programming: Management's Math Model." *Industrial Engineering,* February 1971, pp. 30-35.

McGarrah, R. E. *Production and Logistics Management.* New York: John Wiley & Sons, Inc., 1963.

Silver, E. A. "A Tutorial on Production Smoothing and Work Force Balancing." *Operations Research,* vol. 15, no. 5 (1967), pp. 985-1010.

Taubert, W. H. "A Search Decision Rule for the Aggregate Scheduling Program." *Management Science,* vol. 14, no. 3 (1968), pp. 343-59.

# Chapter 9

## Goal Programming for Financial Decisions

This chapter presents two separate studies in order to demonstrate that the goal programming approach can be effectively applied to financial decision analysis. The first study presents a broad financial planning based on goal programming. In this study, it is assumed that management has a set of goals concerning the capital structure, dividend payout policy, and growth of earnings over time. These goals are not only incommensurable, but they are also incompatible. Through goal programming, multiple goals are sought in a multiyear planning horizon.

The second study presents a goal programming model for mutual fund portfolio selection. During the past two decades, few topics in finance have been the subject of so much controversy as those of security valuation and portfolio analysis. Although the Markowitz model is ideal in its theoretical content, its full implementation by portfolio managers presents some practical difficulties. This study presents a goal programming model that offers a simplified but also a reasonable compromise between the theoretical impeccability of Markowitz's full covariance model and the dictates of practicality. If a model can be formulated that will meet these requirements without necessitating the computation of all-efficient full-covariance portfolios, the fund manager's selection decisions may be made somewhat easier. It is hoped that this study will present at least the beginning of such a model, which can be equally applicable to growth and income-type mutual funds. For purposes of exposition and simplification of the model, the primary emphasis will be placed on common stocks as the investment medium in this study.

## OPTIMIZATION OF FINANCIAL PLANNING

### THE PROBLEM

The financial planning problem to be discussed in this study is adopted from a textbook case in order to obtain necessary data.[1] In this case problem, the company is engaged in the business of leasing railway cars to individual shippers. Management has multiple financial goals as to the capital structure, dividend payout policy, and growth of earnings over a certain planning horizon. These multiple goals are sought while meeting various short-term financial requirements.

In the case problem, the growth of earnings and increase in dividends can be achieved only through investment in new cars. Consequently, a single car represents only a small fraction of the total investment, and therefore, indivisibilities will not distort the solution. Hence, this case is particularly suitable for the application of goal programming.

It is assumed that a specific minimum dividend must be paid every year for the present management to satisfy the stockholders. This target is, therefore, set as the most important goal of the financial planning. The goal with the second highest priority is growth in earnings. The top management has established the investment and financing policies that would maintain the annual target rate of growth of 20 percent in net earnings on the common stock. Finally, the third goal is to follow the dividend payout policy of 50 percent of current earnings as closely as possible. Of course, the underachievement of this goal is allowed if it is necessary to achieve the higher-order goal of earnings growth.

The growth of earnings is possible only when sufficient investment is maintained. There are three sources of investment funds: equipment trust loans, subordinated debt, and internally generated funds. The issue of new stocks is not contemplated in the planning horizon under analysis.

A part of necessary investment funds are generated through depreciation and income tax reserves. These funds are, however, not sufficient, and borrowing is needed to supplement them. The equipment trust loans are available up to 80 percent of the value of new investment. These loans are to be paid back over the period of 15 years. There is a second restriction on the equipment trust loans. The total amount of outstanding loans cannot exceed the upper limit set as twice the sum of net worth and outstanding subordinated debt in any period. This requirement on capital structure is stipulated by the creditors. Therefore, this constraint must be observed in the model. Hence, it is necessary to retain earnings to increase the borrowing base, and this causes the incompatibility between the earnings growth and the high dividend payout.

It is finally assumed that in the fourth year of the financial planning period the firm can issue new subordinated debt. The limit of this debt is set at 43 percent of the net worth at the beginning of the fourth year less any existing subordinated debt.

## THE MODEL

To formulate the goal programming model, the following variables, constants, and preemptive priority factors are to be defined:

## A. Variables

$x_{1i}$ = sales units available in the $i$th period (since the sales revenue is estimated as a percentage of the acquisition value of the fleet, the sales units are simply the total dollar acquisition value of the existing fleet)

$x_{2i}$ = dollars invested in new fleet in the $i$th period

$x_{3i}$ = annual depreciation of the fleet in the $i$th period

$x_{4i}$ = annual interest payments to all outstanding loans in the $i$th year

$x_{5i}$ = profit before the corporate income tax in the $i$th year

$x_{6i}$ = new equipment trust loans taken in the $i$th year

$x_{7i}$ = annual minimum dividends that are to be paid with the highest priority in the $i$th year.

$x_{8i}$ = annual extra dividends that management strives to pay in the $i$th year in order to bring the payout ratio to fifty per cent of current earnings (variable $x_{8i}$ can be at a positive level only when variable $x_{7i}$ is at the level of minimum target dividends)

$x_{9i}$ = net worth in the $i$th year

$x_{10,4}$ = new subordinated debt that can be taken in the 4th year

$d_{1i}^-$ = negative deviation from the minimum dividends in the $i$th year ($b_{8i}$)

$d_{2i}^-$ = negative deviation of the earnings on the $i$th year from the growth target, which is 120 per cent of the earnings of the $(i-1)$th year

$d_{3i}^-$   = negative deviation of the sum of minimum and extra dividends from the target of fifty percent of the current earnings

## B. Constants

$a_1$   = annual sales revenue less directly variable operating costs as a percentage of the acquisition value of the car fleet

$a_2$   = annual depreciation percentage of the new car fleet

$a_3$   = percentage amount of the principal of the new subordinated loan that must be repaid annually

$a_4$   = percentage amount of the principal of the new equipment trust loans that must be repaid annually

$a_5$   = rate in interest on new equipment trust loans and on the new subordinated loan that can be taken in the 4th year

$a_6$   = percentage value of new investment up to which new equipment trust loans can be obtained

$b_i$   = acquisition value of the fleet available at the beginning of the planning horizon

$b_{2i}$   = fixed selling, general, and administrative expenses (all these expenses are assumed to be payable in cash) for the $i$th year

$b_{3i}$   = annual depreciation of the fleet that exists at the beginning of the planning horizon

$b_{4i}$   = beginning cash balance of the first year less the target closing cash balance at the end of the $i$th year and less the cumulative fixed cash outlays from the beginning of the analysis to the end of the $i$th year (the cumulative cash outlays include the fixed selling, general, and administrative cash expenses and the annual repayments of both equipment trust loans and subordinated debt that were outstanding at the beginning of the analysis)

$b_{5i}$   = interest payments to all loans that existed at the beginning of the analysis

$b_{6i}$   = twice the amount of the remaining balance of subordinated debt in the $i$th year which was originated from the beginning period of the planning horizon minus the remaining balance of equipment trust loan in the $i$th year which was also originated at the beginning of the analysis

$b_7$   = net worth at the beginning of the planning horizon

$b_{8i}$   = minimum target dividends in the $i$th year

$b_9$   = target earnings of the first year of the planning horizon (i.e., 120 per cent of the earnings of the preceding year)

$b_{10}$   = the remaining balance of the subordinated debt in the 4th year that was originated at the beginning of the analysis

## C. Constraints

### 1. SALE

$$(9.1) \quad x_{1i} - \sum_{j=1}^{i-1} x_{2j} \leq b_1 \; (i=1,2,\ldots,n)$$

Cars available for leasing in the $i$th year must be less than or equal to the fleet at the beginning of the analysis plus new additions to the fleet at periods 1, 2, ..., (i–1).

### 2. PROFIT BEFORE TAX

$$(9.2) \quad -x_{5i} - x_{4i} - x_{3i} + a_1 x_{1i} = b_{2i} \; (i=1, 2, \ldots, n)$$

The sum of the fixed selling and administrative expenses, interest, depreciation, and profit before tax must equal the amount of sales revenue minus directly variable operation costs.

### 3. DEPRECIATION

$$(9.3) \quad x_{3i} - \sum_{j=1}^{i} a_2 x_{2j} = b_{3i} \; (i=1, 2, \ldots, n)$$

Depreciation for the given year is the sum of the depreciation of old fleet and cars acquired over the horizon up to the end of the current period.

### 4. LIQUIDITY

$$(9.4) \quad \sum_{j=1}^{i} (x_{8j} + x_{7j} + 0.12x_{5j} + x_{4j} + x_{2j}) - \sum_{j=1}^{i} x_{6j} + a_4 [(i-1)x_{6,1}$$
$$+ (i-2)x_{6,2} + \ldots + x_{6,i-1}] - a_1 \sum_{j=1}^{i} x_{ij} - x_{10,4} + (i-4)a_3 x_{10,4} \leq b_{4i}$$
$$(i=4, 5, \ldots, n)$$

The cumulative cash expenditures for minimum and extra dividends, taxes (only one-fourth of the current income tax is assumed to be paid currently, hence the coefficient $0.25 \times 0.48 = 0.12$), interest, and investments less new equipment trust loans plus the repayments of these loans minus sales revenue must be less than or equal to constant $b_{4i}$. The constant $b_{4i}$ shows the cumulative effects of initial cash balance, target closing balance, fixed cash expenditures, and repayments of loans existing at the beginning of the analysis.

The constraint for years 4, 5, . . ., n also takes into account the possible new subordinated debt in the 4th year and its subsequent repayments.

## 5. INTEREST

$$(9.5) \quad -a_5 \sum_{j=1}^{i} x_{6j} + a_5 a_4 \left[ (i-1)x_{6,1} + (i-2)x_{6,2} + \ldots + x_{6,i-1} \right]$$
$$+ x_{4i} = b_{5i} \qquad (i=1,2,3,4)$$

$$-a_5 \sum_{j=1}^{i} x_{6j} + a_5 a_4 \left[ (i-1)x_{6,1} + (i-2)x_{6,2} + \ldots + x_{6,i-1} \right]$$
$$+ x_{4i} - a_5 x_{10,4} + a_5 (i-4) a_3 x_{10,4} = b_{5i}$$
$$(i=5,6, \ldots n)$$

The total interest payments will be the sum of interests on old loans, $b_{5i}$, and on loans taken up to the current period in the horizon. Equipment trust loans are assumed to be taken at the beginning of each year and the new subordinated debt at the end of the 4th year. Therefore, the first interest payment of the latter loan is assumed to take place in the 5th year.

## 6. NEW EQUIPMENT AND TRUST LOANS

$$(9.6) \quad x_{6i} \leqslant a_6 x_{2i} \ (i=1, 2, \ldots, n)$$

New equipment trust loans must be less than or equal to a fraction $a_6$ of the investment in new cars. If the loans are not utilized to the fullest amount possible in a particular year, it is assumed that the unutilized account cannot be used in subsequent years.

## 7. NEW EQUIPMENT TRUST LOAN AND CAPITAL STRUCTURE

$$(9.7) \quad -2x_{9i} + \sum_{j=1}^{i} x_{6i} - a_4 \left[ (i-1)x_{6,1} + (i-2)x_{6,2} + \ldots + x_{6,i-1} \right] \leqslant b_{6i}$$
$$(i=1, 2, 3)$$

$$-2x_{9i} + \sum_{j=1}^{i} x_{6i} - a_4 \left[ (i-1)x_{6,1} + (i-2)x_{6,2} + \ldots + x_{6,i-1} \right]$$
$$-2x_{10,4} + 2(i-4)a_3 x_{10,4} \leqslant b_{6i}$$
$$(i=4, 5, \ldots, n)$$

The total amount of equipment trust loan should not exceed the upper limit set as twice the sum of net worth and subordinated debt.

## 8. NET WORTH

$$(9.8) \quad \sum_{j=1}^{i-1} (x_{8j} + x_{7j}) + x_{9i} - 0.52 \sum_{j=1}^{i-1} x_{5j} = b_7 \ (i=1, 2, \ldots, n)$$

Net worth in the $i$th year is the sum of net worth at the beginning period plus the cumulative profits after tax to the end of the $(i-l)$th year less cumulative dividends to the end of the $(i-l)$th year.

## 9. MINIMUM DIVIDENDS

$$(9.9) \quad y_{1i}^- + x_{7i} = b_{8i} \ (i=1, 2, \ldots, n)$$

The target dividend for the year must equal the sum of minimum dividends and any negative deviation from the target.

## 10. EXTRA DIVIDENDS

$$(9.10) \quad y_{3i}^- + x_{8i} + x_{7i} = 0.5(0.52)x_{5i} \ (i=1, 2, \ldots, n)$$

Management attempts to follow the policy of paying out 50 percent of current earnings. Therefore, the sum of minimum dividend, extra dividend, and any deviation from the above goal should be equal to 50 percent of the current net profit.

## 11. GROWTH OF EARNINGS

$$(9.11) \quad y_{2,1}^- + 0.52x_{5,1} = b_9$$
$$y_{2i}^- - 1.2(0.52)x_{5,i-1} + 0.52x_{5i} = 0 \ (i=2, 3, \ldots, n)$$

The target earnings, which is set as 120 percent of the last year's earnings, must equal the sum of the actual earnings and any negative deviation from the target earnings.

## 12. NEW SUBORDINATED DEBT

$$(9.12) \quad x_{10,4} \leqslant 0.43x_{9,3} - b_{10}$$

New subordinated debt in the $4$th year can be issued up to 43 percent of the net worth in that year less any outstanding subordinated debt.

## 13. OBJECTIVE FUNCTION

$$(9.13) \quad \text{Minimize } Z = P_1 \sum_{i=1}^{n} d_{1i}^- + P_2 \sum_{i=1}^{n} d_{2i}^- + P_3 \sum_{i=1}^{n} d_{3i}^-$$

The objective of the solution is to minimize deviations from various goals. The highest priority is assigned to the goal of minimum dividends. Therefore, the highest preemptive priority factor $P_1$ is associated with the variables that represent deviation from this goal. The growth of earnings is the second goal and, therefore, $P_2$ is assigned to deviational variables from this goal. The last goal is the achievement of the 50 percent dividend payout ratio. Hence, $P_3$ is associated with negative deviational variables from this goal.

## A NUMERICAL EXAMPLE

Next, we consider a numerical example that involved the planning horizon of five years. The following constants were assumed in the model. These values were derived from the case cited above, which served as the basis of our model. However, no attempt was made to use the exact figures in the case.

$a_1$ = 14 percent, i.e., the leasing contracts are written to provide an income of 14 percent over directly variable operating costs

$a_2$ = 4 percent, i.e., the fleet is depreciated over a period of 25 years using the straight-line method

$a_3$ = 2.8 percent, i.e., the new subordinated debt is to be paid back in equal installments over a period of 35 years

$a_4$ = 6.7 percent, i.e., the new subordinated debt is to be paid back in equal installments over a period of 15 years

$a_5$ = 5 percent—this rate of interest is assumed to apply to both equipment trust loans and subordinated debt

$a_6$ = 80 percent, i.e., equipment trust loans can be obtained up to a limit that is eighty percent of the value of the new cars

$b_1$ = the acquisition value of the existing fleet, assumed to be $607 million

$b_7$ = the net worth of the firm at the beginning of the analysis, $217 million

$b_9$ = the target earnings for the first year (i.e., 120 percent of the earnings of the immediately preceding year), assumed to be $20 million.

$b_{10}$ = the remaining balance in the fourth year of the subordinated debt that exists at the beginning of the analysis, assumed to be $50 million.

| | Periods | | | | |
|---|---|---|---|---|---|
| | 1 | 2 | 3 | 4 | 5 |
| $b_{2i}$ = fixed cash expenditure ($ million) | 10 | 11 | 12 | 13 | 14 |
| $b_{3i}$ = depreciation, old fleet | 24 | 22 | 21 | 20 | 19 |
| Liquidity constraints: | | | | | |
| Beginning cash balance | 23 | 25 | 25 | 25 | 25 |
| Less: | | | | | |
| Fixed costs, operations | 10 | 11 | 12 | 13 | 14 |
| Repayments, existing loans | 18 | 18 | 18 | 18 | 17 |
| Target closing cash balance | 25 | 25 | 25 | 25 | 25 |
| Annual Deficit | −30 | −29 | −30 | −31 | −31 |
| $b_{4i}$ = cumulative deficit | −30 | −59 | −89 | −120 | −151 |
| $b_{5i}$ = interest of existing loans | 15 | 14 | 13 | 12 | 12 |
| Capital structure constraint: | | | | | |
| Beginning balance, subordinated debt | 54 | 53 | 51 | 50 | 48 |
| 2 x (subordinated debt) | 108 | 106 | 102 | 100 | 96 |
| Less: existing equipment trust loan | 246 | 231 | 214 | 198 | 181 |
| $b_{6i}$ = right-hand-side constant, capital structure constraint | −138 | −125 | −112 | − 98 | − 85 |
| $b_{8i}$ = target minimum dividends | 6 | 7 | 8 | 8 | 8 |

## Solution

The above goal programming problem was solved through a computer program. The number of variables, including slack variables and deviations from goals, was 82 and the number of constraints was 56. The result of the solution is shown below.

| | Periods | | | | |
|---|---|---|---|---|---|
| | 1 | 2 | 3 | 4 | 5 |
| Sales units $x_{1i}^*$ | 607.0 | 701.7 | 795.1 | 898.4 | 993.1 |
| Sales − variable operation costs $0.14x_{1i}^*$ | 84.9 | 98.1 | 111.3 | 125.7 | 139.0 |
| Less: | | | | | |
| Depreciation $x_{2i}^*$ | 28.1 | 30.4 | 32.7 | 35.4 | 34.3 |
| Interest $x_{4i}^*$ | 18.8 | 21.2 | 22.1 | 21.9 | 22.2 |
| Fixed operating costs $b_{2i}$ | 10.0 | 11.0 | 12.0 | 13.0 | 14.0 |
| Profit before tax $x_{5i}^*$ | 28.0 | 35.3 | 44.5 | 55.4 | 68.4 |

| | | | | | |
|---|---|---|---|---|---|
| Minimum dividends $x_{7i}^*$ | 6.0 | 7.0 | 8.0 | 8.0 | 8.0 |
| Extra dividends $x_{8i}^*$ | 0.0 | 0.0 | 0.0 | 0.0 | 10.8 |
| Negative deviation of total dividends from 50% payout $d_{3i}^-$ | 2.3 | 3.2 | 4.6 | 7.4 | 0.0 |
| Negative deviation of earnings from the growth target $d_{2i}^-$ | 5.4 | 0.0 | 0.0 | 0.0 | 0.0 |
| Investment in new cars $x_{2i}^*$ | 103.5 | 106.1 | 81.8 | 94.7 | 0.0 |
| New equipment trust loans $x_{6i}^*$ | 76.6 | 74.6 | 46.1 | 12.9 | 0.0 |
| New subordinated debt $x_{10,4}$ | | | | 19.7 | |
| Slack of equipment trust loans versus net worth | 39.2 | 0.0 | 0.0 | 27.6 | 136.8 |
| Slack of equipment trust loans versus investment in new cars | 7.1 | 11.3 | 20.3 | 46.8 | 1.0 |
| Net worth $x_{9i}^*$ | 127.0 | 135.6 | 147.0 | 162.2 | 183.0 |
| Slack of liquidity constraint i.e., excess cash | 0.0 | 0.0 | 0.0 | 0.0 | 43.1 |
| Slack of sales constraint i.e., idle cars | 0.0 | 8.7 | 21.4 | 0.0 | 0.0 |

## ANALYSIS OF THE OPTIMAL SOLUTION

### A. Goals

The optimal solution indicates that only the most important goal of paying the minimum dividends can be achieved completely. The minimum dividends recommended by the model are exactly those of the required dividends $b_{8i}$. Therefore, the negative deviational variables from this target, $d_{1i}^-$, are at the zero level.

The second goal, i.e., the growth of earnings of 20 percent, cannot be accomplished completely. The negative deviational variable, $d_{2i}^-$, at the end of the first year was 5.4. This resulted partially from the assumption made in the model that investment in the $i$th year produces revenue beginning in the $(i + 1)$ year. Since sales are at the maximum level ($x_{1,1} = 607.0$), more revenue cannot be generated in the first year. However, there are more subtle reasons for not being able to achieve this goal completely. These reasons are tied to the fact that the target earnings growth can be met in subsequent years. This will be apparent from the discussion below.

The third goal of the financial planning was to pay 50 percent of current earnings as dividends. The negative deviations of total dividends from the 50 percent payout target are positive in all of the first four years. Consequently, the variables representing extra dividends are all zeros in these years. It should be noted that there is a very special reason for the positive extra dividend in the fifth year. This resulted primarily because it is the last year of the planning horizon. Were we to extend the horizon farther in the future, funds would be required for continued growth to such an extent that extra dividends would not have been possible in the fifth year. If it is desired to control the extra dividend at the end of the planning horizon, we can set up a new goal constraint. The new goal concerning the maximum extra dividend at the end of the planning horizon may spread out the dividend payout throughout the planning horizon.

As indicated by the slack variable, which is associated with the liquidity constraint for the fifth year ($43.1 million), there is plenty of idle cash available in the fifth year. This again is the result of the investment at the zero level in the fifth year. In the present model, there would be no investment in the last year of the planning horizon because of the assumption that investment produces earnings only in subsequent years, and these years are beyond the cutoff point. Of course, these weaknesses can be avoided if a continuous planning is assumed in the model or an extra goal constraint is introduced to control such results.

## B. Other variables

New subordinated debt is taken up to the maximum ceiling in the fourth year. Partially, this is a result of the fact that interest for this loan is assumed to be payable from the following year. Therefore, this loan is better than the equipment trust loan in terms of meeting the requirement of earnings growth in the fourth year. As can be seen from the value of slack variables associated with the upper limits of new ETO-loans for the fourth year, these loans are not taken up to their maximum levels in that year. However, should the planning horizon be extended over five years, it is possible that the constraints on capital structure might force new subordinated debt to its maximum level even if interest is payable from the fourth year. This can be conjectured on the basis of the fact that the slack variables associated with the constraints relating new ETO-loans to the capital structure are at zero level in the second and third year.

It is interesting to note positive slack variables for sale constraints in the second and third year. This implies that we are actually investing more than what is necessary to maintain the required growth of earnings in these years. It appears as though, even if we reduced investment in the first year by $8.7 million (the value of cars standing idle in the second year), we could still achieve the required earnings growth in the second year. However, this is not the case. A reduction in investment in the first year would reduce depreciation and interest charges and would increase the profit in that year, granted that other things remain the same.

Since we are aiming at the growth of earnings, which is computed relative to earnings of the preceding year, it obviously is better to have a low starting point. If the profit figure in the first year is actually higher than what is scheduled by the model, higher profits are necessary to meet the goal of earnings growth in the subsequent years. Larger profits would require larger sales, and these would again require larger investments in the subsequent years. The investment requirements would grow so fast that they would not be met in the third year. As indicated by the slack values of variables associated with the constraints that relate new equipment trust loans to the capital structure, we reach the upper limit for new loans in the third year. This was inevitable under the present investment policy no matter how we reshuffle the investment in the first two years.

The model enables us to achieve the goal of 20 percent annual growth in earnings. This was accomplished by pressing the profit in the first year to a level that is large enough to allow payment of minimum dividends, yet small enough to allow us to meet the goal of earnings growth in the subsequent years. The "beauty" of the optimal solution is that the goal is achieved by investing more heavily in the first year than is necessary for the sales increase in the second year but *not* by reducing the first year's sales from the maximum level.

## CONCLUSION

We have discussed a goal programming model for financial planning when management has multiple, conflicting goals. The goal for a satisfactory capital structure was stated in the form of a constraint that must always be satisfied. The goals of maintaining specified dividend policies and a target rate of growth in earnings were treated in such a way that negative deviations from these goals were assigned specific priorities.

The set of goals considered in this study is clearly not the only one that is conceivable. In the above model, the operation was so scheduled that the profit in the first year was less than profits in the year immediately preceding the planning horizon. By so doing the starting base was kept low, and it allowed the achievement of the percentagewise growth target in earnings in later years. Of course, the model can be readily modified in such a way that the most important goal would be to avoid any shortage in the absolute amount of profit from one year to the next.

The model can be also modified to have finer distinctions between policies for increased dividends and increased growth. For example, management might be willing to pay some extra dividends when the 15 percent earnings growth has been achieved, instead of going immediately to the 50 percent payout policy. This kind of preference of management could easily be worked into the model.

Finally, it is not always necessary to treat the debt-equity ratio as a constraint to be adhered. This ratio could be a self-imposed security measure instead of a constraint strictly imposed by the creditors. If this is the case, we could formulate a capital structure that contains less than 60 percent debt. However, if the achievement of higher-order goals requires it, we let that structure deteriorate. Of course, we try to minimize this deterioration with a certain preemptive priority factor. In no case would we allow the deterioration of the debt-equity ratio to go into the danger zone, i.e., beyond 70 percent limit.

As can be expected from the discussion thus far, conventional projected income statements and balance sheets could readily be developed on the basis of the optimal solution. The greatest asset of the goal programming model is its great flexibility, which enables us to handle variations of constraints and goals without difficulty. Indeed, goal programming is an effective tool for financial planning when management has incompatible multiple goals over the planning horizon.

## OPTIMIZATION OF
## MUTUAL FUND PORTFOLIO SELECTION

### INTRODUCTION

Within the past two decades, few topics in investments analysis have been the subject of so much controversy as portfolio analysis. Widely differing opinions, models, and empirical tests have been offered by theoreticians and practitioners alike. It seems that for each different point of view, there is an opposite point of view, both propounded with equal vigor. This situation may in fact be merely indicative that in quasi-efficient markets such as our security exchanges, there is essentially a buyer for every seller—opposing viewpoints of more or less equivalent strength.

One of the theories of investment management that has received relatively little ciriticism on theoretical grounds is that of Markowitz on portfolio selection and diversification.[2] This theory will be the starting point in developing the portfolio selection models presented in this study. Unfortunately, whereas the Markowitz model is ideal in its theoretical content, its full implementation by portfolio managers presents some practical difficulties. Whereas the Markowitz model requires the input of voluminous data that is time-consuming and difficult to obtain, it does not specify any one particular portfolio. Rather, the output consists of an efficient set of many portfolios. The final decision on the choice of a particular portfolio must still be made by the portfolio managers based on their operating goals and preferences. In this case of mutual funds, the legal constraints to be met may make the attainment of a portfolio exactly on the efficient boundary a practical impossibility. Ideally, however, the fund managers should be able to select a portfolio of securities which is as close as possible to the efficient boundary while meeting the goals of the fund and the legal restrictions placed upon it.

The purpose of this study is to offer a simplified but hopefully reasonable compromise between the theoretical impeccability of the full-covariance model and the dictates of practicality. If a model can meet these requirements without the burden of computing all efficient full-convariance portfolios, the fund manager's selection decisions may be made somewhat easier.

## THEORY OF PORTFOLIO SELECTION

The basic philosophy underlying investor portfolio behavior stems from economic utility theory. Given a choice, an investor will derive more satisfaction from greater expected returns on an investment with a certain degree of risk. Conversely, given investment opportunities with the same value of expected return, an investor will prefer those investments having the least risk. This behavior is portrayed graphically in Figure 9.1.[3] Given an investor who is considering a change in his portfolio with the goal of improving his satisfaction, he will select an asset if it: (1) causes a net increase in the total expected return of the portfolio; (2) causes a net decrease in the total risk of the portfolio; of (3) causes a trade-off between change in risk and change in expected return of the portfolio that falls somewhere on or to the right of his subjective risk-return utility curve. Thus, the curve shown in Figure 9.1 represents one of the investor's risk-return indifference curves.

Unfortunately, these generalizations do not explain how the risk and return characteristics of a particular security affect the risk-return parameters of a portfolio. Up until the time that Markowitz provided his comprehensive work, the existing theoretical models failed adequately to explain the influence of diversification on the risk character of a portfolio.[4]

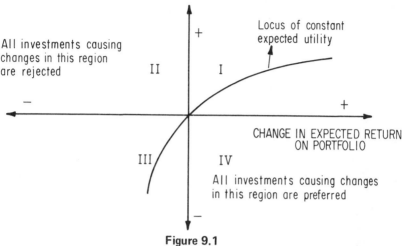

Figure 9.1

In evaluating securities for inclusion in a portfolio there are two types of information that can be utilized. First, the past performance of individual securities may be used as a basis for selection. Second, the subjective beliefs of security analysts about the future performance of a security may be used. This objective or subjective information forms the input to a portfolio analysis. The outputs of this analysis are either portfolios that have performed well in the past or portfolios that are subjectively expected to exhibit better performance in the future. The transformation method of inputs into outputs depends on the purposes for which the portfolio is to be assembled and the needs of the investor. Unfortunately, the only type of accurate information on securities is the past performance. The future behavior of security prices and dividends can be dependent upon many factors other than those that could be encompassed by the subjective evaluation of a security analyst. The international situation, the national economy, industry factors, natural phenomena, investor psychology, and probably numerous other unknown, nonquantifiable variables affect security performance in varying degrees. For these reasons, a portfolio cannot be analyzed as a closed system without some consideration of the responsiveness of securities to external factors as well as company performance.

The motivation for turning over one's funds to be invested by another is assumed to be that the other party is able to generate better results than oneself over some period of time. Certainly, in the event that the mutual fund exhibited poor performance, blaming the failure on faulty guesswork would hardly be the way to remain in business for long. In a similar sense, though, to put sole reliance and blame on a mathematical model would be equally irresponsible. As Markowitz has emphasized: "Carefully and expertly formed judgments concerning the potentialities and weaknesses of securities form the best basis upon which to analyze portfolios."[5] Since very little is known for certain about the future, and since a security's responsiveness to external factors is important, perhaps a useful aid to making expert judgments about the future behavior of a security would be some measure of how that security has reacted to overall conditions in the past. The factors of risk and return thus far mentioned may be used as representative measures of how a security has behaved with respect to both company factors and external factors in the past. These may thus serve as part of the basis for making judgments about the possible or expected future performance of a security.

The return on a security in a given time period in the past may be simply defined as follows:[6]

$$\text{Return} = \frac{(\text{Price, end of period}) - (\text{Price, beginning of period}) + (\text{dividend in period})}{(\text{Price at beginning of period})}$$

This equation does not take the tax situation of the investor into consideration. If it could be assumed that he were taxed at 40 percent of income and 25 percent of capital gains, then the price difference could be multiplied by .75 and the dividends by .60 to reflect a true value of yield. However, these complications will not be considered in the model to be developed since we shall be dealing with a mutual fund and not an individual. If we wished to derive a measure of the performance of a security over several periods of time using consecutive R-values, there could be two approaches taken. First, a simple arithmetic mean of the values could be selected. However, this would not be truly representative of performance as such—particularly when the measurement of growth is desired. Suppose we wished to measure the performance of a non-dividend-paying security over two time periods. The prices at the beginning and end of the first period are 100 and 200 respectively. The price at the end of the second period is 100. The return in the first period is 1.00, but the implied growth is 2.00 or 100 percent. Similarly, the return in the second period is −0.50, but the implied growth is .50 or 50 percent. Obviously, the security has not realized any net growth at the end of the second time period, but the average of the growth multiples is 1.25. This is, of course, an inaccurate measure. If a true measure of the compound growth rate is desired, the geometric mean of the growth multiples should be used.[7] For the above example, it will be $\sqrt{(e_1)(e_2)} = 1.00$, where $e = R+1$. This approach assumes that all returns are reinvested, and, in the case of a growth portfolio, it is a valid assumption. Over a span of n periods, the geometric mean growth rate of a security i may be given by $E_i = \sqrt[n]{(e_1)(e_2)\dots(e_n)} - 1.00$. If, in each of the n periods, the difference between the actual rate of return and the geometric mean is very small or zero, we may view the geometric mean as the expected rate of return for that security. The square root of the sum of the squares of these differences is the standard deviation about the geometric mean. The sum of the squares alone is termed the variance, and it is used in security analysis as a measure of the risk, or variability of return.[8] Obviously, if this variance is very low, the security will have low risk. The higher this measure of risk, the more probable it becomes that the actual return will vary from the mean, or expected return.

In the case of a portfolio of n securities, the expected return will be the weighted sum of the expected returns of each security i as follows:[9]

$$(9.14) \quad E_p = \sum_{i=1}^{n} x_i E_i$$

where $x_i$ is the proportion of the portfolio invested in each security.

Whereas this parameter of a portfolio is relatively uncomplicated, the matter of portfolio variance is not such a simple matter. Given two securities with identical expected returns and identical variances, if placed together in a portfolio, the resulting portfolio variance may differ greatly from their individual variances. The critical factor here is the correlation between the movements of the two security prices. If the two prices tend to move closely together, then the variance of the portfolio may be high. If, on the other hand, the prices have tended to move in opposite directions, the portfolio variance may be lower. Thus, diversification in the sense of selecting securities whose variablities of return are highly uncorrelated should result in a portfolio whose variance is lower than that of any of its component securities. The extent to which diversification will lower portfolio variance depends upon the covariance between each pair of securities in the portfolio. The covariance between two securities, 1 and 2, is calculated as follows:[10]

$$(9.15) \quad Cov_{12} = r_{12} s_1 s_2$$

where $r_{12}$ is the correlation coefficient between the returns of the two securities, and $s_1$ and $s_2$ are their respective standard deviations. For the two asset portfolio, the portfolio variance is given by the following:

$$(9.16) \quad s_p^2 = x_1^2 s_1^2 + x_2^2 s_2^2 + 2x_1 x_2 Cov_{12}$$

Obviously, the more negatively correlated the individual returns (i.e., the closer

$r_{12}$ is to $-1$), the more the portfolio variance will be reduced. The general formula for the variance of a portfolio of n securities is:[11]

$$(9.17) \quad s_p^2 = \sum_{i=1}^{n} x_i^2 s_i^2 + \sum_{i=1}^{n} \sum_{j=1}^{n} x_i x_j Cov_{ij}, \, i \neq j$$

If the values of $E_p$ and $s_p^2$ are calculated using all possible combinations and values of $x_i$ through $x_n$, and if one is very patient and has access to a computer, a two-dimensional boundary may be mapped representing all available portfolios. This set of portfolios may look something like the bounded space in Figure 9.2. Within this space lie the $E_p$ and $s_p^2$ values for every possible portfolio. But not all of these portfolios will provide the most investor satisfaction. The portfolios that will accomplish this most efficiently will lie as far to the right as possible for any value of $s_p^2$ and as far down as possible for any value of $E_p$. Thus, the efficient set is found along the parabolic curve to the right. Any other portfolio is inefficient because it could be replaced by one with either higher return or lower risk. In order to determine the exact composition of any efficient portfolio, a quadratic programming problem must be solved maximizing the function $Z = \lambda E_p - s_p^2$ where $\lambda$ moves from zero to infinity, subject to the constraint that the objective function does not depart from the efficient boundary.[12] It follows that an investor whose objective is to maximize expected return or growth at the sacrifice of safety will prefer a portfolio far up on the efficient boundary. Conversely, an investor seeking stability of return at the sacrifice of growth will be satisfied with a portfolio lower on the curve. This same rationale applies to mutual funds and has been borne out empirically by Farrar.[13] His findings for both high-risk and low-risk mutual funds may be seen in Figure 9.3.

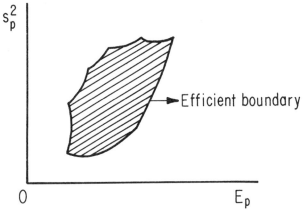

Figure 9.2

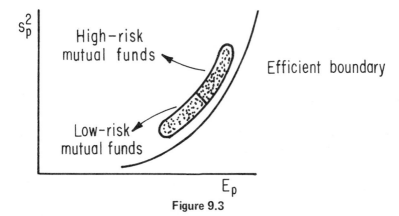

**Figure 9.3**

Although the Markowitz model provides an excellent criterion for portfolio selection, it is complicated and difficult for the practitioner to apply. The data requirements for selecting large portfolios are enormous. In addition, before the input data can be computed, a tremendous volume of initial data processing is required. Finally, the portfolio manager must still make a judgment as to which of many portfolios to select.

In an effort to lighten some of the computational load involved in the Markowitz full-covariance analysis, Sharpe developed a model based on a single market index.[14] This index could be the gross national product or some stock market index. The reasoning behind the use of such an index is that security prices seem to have a general tendency to rise and fall at the same time with some random variability among them. This being the case, then, much of the covariance between securities might be explained by their relationship to some common index, thereby eliminating the need to compute all of the covariances. This approximation could not be expected to yield the same results exactly as the full-covariance model, but any loss in accuracy would be offset by a considerable cost saving. In a subsequent paper, Sharpe applied this approach to the problem of mutual fund portfolio selection and developed some further approximations.[15] Given a mutual fund portfolio, $\Sigma x_i = 1$ assuming that the portfolio consists entirely of securities. But according to the Investment Company Act of 1940, the legal maximum that may be invested in any one security is 5 percent of the fund's assets. Thus, $x_i \leq 1/20$ for all i. Applying the single index relationship, the predicted return on a security i is given by:

$$(9.18) \quad R_i = E_i + B_i [I - E(I)] + C_i$$

where $E_i$ is the expected rate of return for the security, I is the actual level of the index at some specified future date, $E(I)$ is the expected value of the index, $B_i$ is a measure of the security's responsiveness to changes in the index (i.e., the regression coefficient of index vs. security price), and $C_i$ is a random variable with a mean of zero and standard deviation of $s_i$ . The random variables for all pairs of securities are assumed to have zero covariance, and they are assumed to be uncorrelated with I. In other words, this particular component of security price behavior is completely random, and all securities share some common relationship with I. The response of the portfolio to changes in I is a weighted average sum of all the component $B_i$ values:

$$(9.19) \quad B_p = \Sigma x_i B_i$$

The variance of the portfolio under these assumptions is composed of the term related to the variance of the market index plus the weighted average sum of the random variances of the component securities:

$$(9.20) \quad s_p^2 = B_p^2 s_I^2 + \Sigma x_i^2 s_i^2$$

where $s_I$ is the standard deviation of the market index. But Sharpe has demonstrated that given the diversification constraint placed on mutual fund portfolios, as the number of securities increases beyond 20, the closer to an equality becomes the approximation, $s_p \approx | B_p s_I |$. He finally concludes that $B_p$ may be used as a surrogate for $s_p$ or $s_p^2$, and that the efficient set of portfolios may be plotted in the $B_p$, $E_p$ plane. Thus, we now have a linear system of equations that may be used to approximate Markowitz's efficient set for mutual funds with considerably less complexity than that of the full-covariance model. The objective function to be maximized in the Sharpe model is $Z = (1 - \lambda) E_p - \lambda B_p$. Solving for values of $\lambda$ between zero and one defines an efficient boundary made up of linear segments rather than a paraboia. In an empirical test of 30 portfolios composed of up to 63 securities selected using both the full-covariance method and the linear approximation, Sharpe found that for any given average $E_p$, the approximated portfolio had an $s_p$ less than 1 percent greater than the full-covariance portfolio. However, for portfolios with lower variance, the difference between the two models was greater because the linear approximation gives an imperfect measure of individual security variance.

Whereas the linear approximation brings a powerful analytical model closer to practical implementation by the fund manager, he must still go through the derivation and evaluation of numerous efficient portfolios before he can choose the portfolio that meets his objectives. It is proposed in this study that the goal programming extension of linear programming will provide him with a much more versatile and faster tool for selecting one approximately efficient portfolio based on his goals and their priorities.

## THE MODEL

The previous discussion of a linear approximation model for selecting efficient mutual fund portfolios has described two linear equations that may serve as primary goals. The first of these is the equation for $E_p$, the expected return of the portfolio. This is a weighted average sum of all $E_i$'s where $E_i$ may be the geometric mean rate of return of a security over some reasonable number of years, and the $x_i$'s are unknowns to be solved for. The second goal is the surrogate equation for variance, the $B_p$ or market volatility of the portfolio. This is a weighted average sum of all $B_i$'s that are slopes of the regression lines of security prices vs. a general market index. If there were no other constraints on the portfolio and maximum expected return was desired with no consideration of volatility, then we could invest 100 percent of assets in the security with the highest expected return. This procedure was used to establish an upper bound for the first goal. This value also falls on the apex of the efficient boundary.

For the second major goal, if we desired the minimum possible volatility, we could invest 100 percent of assets in the security exhibiting the lowest $B_i$ value. This value serves as a lower bound for the goal as well as the lowest point on the efficient boundary. Both of these goals thus far serve to establish the end points of the efficient boundary. Obviously, it will be impossible to achieve both goals, and the portfolio chosen would have to fall somewhere in between the two end points. But if a fund with a strict growth objective were using the model, the highest priority would be placed on the achievement of the maximum possible rate of return, rather than on the minimum volatility. Conversely, a fund with an income objective would place the highest priority on achieving minimum volatility.

As we have stated previously, however, the linear approximation model gives an imperfect measure of individual security variance; and it is assumed that one of the objectives of an income portfolio would be to maintain reasonable stability of principal while seeking high current income. If we were completely proper about solving this problem, we would have to put back in the model the term representing non-market-related variance for each security. This in turn would reestablish the model as a quadratic programming problem with full covariances, obviously putting us back where we started. Therefore, a reasonable substitute will have to be put into the model to help us compensate for having the income fund suffer some unaccounted-for risk. Since one of the major goals of this type of fund is assumed to be high current income, we shall include an equation similar to the other two that will establish a goal for the highest current dividend yield obtainable. This would be given by 100 percent investment in that stock with the highest current yield.

It was assumed for this study that the models were used to select portfolios at the end of 1968, which would be held through 1969 in order to compare actual versus expected performance. Thus, the current dividend yields used as inputs were computed on the basis of 1968 dividends and year-end 1968 prices. The expected rates of return were computed based on prices over a 10-year period from 1959 through 1968 where the prices were adjusted for stock splits and stock dividends, and the cash dividends paid in each year. The $B_i$ data was obtained from the work of Treynor.[16] It was based on periods of up to 40 years where applicable; however, the periods of study for each security are not known. These data were computed as slopes of regression lines of security price relatives versus market price relatives where the price relative equals the price at the beginning of each month. It should be emphasized here that the limited availability of such data had a significant bearing on the securities chosen for potential inclusion in the growth and income portfolios.

In addition to the three major goals already mentioned, three more series of equations were included representing physical and legal constraints. The physical constraint was that the total composition of the portfolio equaled 100 percent of assets. Included in this equation were two variables in addition to the variables representing the common stocks. The first of these was a variable representing the proportion of cash and equivalent liquidity to be set aside for redemption, etc. The second variable was the proportion of the fund allocated to bonds and preferred stocks. The next set of constraints was for the legal limitation of 5 percent mentioned in discussing the Sharpe model. One final set of constraints represented the maximum allowable concentration of the portfolio in any industry. This was set at 25 percent.

We have mentioned that the goal programming model would select portfolios somewhere in between the two extremities of the efficient boundary. The closeness of this point for each fund to the efficient boundary is dependent upon a trade-off between the effects of the number of securities in the portfolio and the constraints placed upon the portfolio. Increasing the number of securities will have a beneficial effect. The constraints have a detrimental effect. For example, Figure 9.4 represents both the efficient boundary and the goal programming solution for a portfolio of three securities. Without any constraints on concentration, the solutions of both the income model and the growth model

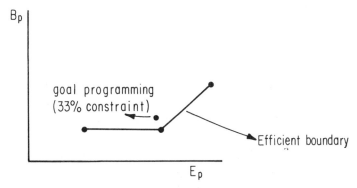

**Figure 9.4**

fall on the efficient boundary, at their respective extremities. However, as soon as a concentration constraint of 1/3 or 33 percent is placed on the portfolio, the solution no longer falls on the efficient boundary. The solution shown in this case is for the growth model. Of course, three securities is too small a portfolio to make Sharpe's linear assumption valid. But the example does show how the relaxing of constraints or the increase in securities allows the goal program more closely to attain an efficient portfolio.

A goal programming model to describe the above three-security analysis may be described by the following linear system of equations:

$$(9.21) \quad E_1 x_1 + E_2 x_2 + E_3 x_3 + d_1^- = E_i \text{ max.}$$
$$B_1 x_1 + B_2 x_2 + B_3 x_3 + d_2^- - d_2^+ = B_i \text{ min.}$$
$$x_1 + x_2 + x_3 + S_1 = 1.00$$
$$x_1 \qquad\qquad + d_3^- = .33$$
$$x_2 \qquad + d_4^- = .33$$
$$x_3 + d_5^- = .33$$
$$x_1, x_2, x_3, d_1^-, d_2^-, d_3^-, d_4^-, d_5^-, d_2^+, S_1 \geqslant 0$$

where $x_i$ represents the proportion of the fund's assets invested in security i, $E_i$ is the geometric mean rate of return, $B_i$ is the market volatility factor, and $d_i$'s represent the deviations from the goals specified on the right-hand side, and $S_1$ is an artificial slack variable that indicates that no deviation will be allowed. The first equation in the system represents the goal of achieving the highest possible expected return for the portfolio. The second equation is the goal for achieving the lowest possible market volatility for the portfolio. Obviously these two goals are incommensurate and cannot both be attained. Because of this, fund management may resolve this conflict by placing a higher priority on the achievement of one goal than on the other. If the main objective of the fund is high capital growth with little regard for volatility, then the highest priority may be placed on the attainment of the first goal with a low priority assigned to the second goal.

In the present case, $d_1^-$ is the negative deviation or underachievement of the growth goal. In assigning the highest priority to this deviation, we are reflecting the objectives of the fund in the model. But we cannot ignore the other equations in the system, which act as constraints. The third equation in the system represents a budget constraint, i.e., the whole can be no larger than the sum available. The remaining three equations we shall term legal constraints on concentration of the fund's assets in any one security. In this simple model, the fund can have no more than 33 percent of its portfolio invested in any one security. However, in the case of an actual mutual fund, this constraint is set by law at no more than 5 percent. The priorities to be assigned to these four constraints may be viewed in two ways. First, considering the legal nature of three of these constraints, one would be tempted to assign the highest priority to achieving these. But we must be concerned with the priority of minimizing the deviations in each of these equations. If meeting the legal constraints receives the highest priority, we might end up selecting securities which do not belong in an efficient portfolio. This would have the effect of "locking up" the proper functioning of the model in sorting out efficient portfolios from inefficient ones. What is just as important to observe is that as the equations are written, the $x_i$ values can never be any greater than the legal limit, regardless of the priority level assigned. Therefore, if the lowest priority level is assigned to these constraints, the model will not be forced to include securities in the portfolio that would make it inefficient. The budget constraint, on the other hand, must have the highest priority assigned to it since the artificial slack variable must be reduced to zero. This priority should be even higher than that assigned to the most desirable goal. Thus, in the three-security model we have just described, we will have four priority levels. The highest level will be assigned to $S_1$, the next level to $d_1^-$, the next level to $d_2^+$, and the lowest level to $d_3^-$, $y_4^-$ and $d_5^-$. The objective function for this particular model may then be written as:

$$(9.22)\quad \text{Min } Z = P_1 S_1 + P_2 d_1^- + P_3 d_2^+ + P_4(d_3^- + d_4^- + d_5^-)$$

The goal programming portfolio selection model developed here is similar to the simplified example we have just described except for two additions. The first addition is an equation representing the goal of maximum current dividend yield described earlier. The second is the series of constraints limiting the protfolio concentration in any one industry to 25 percent, also described earlier. A detailed algebraic description of the model is given in Table 9.1. The only

Table 9.1

The Goal Programming Model for Portfolio Selection

Objective Functions

Growth Portfolio:  $\text{Min } Z = P_1 d_1^- + P_2 d_2^+ + P_3 d_3^- + P_4 \sum_{i=5}^{69} d_i^-$

Income Portfolio:  $\text{Min } Z = P_1 d_3^- + P_2 d_2^+ + P_3 d_1^- + P_4 \sum_{i=5}^{69} d_i^-$

Constraints and Goals

$\sum_{i=1}^{53} E_i x_i + d_1^- = 1.391$ (the goal of maximum expected growth)

$\sum_{i=1}^{53} B_i x_i + d_2^- - d_2^+ = 0.403$ (the goal of minimum market volatility)

$\sum_{i=1}^{53} D_i x_i + d_3^- = 0.048$ (the goal of maximum current income)

$\sum_{i=1}^{55} x_i + S_1 = 1.00$ (the budget constraint)

$x_i + d_{i+4}^- = 0.05 \quad 1 \leqslant i \leqslant 53$ (up to 5% legal constraint)

$\qquad\qquad\qquad 54 \leqslant i \leqslant 55$ (up to 5% in cash and bonds and preferred)

$\sum_{i=1}^{4} x_i + d_{60}^- = 0.25$ (diversification constraint, Automation & Business Services Industry, 25% limit)

$\sum_{i=5}^{12} x_i + d_{61}^- = 0.25$ (Electronics & Instrument Industry)

$\sum_{i=13}^{20} x_i + d_{62}^- = 0.25$ (Cosmetics & Drugs Industry)

$\sum_{i=21}^{26} x_i + d_{63}^- = 0.25$ (Air Transport Industry)

$\sum_{i=27}^{31} x_i + d_{64}^- = 0.25$ (Industrials)

$$\sum_{i=32}^{34} x_i + d_{65}^- = 0.25 \text{ (Chemicals)}$$

$$\sum_{i=35}^{37} x_i + d_{66}^- = 0.25 \text{ (Aluminum)}$$

$$\sum_{i=38}^{43} x_i + d_{67}^- = 0.25 \text{ (Oils)}$$

$$\sum_{i=44}^{46} x_i + d_{68}^- = 0.25 \text{ (Financial)}$$

$$\sum_{i=47}^{53} x_i + d_{69}^- = 0.25 \text{ (Utilities)}$$

$$x_i, d_i^-, d_i^+ \geqslant 0$$

difference between using the model for a growth fund and using it for an income fund is in the assignment of the four priority levels. We only deal with real priority levels here. The objective function of the growth model places the highest priority on the expected return goal, the next priority on the market volatility goal, the next priority on the current dividend yield goal, and the lowest priority on the concentration and diversification constraints. With the simple exchange of priority levels between the dividend yield goal and the growth goal, we can establish the objective function for the income model.

The input data for these models was derived for 53 potential securities classified by industry. These data are listed in Table 9.2. The values for $E_i$ were

Table 9.2

Input Data—Security by Industry Group*

| Industry Group | Security # $i$ | 1959 1968 $E_i$** | $B_i$*** | 1968 $D_i$**** |
|---|---|---|---|---|
| A.   Automation & | 1 | 1.095 | 1.072 | .004 |
| Business Services | 2 | 1.391 | 1.500 | 0 |
| | 3 | 1.311 | .757 | .008 |
| | 4 | .881 | 1.352 | .006 |

| | | | | | |
|---|---|---|---|---|---|
| B. | Electronics & | 5 | 1.217 | 1.378 | 0 |
| | Instrumentation | 6 | 1.014 | 1.296 | .015 |
| | | 7 | .930 | 1.262 | .002 |
| | | 8 | .923 | 1.575 | .008 |
| | | 9 | .953 | 1.717 | .006 |
| | | 10 | 1.188 | 1.496 | .014 |
| | | 11 | .124 | 1.306 | .029 |
| | | 12 | .083 | 1.290 | .009 |
| C. | Cosmetics & Drugs | 13 | .316 | 1.589 | .013 |
| | | 14 | .092 | 1.133 | .019 |
| | | 15 | .064 | 1.092 | .017 |
| | | 16 | .201 | 1.022 | .023 |
| | | 17 | .150 | .851 | .022 |
| | | 18 | .154 | .850 | .020 |
| | | 19 | .161 | .819 | .023 |
| | | 20 | .146 | .593 | .014 |
| D. | Air Transport | 21 | .296 | 1.687 | .009 |
| | | 22 | .203 | 1.517 | .014 |
| | | 23 | .243 | 1.378 | .008 |
| | | 24 | .391 | 1.302 | .011 |
| | | 25 | .207 | 1.277 | .024 |
| | | 26 | .103 | 1.524 | .023 |
| E. | Industrials | 27 | .257 | 2.088 | .032 |
| | | 28 | .206 | 1.303 | .014 |
| | | 29 | .120 | 1.157 | .004 |
| | | 30 | .143 | .985 | .028 |
| | | 31 | .093 | .936 | .035 |
| F. | Chemicals | 32 | .199 | .945 | .011 |
| | | 33 | .017 | .738 | .033 |
| | | 34 | .077 | .649 | .032 |
| G. | Aluminum | 35 | −.004 | 1.394 | .021 |
| | | 36 | .017 | 1.191 | .026 |
| | | 37 | −.006 | 1.011 | .025 |

| | | | | | |
|---|---|---|---|---|---|
| H. | Oils | 38 | .120 | .727 | .035 |
| | | 39 | .302 | 1.098 | .014 |
| | | 40 | .478 | 1.114 | 0 |
| | | 41 | .208 | .850 | .022 |
| | | 42 | .151 | .752 | .026 |
| | | 43 | .077 | .587 | .038 |
| I. | Financial | 44 | .414 | 1.506 | 0 |
| | | 45 | .439 | 1.504 | 0 |
| | | 46 | .119 | .565 | .033 |
| J. | Utilities | 47 | .076 | 1.019 | .041 |
| | | 48 | .058 | .888 | .028 |
| | | 49 | .101 | .610 | .035 |
| | | 50 | .070 | .607 | .039 |
| | | 51 | .095 | .549 | .034 |
| | | 52 | .066 | .451 | .048 |
| | | 53 | .078 | .403 | .045 |

[*]Price and dividend data used in the calculation of $E_i$ and $D_i$ were obtained from charts adjusted for stock splits and stock dividends published by Securities Research Company. $B_i$ values were obtained from Treynor *et al*. The names of the firms are not listed because this study is intended only to demonstrate the applicability of goal programming to portfolio selection rather than analyzing the portfolio components.

[**]$E_i = 10 \sqrt{(e_{i1})(e_{i2})(e_{i3}) \ldots (e_{i10})} - 1.00$, where

$$e_i = \frac{(\text{year-ending price}) - (\text{year-beginning price}) + (\text{cash dividend for year})}{(\text{year-beginning price})}$$

[***]$B_i$ represents the slope of the regression line of a security price relative vs. market price relatives.

$$\text{Price relative} = \frac{\text{value at end of month} + \text{any intervening dividends}}{\text{value at beginning of month}}$$

[****]$D_i = \frac{\text{dividend paid in cash}}{\text{year-ending price}}$

computed by a separate computer program for a period of ten years from 1959 through 1968. The closing prices for each of these periods were read from charts of monthly prices adjusted back for any stock dividends or stock splits.[17] The dividend data were also obtained from these charts. The values for $B_i$ were taken from the Treynor study. Although it would have been desirable to include more securities in many of the industries, being dependent upon the $B_i$ data that was available, it was not possible.

## RESULTS AND CONCLUSIONS

The portfolio selected by the goal programming models is exhibited in Table 9.3 along with the actual rates of return and dividend yields based on January 1969 closing prices.[18] As may have been predicted from the discussion of the models, the securities selected were "in" to their legal limit of 5 percent. With a diversified list of securities, a security selected at less than 5 percent could probably be replaced by a security having a higher contribution to the goals of the model. Therefore, there would be no reason for having anything less than 5 percent invested in a security. Sharpe also observed this phenomenon in his linear mutual fund model.[19]

Table 9.3

Computed Portfolios and Comparative Data

| The Growth Portfolio | Security # $i$ | $x_i$ | 1969 $e_i-1$ | 1969 $d_i$ |
|---|---|---|---|---|
| Industry Group | | | | |
| Automation & | 1 | .050 | .35 | .005 |
| Business Services | 2 | .050 | −.18 | 0 |
| | 3 | .050 | .21 | .014 |
| | 4 | .050 | .20 | .007 |
| Electronics & | 5 | .050 | −.15 | 0 |
| Instrumentation | 6 | .050 | .11 | .018 |
| | 7 | .050 | .27 | .002 |
| | 9 | .050 | .20 | .006 |
| | 10 | .050 | −.18 | .015 |
| Cosmetics & Drugs | 13 | .050 | .34 | .014 |

| Air Transport | 21 | .050 | −.34 | .010 |
|---|---|---|---|---|
| | 23 | .050 | −.48 | .009 |
| | 24 | .050 | −.18 | .011 |
| | 25 | .050 | −.56 | .012 |
| | | | | |
| Industrials | 27 | .050 | −.41 | .030 |
| | | | | |
| Oils | 39 | .050 | −.33 | .015 |
| | 40 | .050 | .23 | 0 |
| | 41 | .050 | −.31 | .028 |
| | | | | |
| Financial | 44 | .050 | −.42 | 0 |
| | 45 | .050 | .03 | 0 |

$E_p = 0.677$       $D_p = 0.009$       $d_p = 0.0098$

$B_p = 1.361$       $e_p = -0.08$

Where   $E_p$ = expected portfolio growth ($\Sigma E_i x_i$)
  $e_p$ = actual 1969 portfolio yield ($\Sigma(e_i - 1)x_i$)
  $D_p$ = 1968 year-ending portfolio dividend yield ($\Sigma D_i x_i$)
  $d_p$ = actual 1969 portfolio dividend yield ($\Sigma d_i x_i$, where

$$d_i = \frac{1969 \text{ cash dividends}}{\text{closing price, Jan. 1969}})$$

  $B_p$ = the market volatility of the portfolio ($\Sigma B_1 x_i$)

| *The Income Portfolio* | Security #<br>$i$ | $x_i$ | 1969<br>$e_i - 1$ | 1969<br>$d_i$ |
|---|---|---|---|---|
| Industry Group | | | | |
| Electronics & | | | | |
| Instrumentation | 11 | .050 | −.23 | .029 |
| | | | | |
| Cosmetics & Drugs | 16 | .050 | .16 | .023 |
| | | | | |
| Air Transport | 25 | .050 | −.56 | .012 |
| | 26 | .050 | −.49 | .011 |

| | | | | |
|---|---|---|---|---|
| Industrial | 27 | .050 | $-.41$ | .030 |
| | 30 | .050 | $-.06$ | .026 |
| | 31 | .050 | $-.33$ | .034 |
| Chemicals | 33 | .050 | $-.27$ | .034 |
| | 34 | .050 | $-.29$ | .034 |
| Aluminum | 36 | .050 | $-.04$ | .025 |
| | 37 | .050 | $-.08$ | .023 |
| Oils | 38 | .050 | $-.24$ | .035 |
| | 42 | .050 | $-.30$ | .030 |
| | 43 | .050 | $-.18$ | .040 |
| Financial | 46 | .050 | 0 | .035 |
| Utilities | 47 | .050 | $-.05$ | .043 |
| | 49 | .050 | $-.11$ | .036 |
| | 50 | .050 | $-.01$ | .042 |
| | 52 | .050 | $-.10$ | .053 |
| | 53 | .050 | $-.04$ | .048 |

$E_p = .105$    $D_p = .033$    $d_p = 0.032$

$B_p = .922$    $e_p = 0.181$

Whereas the models performed well in choosing the desired $E_p B_p$ efficient portfolios within the existing constraints, the actual performances as measured by $e_p$ for 1969 were poor. There may be several reasons given for this unfortunate result. In the first place, anyone who is even remotely aware of 1969 market conditions would admit that they were not good at all. In fact, measuring this model's performance on the basis of 1969 results might be described as an "acid test." However, the models may not have performed as poorly relative to the total market as it would appear. The $B_p = 1.361$ for the growth model indicates that the asset value for the portfolio can be expected to move 1.361 times the amount that the general market moves. Thus, if some market average dropped by 1, our portfolio asset value may be expected to drop by approximately 1.36. In 1969, the Dow-Jones 30 Industrials dropped 15.2 percent and mutual funds on the average lost 14.5 percent.[20] However, our growth portfolio lost only 10.5 percent. Granted, this is not amazingly better performance than the average, but it provides a stimulus for further exploration. Of course, our model does not take into consideration distributions and expenses paid out, and this could account for the difference. For the income model, the drop in return was even worse at 17.7 percent. This is certainly not meeting a goal of stability of principle; however, it can be expected to be more closely related to the 30 Industrials on the basis of the type of securities held. Nevertheless, this weakness in the model warrants improvement.

Another difficulty of the model is that holding a portfolio dormant for one year is unrealistic. Managerial judgment would certainly have called for replacing those securities in the portfolios that exhibited poor performance with more defensive issues before substantial losses were suffered. In addition, with the prospect of a long bear market ahead, management could place higher priorities and weights on holding cash or investing in defensive issues. It is also assumed that fund management would be able to place more weight on currently attractive issues or industries that would tend to offset the vicissitudes of the market. Although the parameters of $B_i$ and $E_i$ used in the models reflect past performance and a reasonable estimate of future behavior in the long run, there is no substitute for managerial judgment based on current information. This type of information may form the basis for selecting an attractive potential set of securities as inputs to the models and also for establishing appropriate weights or priorities. The use of the goal programming models implicitly assumes that the potential list of securities is composed of those issues that have been screened on the basis of current attractiveness. They are to be viewed as facilitating devices for selecting approximately efficient portfolios only. They are not a magical system for success. There is no substitute for astute managerial judgments, but after these judgments have been made, the models provide work-saving methods for selection.

The major benefits of the models are their tremendous flexibility in incorporating managerial tastes and value judgments and their relative simplicity of data and computational requirements.

The disadvantages of the present models have been made known. The failure to include sufficient and necessary measures of judgments about current investment potentials has placed major reliance upon past data. The only excuse to be offered in this area is that the author is not a seasoned investment analyst with the ability and information to make valid judgments about current investment prospects. If this were the case, then more weight would have been placed on securities and industries that were felt to promise more defensive behavior under the prospect of a long bear market in 1969. However disappointing the present results, the usefulness and flexibility of the models remains promising. Perhaps, with further development, more measures of security behavior under different conditions may be added to the models to improve their sophistication and reliability. Nevertheless, it is believed that the objective of this study in providing a starting point for such a model has been achieved.

## REFERENCES

"The Markets." *Business Week,* January 10, 1970, pp. 86-87.

Farrar, D. E. *The Investment Decision Under Uncertainty.* Englewood Cliffs, N. J.: Prentice-Hall, Inc., 1965.

Lee, S. M. and Lerro, A. J. "Optimizing the Portfolio Selection for Mutual Funds." Working paper, Virginia Polytechnic Institute and State University, 1971.

Levy, R. A. "Measurement of Investment Performance." *Journal of Financial and Quantitative Analysis,* vol. 3, no. 1 (1968), pp. 35-57.

Markowitz, Harry M. *Portfolio Selection: Efficient Diversification of Investments.* Cowles Foundation Monograph 16. New York: John Wiley & Sons, Inc., 1959.

Renshaw, E. F. "Portfolio Balance Models in Perspective: Some Generalizations That Can Be Derived from the Two-Asset Case." *Journal of Financial and Quantitative Analysis,* vol. 3, no. 2 (June 1967), pp. 123-49.

Renwick, F. B. "Asset Management and Investment Portfolio Behavior." *Journal of Finance,* vol. 24, no. 2 (May 1969), pp. 181-206.

Securities Research Company. *3-Trend Cycli-Graphs.* 109th ed. Boston: Securities Research Co., January 1970.

Sharpe, W. F. "A Linear Programming Algorithm for Mutual Fund Portfolio Selection." *Management Science,* vol. 13, no. 7 (March 1967), pp. 499-510.

_____. "A Simplified Model for Portfolio Analysis." *Management Science,* vol. 9, no. 5 (January 1963), pp. 277-93.

Soldofsky, R. M. "Yield-Risk Performance Measurements." *Financial Analyst Journal,* vol. 24 (September-October 1968), pp. 130-39.

Treynor, J. L., *et. al.* "Using Portfolio Composition to Estimate Risk." *Financial Analysts' Journal,* vol. 24 (September-October 1968), pp. 93-100.

Wallingford, B. A. "A Survey and Comparison of Portfolio Selection Models." *Journal of Financial and Quantitative Analysis,* vol. 3, no. 2 (June 1967), pp. 85-106.

Weston, J. F., and Brigham, E. F. *Managerial Finance.* 3rd ed. New York: Holt, Rinehart & Winston, 1969.

# Chapter 10

## Goal Programming for Marketing Decisions

Goal programming is an effective technique for many marketing decisions. It can be applied to new product decisions, marketing research, advertising media planning, multi-stage price determination, sales effort allocation, and many other areas. In this chapter, application of goal programming is explored for the optimization of sales effort allocation and advertising media scheduling.

### OPTIMIZATION OF SALES EFFORT ALLOCATION

One of the most difficult decision problems in marketing is the determination of the optimum allocation of sales effort among the various market elements. Because of the multiplicity of factors and the complex relationships among these factors, it is often beyond the ability of the sales manager to identify the optimum allocation alternative. Thus it has become important for the retail manager to understand the capabilities and applications of the various quantitative or management science techniques so that he can thoroughly evaluate the alternative allocation opportunities.

Wages generally represent more than half the total expenses in retailing. This alone would single out effective manpower scheduling as one of the most important jobs for the retail manager. In the 1930's, the daily and weekly hours of a full-time salesman were almost identical with the hours of store operation. Since then store hours have increased tremendously while working hours have decreased. The average weekly store hours were 48 hours in the 1940's and 58 hours in the 1950's, and now they are over 68 hours.[1] To compound the problem, not only has the average employee decreased his working hours but his salary has increased at an even greater rate. The increased store time has also resulted in greater valleys and peaks in the pattern of daily and hourly sales. The manager's response to fluctuations in demand is limited because of either the maximum or minimum daily working hours of employees (resulting from unionization or increased legal pressures) and because of the increased costs associated with overtime. The combination of these developments has pointed to the need for some efficient manpower scheduling as a means to the optimum sales effort allocation.

This chapter reviews several different techniques that have been used and presents the goal programming approach as the new optimization model for sales effort allocation.[2]

## TECHNIQUES OF SALES EFFORT ALLOCATION

### A. The Intuitive Approach

This approach to allocating sales effort has enjoyed the greatest application to date of all the methods to be discussed. This is due to its ease of application and understanding. However, with the development of more scientific procedures the value of the intuitive approach will be diminished. To begin with, management takes a survey to determine the pattern of distribution of daily and hourly sales volume by sales area. This will determine the normal hourly requirements for sales floor coverage. Having determined the sales requirements, management allocates the sales personnel through a work schedule that it "feels" will best satisfy the customers needs.[3] Management depends primarily upon past experience and an intuitive awareness for the ability of the sales force to handle the work load through proper scheduling. This method suffers from a lack of standards for allocating effort, an inability to handle multiple goals, and a dependence upon the estimation of the individual manager in charge. However, for an extremely small retailing firm this approach may be the only feasible alternative; the larger the establishment, the less applicable it becomes.

## B.  The Systems Approach

Basically the systems approach attempts to describe the types of information that management needs for sales force scheduling and to explain some of the systems of concept and tools available to aid in this task. The systems concept pictures the selling unit as composed of three basic elements — inputs (customer needs), service facility (sales force, physical assets, inventories), and output (the served customer) — related as shown in Figure 10.1.[4]

**Figure 10.1**

Each element in the system has certain characteristics that are measurable and quantifiable.  For example, random samplings of a number of customers within a selling area over any particular length of time plus the normal sales records provide predictions and measures of sales volume. By utilizing work sampling techniques it is possible to obtain an average service time. This information is then used to determine sales standards. Combining the sales standards and transaction data, a standard for time sales transactions can be set.

Once the scheduling standards and customer arrival patterns are computed, staffing schedules can be completed. These schedules should be based on predetermined sales levels for each day of the week with provisions for adding extra personnel or personnel hours in accord with sales trend needs. The schedule should specify floor coverage by the hour and the number of personnel assigned along with scheduled working hours. An example is illustrated below in Table 10.1.[5]

The control section acts as a self-checking device. As can be seen, the schedule portion is derived easily from the assigned hours column while the base is determined as previously discussed. The variance section is the actual implementation of control. After checking for extenuating circumstances that may have caused the variance, the manager has a basis for remedial action. The main objective with this scheduling method is that it can handle the objective of minimizing cost and it treats all the employees as being equal in their ability to make sales when in reality this is often not true.[6]

Table 10.1

| Employee | Assigned Hours | Floor Coverage Control (No. of Sales People) | | | |
|---|---|---|---|---|---|
| | | Time | Base | Schedule | Variance |
| A | 9-5 | 9-10 | 4 | 4 | — |
| B | 11-3 | 10-11 | 4 | 4 | — |
| C | 9-5 | 11-12 | 5 | 7 | +2 |
| D | 9-5 | 12-1 | 7 | 8 | +1 |
| E | 11-3 | 1-2 | 7 | 8 | +1 |
| F | 9-5 | 2-3 | 7 | 8· | +1 |
| G | 11-3 | 3-4 | 4 | 5 | +1 |
| H | 12-4 | 4-5 | 4 | 4 | — |

| Customer (input) | → | Assigned hours (service facility) | → | Base section (control section) | → | Satisfied customer (output) |
|---|---|---|---|---|---|---|

## C. Staff Engineering

The staff engineering procedure is much like the systems approach except for the methodology used in determining the basis of customer needs. Instead of positioning a certain selling expense level and hoping that service will be acceptable, the aim is to start with a service level that will maximize customer satisfaction and work from this in forecasting a selling expense. The hope is that by maximizing customer satisfaction the firm will also be minimizing costs while maximizing profits.[7]

Basically allocating sales personnel is a problem dealing with the efficient allocation of a scarce resource. The idea is to apply these resources (sales personnel) where they will produce the most return (reconcilable with the firm's service requirements). This is a function of systems, staffing, and environment, which in turn are affected by merchandise type, physical plant, salesclerk caliber, etc. It is first necessary to quantify certain data.[8]

$$\text{Number of arrivals: } A = \frac{\overline{N} \times 60}{\overline{E}}$$

where $\overline{N}$ = average number of customers in store at a given point in time

$\overline{E}$ = length of time average customer remains in the store.

$$\text{Salesforce capacity: } Sc = \frac{Ns \times 60}{Ts}$$

where Ns = total number of salesmen

Ts = average time to wait on each customer.

The number of arrivals per hour, waiting time per customer, and the service time can be computed through the analysis of time series observations. These variables are used to determine the probability of a person purchasing, given a certain length of waiting time before service as shown in Figure 10.2

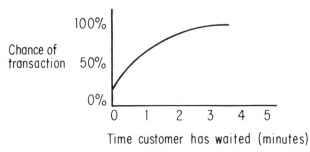

**Figure 10.2**

The object of this analysis is to staff the store with that number of employees that will cause the average waiting time to equal $T_o$. $T_o$ is determined by combining the two waiting-time graphs shown Figures 10.2 and 10.3. It is that waiting time that will allow for the minimum number of sales personnel and provide for a satisfactory level of customer satisfaction as shown in Figure 10.4.

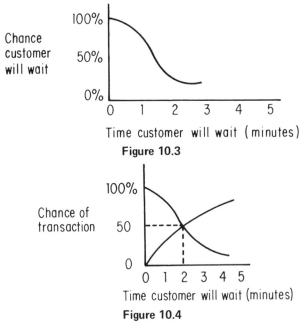

**Figure 10.3**

**Figure 10.4**

The total number of hours required by the sales staff would then be computed:

$$\frac{Sc \times A}{To}$$

The scheduling approach used to allocate the individual salesman's working hours would be identical to that used under the systems approach to allocating sales effort. Excessive demands by customers will be balanced with excessive staffing costs and the customer will receive the greatest satisfaction while the retailer will be able to operate at a minimum cost. The disadvantages of this type of system is that it only has the objective of determining the minimum cost at which to operate efficiently, and it does not take into account the individual capacity to satisfy customer demand (all calculations are on an aggregate basis and the approach does not provide for differences in individual salesmen.)

## D.  Linear Programming

The linear programming approach to sales effort allocation attempts to minimize cost or to maximize sales revenues.[9] It is possible to evaluate the achievement of only one goal under this technique. A linear program is a mathematical program in which the constraints (limiting factors) and the objective function (that goal which management has decided their sales effort scheduling should strive toward) involve the choice variables (sales personnel) in a mathematically linear way. A solution of a linear program is a particular set of values for the choice variables, for which all constraints are satisfied. More than one solution may exist, if any exist at all given the limits of the constraining functions. The optimum solution is a solution for which the computed value of the objective functions is as large (or as small, depending upon whether the objective is to maximize return or minimize investment) as possible. Since there exists an abundance of linear programming applications in the literature, its property will not be discussed here.

## THE GOAL PROGRAMMING APPROACH

To demonstrate the capabilities of the goal programming model for sales effort allocation, a simple real-world problem will be presented as an illustration.

To gather data for the goal programming model formulation, a small specialty shoe store in a medium size city (population 100,000) in southeastern Virginia is selected. The store employs five salesmen: the owner, the manager, two full-time salesmen, and a part-time sales clerk. The basic sales effort allocation problem of the store is how best to employ the time of the full-time and part-time salesmen in order to serve the objectives of the store.

Through time-series analysis the owner has determined that his contribution per hour on the sales floor is $24. The sales manager's contribution is $16 per hour, the first full-time salesman's contribution is $9, the second full-time salesman's contribution is $5 and the part-time salesman contributes only $1.50.

The scheduled working hours for each employee are predetermined. The manager and the owner are supposed to work 200 hours a month. The first full-time salesman is committed to 172 hours, the second full-time salesman to 160 hours, and the part-time salesman is scheduled to work 100 hours a month. As an incentive to the full-time employees the owner is trying to implement a commission-based salary structure. The manager and two full-time salesmen receive the commission of 5.5 percent of their total sales. Through scheduling and the differences in sales volume the manager should make at least $170 from his commission in a month. The first full-time salesman should make $87, while the second full-time salesman should make $52.

To ensure that the sales personnel retain some degree of sales effectiveness and that their morale does not suffer from too much work the owner is attempting to limit overtime. He limits himself and the manager to 24 hours a month. The first full-time salesman has a great deal of ambition and energy, so his limit is 52 hours a month, and the second full-time salesman is limited to 32 hours a month. The owner has set 32 hours as the maximum overtime hours for the part-time worker.

The owner is strongly sales-oriented and unable to understand the idea that he should value profits more than absolute sales. For this reason all his return objectives are set in terms of sales. He had decided that it was not worthwhile to introduce a constraint on sales costs even though it would be relatively easy to introduce a restraint that would attempt to limit the total sales expenditure by assigning a cost per hour to each salesman and limiting the total to some percentage of anticipated sales. In essence this would incorporate a return-on-investment restraint; the cost constraint could be expanded to include all costs for operating the store and thus it becomes a return-on-investment criterion.

## A. The First Run

After a considerable amount of deliberation, the owner of the store lists the following set of goals in the ordinal ranking of importance:

1. Achieve the sales goal of $14,500 for the month.
2. Provide job security to all employees by allowing them to work at least their regular working hours.

3. Make sure that the manager receives at least $170 commission.
4. Do not exceed the overtime ceiling set for the owner, manager, and the first full-time salesman.
5. Do not exceed the maximum overtime allowed for the second full-time and the part-time salesmen.
6. Make sure that the two full-time salesmen get their commission goals of $87 and $52 respectively.

To formulate the model, the following variables must be defined:

$x_1$ = owner's working hours for the month
$x_2$ = manager's working hours for the month
$x_3$ = the first full-time salesman's working hours for the month
$x_4$ = the second full-time salesman's working hours for the month
$x_5$ = part time salesman's working hours for the month

Now, the goal programming model can be formulated as follows:

(10.1) $\text{Min } Z = P_1 d_1^- + P_2 (d_2^- + d_3^- + d_4^- + d_5^- + d_6^-) + P_3 d_7^- + P_4 d_{21}^+ + P_4 d_{31}^+$
$\qquad\qquad + P_4 d_{41}^+ + P_5 d_{51}^+ + P_5 d_{61}^+ + P_6 d_8^- + P_6 d_9^-$

s.t. $\quad 24x_1 + 16x_2 + 9x_3 + 5x_4 + 1.5x_5 + d_1^- - d_1^+ = 14{,}500$
$\qquad x_1 + d_2^- - d_2^+ = 200$
$\qquad x_2 + d_3^- - d_3^+ = 200$
$\qquad x_3 + d_4^- - d_4^+ = 172$
$\qquad x_4 + d_5^- - d_5^+ = 160$
$\qquad x_5 + d_6^- - d_6^+ = 100$
$\qquad .055 (16x_2) + d_7^- - d_7^+ = 170$
$\qquad .055 ( 9x_3) + d_8^- - d_8^+ = 87$
$\qquad .055 ( 5x_4) + d_9^- - d_9^+ = 52$
$\qquad d_2^+ + d_{21}^- - d_{21}^+ = 24$
$\qquad d_3^+ + d_{31}^- - d_{31}^+ = 24$
$\qquad d_4^+ + d_{41}^- - d_{41}^+ = 52$
$\qquad d_5^+ + d_{51}^- - d_{51}^+ = 32$
$\qquad d_6^+ + d_{61}^- - d_{61}^+ = 32$

The above problem is solved by a computer program written by the author. The solution indicates the following values for the variables: $x_1 = 224$; $x_2 = 224$; $x_3 = 224$; $x_4 = 665$; and $x_5 = 132$. The total sales for the month are $14,500. As is apparent from the solution, all the goals of the owner have been achieved except the fifth goal. The second full-time salesman has to work 505 hours of overtime in the month. This result clearly demonstrates that the owner's sales goal is unrealistic.

## B. The Second Run

After reviewing the model result, the owner decides to modify his goals to be more realistic. The owner now realizes that he cannot achieve the sales goal of $14,500 with the existing sales force. Therefore, he decides to introduce some advertising and also reduce his sales goal to a reasonable level.

According to the owner's past experience, advertising in the local paper and radio has brought the sales increase of 2 percent for every $100 of advertising expenditure. This index of advertising effectiveness is relatively constant up to $500 advertising expenditure, and it declines rather sharply after that point.

The owner's record indicates that sales in the same month last year were $11,000. Realizing that the sales of shoes in the city have increased approximately 9 percent, the owner feels that in order to maintain his market share he must achieve at least $12,000 sales in the month. With such additional information, the owner set a list of new goals as follows:

1. Provide job security to all employees by allowing them to work at least their regular working hours
2. Maintain current market share ($12,000)
3. Make sure that the manager receives at least $170 commission
4. Limit the advertising expenditure to $450
5. Do not exceed the maximum allowed overtime for all employees
6. Achieve an increase of market share by 11 percent
7. Make sure the two full-time salesmen receive their commissions of $87 and $52 respectively

The goal programming model of the second run is formulated as follows:

$$(10.2) \quad \text{Min } Z = P_1 (d_2^- + d_3^- + d_4^- + d_5^- + d_6^-) + P_2 d_1^- + P_3 d_8^- + P_4 d_7^-$$
$$+ P_5 (d_{21}^+ + d_{31}^+ + d_{41}^+ + d_{51}^+ + d_{61}^+) + P_6 d_{11}^- + P_7 (d_9^- + d_{10}^-)$$

$$\text{s.t.} \quad 24x_1 + 16x_2 + 9x_3 + 5x_4 + 1.5x_5 + 2.4x_6 + d_1^- - d_1^+ = 12{,}000$$
$$x_1 + d_2^- - d_2^+ = 200$$
$$x_2 + d_3^- - d_3^+ = 200$$
$$x_3 + d_4^- - d_4^+ = 172$$
$$x_4 + d_5^- - d_5^+ = 160$$
$$x_5 + d_6^- - d_6^+ = 100$$
$$x_6 + d_7^- - d_7^+ = 450$$
$$.055 (16x_2) + d_8^- - d_8^+ = 170$$
$$.055 ( 9x_3) + d_9^- - d_9^+ = 87$$
$$.055 ( 5x_4) + d_{10}^- - d_{10}^+ = 52$$
$$d_1^+ + d_{11}^- - d_{11}^+ = 1{,}320$$
$$d_2^+ + d_{21}^- - d_{21}^+ = 24$$

$$d_3^+ + d_{31}^- - d_{31}^+ = 24$$
$$d_4^+ + d_{41}^- - d_{41}^+ = 52$$
$$d_5^+ + d_{51}^- - d_{51}^+ = 32$$
$$d_6^+ + d_{61}^- - d_{61}^+ = 32$$

In the above model, $x_6$ represents the amount of advertising expenditure. The coefficient of 2.4 for $x_6$ in the first constraint was derived through the following calculations:

$$(.02) \times 4.5 \times 12{,}000 = 1{,}080$$
$$1{,}080/450 = 2.4$$

The solution of the second run derived the following results: $x_1 = 224$, $x_2 = 224$, $x_3 = 224$, $x_4 = 192$, $x_5 = 132$, and $x_6 = 450$. The total sales of the store resulted were \$13,214. With the above solution, all goals are achieved, except the sixth goal (sales of \$13,320). However, the underachievement of the sales goal is only \$106. The most important goals of the owner are achieved by restructuring goals and their priorities and by providing advertising expenditure of \$450. As can readily be seen, the goal programming model allows the violation of lower-order goals when it is necessary to achieve goals with high priorities. The goal programming model provides management an opportunity for a critical evaluation of the soundness of its goal structure in view of the goal achievements and trade-offs among the marketing goals presented by the solution.

## CONCLUSION

Changes in the retail marketing structure have made it necessary for the marketing manager to utilize the various decision aids available to him. It is especially important for him to be able to exert control over the elements of the sales force. This is not only because they constitute one of the retailer's largest expenses, but also because they are typically his greatest asset. As is true with any valuable and scarce resource, there is a great need for efficiency in the allocation of the sales effort to ensure that there are enough sales personnel on hand to meet the organization's marketing goals but not so many as to lose market effectiveness. Typically, the larger the retailer the better suited the organization is to utilize the more scientific management tools. However, this does not negate the need for efficiency on the smaller-scale retail level; in fact the small operator can least afford inefficiency.

The shoe-shop illustration demonstrated one example of solving sales effort allocation problems with goal programming. Goal programming should also be applicable to a wide range of other marketing problems. This is easily conjectured in view of other studies reported recently.[10] With recent developments in the marketing area, especially in the fields of international marketing and consumerism, it appears that goal programming will find a wide range of applications to managerial decisions in marketing. In fact, marketing promises to be an area where goal programming will play a vital role in decision analysis.

## OPTIMIZATION OF ADVERTISING MEDIA SCHEDULING

One of the most difficult and most heavily subjective decisions that face an advertising manager is that of specific media selection. If the choices are limited to only two or three individual media, an experienced advertising man can probably do a fairly decent job of approaching the optimum mix for his campaign by using his own judgment. In the highly differentiated milieu of the advertising field, however, where specific media choices reach into the hundreds, and different combinations reach the millions, it becomes physically impossible for even a large team of experienced men to scan every possibility by hand and select an optimum combination alternative. It has become necessary to rely on the assistance of management science techniques in order to optimize advertising media scheduling. Thus far there have been various studies in which a number of different approaches have been attempted for advertising media planning. However, most of these studies have found very limited practical applications, primarily because the models utilized often neglected the multiple conflicting goals of the advertising campaign. In this study we explore the possibility of goal programming application for the optimization of the advertising campaign. More specifically, this study presents a highly simplified hypothetical model in order to demonstrate the potential of goal programming for advertising media scheduling.

## MANAGEMENT SCIENCE FOR MEDIA PLANNING

The application of the computer and management science to actual advertising was first pioneered in 1961 by the advertising agency Batten, Barton, Durstine, and Osborne (BBDO). Since then a great number of studies using various management science techniques have been published.[11] However, the most widely used technique appears to be linear programming. The model that was used by BBDO, and which since has received a great deal of attention both laudatory and derogatory, was also based on linear programming. Its basic purpose was to allocate available advertising budget to selected media by scanning all the feasible media in such a way as to maximize the effectiveness for available resources or to minimize the costs for a desired level of effectiveness.

Since the BBDO linear programming model, numerous studies have been published concerning the use of improved linear programming models for media scheduling. We will review several important studies here. A study by Day reviewed the feasibility of linear programming applications to advertising field.[12] The study also emphasized the importance of sensitivity analysis of the optimal solution, because there usually exists some degree of uncertainty in the model parameters. Day pointed out three major problems in applying linear programming to advertising decisions:

1. Selecting candidate advertising units, which requires expert knowledge of the target audience, media audience, etc.

2. Forming equalities and inequalities to reflect real-world conditions, either objectively or subjectively, but subjective constraints should be minimized if possible.

3. Establishing effectiveness measures. This is an extremely difficult criterion to analyze, yet it is used as the objective criterion. In an attempt to extend Day's article, Engel and Warshaw described the method by which effectiveness measures of media could be determined.[13] The study proposed the inclusion of the audience dimensions in the objective function so that an optimal solution not only maximizes the number reached but also readers who are likely to be prospects. The audience dimension is called the audience profile match. When this rating is included with the qualitative factor based on editorial climate and media effectiveness and multiplied by the total audience, the effective audience can be determined.

A study by Kotler attempted to adjust the programming algorithm to include information that previous models did not include.[14] Kotler's study included quantity discounts and the "decay" of successive insertions. This approach was intended to help design a more realistic model.

A similar and even more extensive study was presented by Brown and Warshaw.[15] This study demonstrated how the objective function could be altered by a step-wise analysis to include most of the realistic constraints of the problem. The authors also presented a piecewise linear approximation approach to include some nonlinear-type constraints.

An additional modification of the objective function to include previously neglected conditions was presented by Stasch.[16] He altered the basic model to include space and time considerations. The space is the "where" question that discerns the differentiated markets and assigns different numbers of media inclusions to them. For example, in certain regional areas heavier concentration may be placed on regional magazine editions, local TV and radio, etc. The time is the "when" question and concerns itself with which editions or who should be used for advertisement. For example, if the optimum allocation shows four inclusions in a monthly magazine in a six-month period, then there should also be the determination of which months to be included.

The last study to be discussed here is one by Bass and Lonsdale.[17] The study examined two important aspects of linear programming models for advertising media scheduling. First, it attempted to evaluate whether or not the refined adjustments of audience data is justified—that is, whether they result in a significantly better solution than if a simple gross audience is used. Second, it was the first study that analyzed the model by using empirical data. The study found that it made little, and often no, difference which audience data were used for the model; the results of the model were not significantly different. From these results, the authors concluded that the objection to linear programming because of its arbitrary or judgmental constraints was largely unfounded, because less arbitrary constraints yielded similar results. These arguments, however, have been criticized on the grounds of methodological errors. It appears that greater improvements are needed for a linear programming model to be truly effective in optimizing advertising media scheduling.

## GOAL PROGRAMMING FOR MEDIA SCHEDULING

In order to explore the potential application of goal programming, let us consider a simple hypothetical case.[18] Suppose we are planning an advertising campaign for the coming fiscal year. We have selected 47 media vehicles in seven categories as possible candidates for the campaign. Table 10.2 presents the list of

Table 10.2
Media Vehicles, Cost, and Audience Data Used for the Goal Programming Model

| Media Category & Vehicle #i | Cost[*] | Exposure[**] |
|---|---|---|
| A. General public magazine (4 color, page) | | |
| 1 | $20,100 | 62,000 |
| 2 | 48,000 | 93,000 |

| | | |
|---|---|---|
| 3 | 4,900 | 11,500 |
| 4 | 3,200 | 8,000 |
| 5 | 15,500 | 18,000 |
| 6 | 4,600 | 7,000 |
| 7 | 18,000 | 22,500 |

B. General public magazine (2 color, page)

| | | |
|---|---|---|
| 8 | 2,200 | 10,500 |
| 9 | 4,200 | 14,500 |
| 10 | 1,500 | 6,500 |
| 11 | 1,900 | 7,200 |
| 12 | 3,200 | 18,500 |

C. Ladies' magazines (4 color, page)

| | | |
|---|---|---|
| 13 | 12,100 | 48,500 |
| 14 | 33,000 | 58,000 |
| 15 | 24,000 | 26,000 |
| 16 | 26,000 | 36,000 |
| 17 | 18,000 | 22,600 |

D. Men's magazine (4 color, page)

| | | |
|---|---|---|
| 18 | 12,000 | 15,200 |
| 19 | 10,000 | 14,500 |
| 20 | 9,000 | 10,200 |
| 21 | 18,000 | 26,000 |
| 22 | 14,000 | 19,500 |

E. Spot radio (day, 60 seconds)

| | | |
|---|---|---|
| 23 | 2,000 | 1,800 |
| 24 | 1,800 | 2,150 |
| 25 | 1,500 | 1,600 |
| 26 | 1,800 | 2,200 |

F. Spot radio (prime, 60 seconds)

| | | |
|---|---|---|
| 27 | 2,800 | 2,800 |
| 28 | 2,600 | 2,200 |
| 29 | 3,000 | 3,600 |
| 30 | 2,400 | 2,850 |

| | | | |
|---|---|---|---|
| G. | Day T.V. (30 seconds) | | |
| | 31 | 2,000 | 8,500 |
| | 32 | 2,800 | 8,900 |
| | 33 | 1,600 | 11,000 |
| | | | |
| H. | Day T.V. (60 seconds) | | |
| | 34 | 3,500 | 10,500 |
| | 35 | 6,800 | 18,000 |
| | 36 | 4,200 | 11,500 |
| | | | |
| I. | Spot T.V. (day, 30 seconds) | | |
| | 37 | 21,000 | 13,500 |
| | 38 | 18,500 | 14,000 |
| | 39 | 24,000 | 15,600 |
| | | | |
| J. | Spot T.V. (prime, 60 seconds) | | |
| | 40 | 36,000 | 56,500 |
| | 41 | 42,000 | 58,500 |
| | 42 | 28,000 | 36,000 |
| | 43 | 31,000 | 42,000 |
| | | | |
| K. | Night T.V. (30 seconds) | | |
| | 44 | 6,800 | 12,000 |
| | 45 | 7,500 | 12,500 |
| | 46 | 8,200 | 11,200 |
| | 47 | 5,800 | 8,000 |

*Cost per inclusion

**The exposure represents the weighted exposure unit per inclusion. For the detailed definition see footnote 19.

vehicles, cost per inclusion, and the weighted exposure units per inclusion.[19] The advertising director has multiple conflicting goals to achieve from the campaign. He lists the following goals in ordinal ranking of importance:

1. Allocate at least $2,500,000 for the advertising campaign for the coming year.

2. Achieve at least 7,000,000 weighted exposure units from the entire campaign.

3. Limit the number of exposures per media category as follows:

| Media Category | Maximum Exposures |
|---|---|
| General public magazines (A and B) | 6 |
| Ladies' magazines (C) | 10 |
| Men's magazines (D) | 5 |
| Spot radio-day (E) | 15 |
| Spot radio-prime (F) | 7 |
| Day T.V. (G and H) | 18 |
| Spot T.V. (I and J) | 24 |
| Night T.V. (K) | 12 |

4. The total advertising expenditure should not exceed the budget by more than 5 percent.

5. Limit the advertising expenditures for various media categories as follows:

| Media Category | Maximum Expenditure |
|---|---|
| All magazines (A, B, C, and D) | $500,000 |
| Spot radio (E and F) | 125,000 |
| Day T.V. (G and H) | 600,000 |
| Spot T.V. (I and J) | 400,000 |
| Night T.V. (K) | 700,000 |

6. Assure that there is at least the following number of exposures for certain specific vehicles in order to maintain the current market shares in various market segments:

| Vehicle No. | Exposures |
|---|---|
| 2 | 2 |
| 10 | 3 |
| 14 | 2 |
| 21 | 4 |

7. Only four-color, full-page advertisements must be used in the general public magazines.

Now we can formulate the above advertising scheduling problem as a goal programming model. First, we shall formulate constraints, and next the objective function will be defined.

## A. Constraints

### 1. THE BUDGET

The total advertising budget, which is the sum of cost per exposure times the number of exposure for each vehicle, should come out to the allocated $2,500,000 for the coming year.

$$(10.3) \quad \sum_{i=1}^{47} c_i x_i + d_1^- - d_1^+ = 2,500,000$$

where $c_i$ = cost per exposure for the ith vehicle

$x_i$ = number of exposures for the ith vehicle

$d_1^-$ = underexpenditure of the advertising budget

$d_1^+$ = overexpenditure of the advertising budget.

### 2. DESIRED LEVEL OF WEIGHTED EXPOSURE UNITS

The total weighted exposure units, which is the sum of the weighted exposure units per exposure times the number of exposure for each vehicle, should reach the desired 7,000,000 units.

$$(10.4) \quad \sum_{i=1}^{47} b_i x_i + d_2^- - d_2^+ = 7,000,000$$

where $b_i$ = the weighted exposure units per exposure for the ith vehicle

$d_2^-$ = underachievement of the desired level

$d_2^+$ = overachievement of the desired level.

### 3. MAXIMUM DESIRED EXPOSURES FOR MEDIA CATEGORIES

The advertising director specified the maximum desired number of exposures for eight media categories as described in the third goal.

$$(10.5) \quad \sum_{i=1}^{12} x_i + d_3^- - d_3^+ = 6 \text{ (general public magazines)}$$

$$\sum_{i=13}^{17} x_i + d_4^- - d_4^+ = 10 \text{ (ladies' magazines)}$$

$$\sum_{i=18}^{22} x_i + d_5^- - d_5^+ = 5 \text{ (men's magazines)}$$

$$\sum_{i=23}^{26} x_i + d_6^- - d_6^+ = 15 \text{ (spot radio–day)}$$

$$\sum_{i=27}^{30} x_i + d_7^- - d_7^+ = 7 \text{ (spot radio–prime)}$$

$$\sum_{i=31}^{36} x_i + d_8^- - d_8^+ = 18 \text{ (day T.V.)}$$

$$\sum_{i=37}^{41} x_i + d_9^- - d_9^+ = 24 \text{ (spot T.V.)}$$

$$\sum_{i=42}^{47} x_i + d_{10}^- - d_{10}^+ = 12 \text{ (night T.V.)}$$

## 4. LIMIT ON OVEREXPENDITURE OF THE BUDGET

The director desires to limit the overexpenditure of the campaign to 5 percent of the budget. In other words, the overexpenditure $(d_1^+)$ should be limited to \$125,000.

$$(10.6) \quad d_1^+ + d_{11}^- - d_{11}^+ = 125,000$$

## 5. MAXIMUM EXPENDITURES FOR MEDIA CATEGORIES

The director specified the maximum level of advertising expenditures for the five media categories as explained in the fifth goal.

$$(10.7) \quad \sum_{i=1}^{22} c_i x_i + d_{12}^- - d_{12}^+ = 500,000 \text{ (all magazines)}$$

$$\sum_{i=23}^{30} c_i x_i + d_{13}^- - d_{13}^+ = 125,000 \text{ (spot radio)}$$

$$\sum_{i=31}^{36} c_i x_i + d_{14}^- - d_{14}^+ = 600,000 \text{ (day T.V.)}$$

$$\sum_{i=37}^{43} c_i x_i + d_{15}^- - d_{15}^+ = 400,000 \text{ (spot T.V.)}$$

$$\sum_{i=44}^{47} c_i x_i + d_{16}^- - d_{16}^+ = 700,000 \text{ (night T.V.)}$$

## 6. DESIRED NUMBER OF EXPOSURES FOR SELECTED VEHICLES VEHICLES

The director has determined that there should be at least a specified number of exposures for selected vehicles in order to maintain present market shares in certain market segments as described in the sixth goal.

$$(10.8) \quad \begin{aligned} x_2 + d_{17}^- - d_{17}^+ &= 2 \\ x_{10} + d_{18}^- - d_{18}^+ &= 3 \\ x_{14} + d_{19}^- - d_{19}^+ &= 2 \\ x_{21} + d_{20}^- - d_{20}^+ &= 4 \end{aligned}$$

## 7. USE OF FOUR-COLOR FULL-PAGE ADS

The director has decided that in order to present quality advertisements in magazines it is essential that only four-color full-page ads be utilized. Therefore, this goal should eliminate all two-color ads as long as it does not hinder the achievement of the first six goals.

$$(10.9) \quad \sum_{i=8}^{12} x_i + d_{21}^- - d_{21}^+ = 0$$

## B. The Objective Function

The objective of the model is to minimize deviations from a set of goals by assigning appropriate priority factors. After reviewing the priority structure of goals and the model constraints formulated above, we can derive the following objective function:

$$(10.10) \quad \text{Min } Z = P_1 d_1^- + P_2 d_2^- + P_3 \sum_{i=3}^{10} d_i^+ + P_4 d_{11}^+ + P_5 \sum_{i=12}^{16} d_i^+ + P_6 \sum_{i=17}^{20} d_i^- \\ + P_7 (d_{21}^- + d_{21}^+)$$

In the above objective function, it is of course possible to assign differential weights to deviational variables at $P_3$, $P_5$, and $P_6$ levels. The criterion for the determination of differential weights may be the weighted exposure units per dollar of cost for each of the media categories considered.

## CONCLUSION

In this section of the chapter we explored the application potential of goal programming for advertising media scheduling through a highly simplified hypothetical example. The model neglected many important aspects of advertising for the sake of simplicity. For example, the time dimensions (i.e., time intervals between exposures in the same vehicle, harmonized exposures in various vehicles for special purposes, etc.), weighted exposure index of each vehicle for specific market segments (i.e. women, men, teen-age market, etc.), and advertising goals for certain time periods, products, and market segments have been ignored in the model. However, if such data are available, these can easily be incorporated in the model. It appears that goal programming has a great potential for advertising media scheduling when there are multiple conflicting goals of the advertising campaign.

## REFERENCES

Bass, F. M. and Londsdale, R. T. "An Exploration of Linear Programming in Media Selection." *Jourhal of Marketing Research*, vol. 3 (1966), pp. 179-88.

Beale, E. M. L., Hughes, P. A. B., and Broadbent, S. R. "A Computer Assessment of Media Schedules." *Operational Research Quarterly*, vol. 17 (1966), pp. 381-412.

Bell, R. W. and Paul, R. W. "Quantitative Determination of Manpower Requirements in Variable Activities." *Journal of Retailing*, vol. 43 (1967), pp. 21-27.

Brown, D. B. "A Practical Procedure for Media Selection." *Journal of Marketing Research*, vol. 4 (1967), pp. 262-64.

_____ and Warshaw, M. R. "Media Selection by Linear Programming." *Journal of Marketing Research*, vol. 2 (1965), pp. 83-88.

Charnes, A., Cooper, W. W., and Ferguson, R. "Optimal Estimation of Executive Compensation by Linear Programming." *Management Science*, vol. 1, no. 2 (January 1955), pp. 138-51.

Charnes, A., Cooper, W. W., and Niehaus, R. J. "A Goal Programming Model for Manpower Planning." Management Science Research Report No. 115 (also see No. 188). Carnegie-Mellon University, August 1968.

Charnes, A., *et al.* "A Goal Programming Model for Media Planning." *Management Science*, vol. 14, no. 8 (April 1968), pp. 423-30.

_____. "Note on an Application of a Goal Programming Model for Media Planning." *Management Science*, vol. 14, no. 8 (April 1968), pp. 431-36.

Davis, O. A. and Farley, J. U. "Allocating Sales Force Effort with Commissions and Quotas." *Management Science*, vol. 18, no. 4, III (1971), pp. 55-63.

Day, R. L. "Linear Programming in Media Selection." *Journal of Advertising Research*, vol. 2 (June 1962), pp. 40-44.

Duncan, Delbert J., and Phillips, Charles F. *Retailing: Principles and Methods* (Homewood, Ill.: Richard D. Irwin, Inc., 1963).

Ellis, D. M. "Building Up a Sequence of Optimum Media Schedules." *Operations Research Quarterly*, vol. 17 (1966), pp. 413-24.

Engel, George C. "How You Can Get More Volume from Your Sales Staff." *Stores* (January 1965), pp. 62-63.

Engel, J. F. and Warshaw, M. R. "Allocating Advertising Dollars by Linear Programming." *Journal of Advertising Research*, vol. 4 (September 1964), pp. 42-48.

Horgan, Charles. "A Systems Approach to Manpower Planning in Department Stores." *Journal of Retailing*, vol. 44, no. 3 (Fall 1968), pp. 13-30.

Johnson, Gary. "The Thinking Man's Approach to Sales Force Scheduling." *Stores*, November 1969, pp. 37-38.

Kotler, P. "Toward an Explicit Model for Media Selection." *Journal of Advertising Research*, vol. 4 (March 1964), pp. 34-41.

Lee, A. M. "Decision Rules for Media Scheduling: Static Campaigns." *Operational Research Quarterly,* vol. 13 (1962), pp. 229-42.

_____. "Decision Rules for Media Scheduling: Dynamic Campaigns." *Operational Research Quarterly,* vol. 14 (1963), pp. 355-72.

Lee, S. M. and Bird, M. "A Goal Programming Model for Sales Effort Allocation." *Business Perspectives,* vol. 6, no. 4 (Summer 1970). Also published in *Operations Research: An Introduction to Modern Applications,* ed. William C. House. Philadelphia: Auerbach Publishers Inc.

Little, J. D. C., and Lodish, L. M. "A Media Selection Model and Its Optimization by Dynamic Programming." *Industrial Management Review,* vol. 8 (1966), pp. 15-24.

Montgomery, D. B., Silk, A. J., and Zaragoza, C. E. "A Multiple-Product Sales Force Allocation Model." *Management Science,* vol. 18, no. 4, III (1971), pp. 3-24.

Morgon, W. T. "Practical Media Decisions and the Computer." *Journal of Marketing,* vol. 27 (July 1963), pp. 26-30.

Naert, P. A. "Optimizing Consumer Advertising, Intermediary Advertising and Markup in a Vertical Market Structure." *Management Science,* vol. 18, no. 4, III (1971), pp. 90-101.

Rohloff, A. C. "Quantitative Analysis of the Effectiveness of TV Commercials." *Journal of Marketing Research,* vol. 3 (1966), pp. 239-45.

Sasieni, M. W. "Optimal Advertising Expenditure." *Management Science,* vol. 18, no. 4, III (1971), pp. 64-72.

Stasch, S. F. "Linear Programming and Media Selection—a Comment." *Journal of Marketing Research,* vol. 4 (1967), pp. 205-06.

_____. "Linear Programming and Space-Time Considerations in Media Selection." *Journal of Advertising Research,* vol. 5 (1965), pp. 40-46.

Taylor, C. J. "Some Developments in the Theory and Application of Media Scheduling Methods." *Operational Research Quarterly,* vol. 14 (1963), pp. 291-305.

Weller, R. S. "Don't Get Locked In on Inflexible Employee Work Week." *Stores,* May 1965, p. 22.

Wilson, C. L. "Use of Linear Programming to Optimize Media Schedules in Advertising." In *Proceedings of the Forty-Sixth Conference of the American Marketing Association,* ed. H. Gomez, pp. 178-91. Chicago: American Marketing Association, 1963.

Zangwill, W. I. "Media Selection by Decision Programming." *Journal of Advertising Research,* vol. 5 (September 1965), pp. 30-36.

# Chapter 11

## Goal Programming for Corporate Planning

This chapter presents the use of goal programming for the optimization of top-management corporate planning. There have been a number of studies concerning applications of goal programming to various decision problems, but none of these has dealt with top-level corporate planning. In this study, goal programming is applied to the corporate financial planning for a publicly owned utility company. The study presents optimal solutions to the financial decisions of the firm, given the firm's basic financial constraints, legal obligations, and the long- and short-run goals of its top management.

### TECHNIQUES FOR TOP-LEVEL PLANNING DECISIONS

There have been a great number of studies published concerning the application of various management science techniques to top-level corporate financial decisions. Let us review some of these techniques.

One technique that has been used widely in the analysis of financial decisions is that of setting the various constraints of a decision into a general mathematical form and solving the model through calculus or some other mathematical method. There is no universal format of the model; instead each model is custom-made for the problem at hand, and the techniques that are to be used in analyzing the models are specific for each problem. For example, Morris presented a model for the financial decision of whether or not to diversify a firm.[1] He formulated the outcomes of various decisions concerning the usage of the firm's available resources in the form of second-degree mathematical equations and solved them in order to maximize the utility of a decision. Zangwill derived a method for solving a non-linear pricing decision problem by transforming it into a sequence of unconstrained maximization problems.[2] A penalty cost was imposed in the model if the variables in the unconstrained equations did not satisfy the original constraints of the problem.

A list of examples of this general model technique is virtually endless. The illustrations are presented here simply to show the scope of the use of this technique. Models such as these can be developed for almost any decision situation that can be quantitatively measured. The general problems involved with the use of mathematical conceptualizations of the real world are that (1) there is a question of which is the most efficient mathematical technique to use in analyzing each model, and (2) the models do not always work after they are tried.

A second major technique that is used for financial planning decisions is the direct computer search procedure. This procedure was applied by Jones and by Buffa and Taubert to aggregate planning decisions.[3] The technique involves the sequential examination of a finite set of feasible trial solutions of a certain function. By specifying values for all the independent variables, the technique produces the trial evaluation. If a trial improves the previous trial solution, it is accepted; if not it is rejected. The criterion function represents the costs to be minimized, and it is expressed as a function of employment level and production rates.

Another popular technique is the Lagrange multiplier, which has been used by Eppen and Gould and by Everett to identify the optimum decision in planning resource allocation.[4] The procedure involves the development of a "payoff" function that is the utility that accrues from employing a strategy (also defined as a function) with a given set of resource constraints. This technique has been widely applied to the decision problems of the optimum allocation of multiple resources to multiple independent activities.

Input/output analysis has been also applied to corporate planning by Hetrick and by Farag.[5] This technique involves a deterministic mathematical model in a matrix form. Output—i.e., sales, inventory, etc.—is expressed in the form of equations that use inputs as their variables. Predictions as to production levels, the profits and costs associated with each level, and the degree of utilization of each of the firm's resources can be derived from the model.

Steiner, Ansoff, and Crecine have all advocated the use of the simulation approach for aggregate planning problems.[6] This approach is based on the designing of an abstraction of the real-world problem expressed in mathematical equations to represent interrelationships among system components. Through simulation, the model allows the analysis of a numerical trace of all endogenous variables generated by the model with varying combination of inputs.

Dynamic programming is applied by Petrovic and by Briskin for planning decisions.[7] This technique involves the formulation of relationships that describe the management problem and their association to a criterion function. The criterion function represents the management goal that requires planning and action. The optimization of management actions is viewed as a multistage decision problem.

Applications of the linear decision rule to planning problems have been examined by Bowman, by Jones, and by Holt, Modigliani, and Simon.[8] This technique is a mathematical model designed to determine the optimal aggregate production rates and employment levels throughout the planning horizon. The two decisions rules (one for production rate and one for employment rate) are derived from a cost model that expresses each firm's unique situation. The linear decision rule itself prescribes a course of action that is optimal in terms of management's objective criterion.

Linear programming has found wide applications in various planning problems. Examples of the uses of this technique abound in the literature.[9]

In this study, goal programming is proposed for top management financial planning. It appears that goal programming overcomes many limitations of the techniques mentioned above that have been suggested for management planning problems.

## AN EMPIRICAL STUDY

In order to demonstrate the application of goal programming to top-management-level financial planning, data were obtained from a public utility firm located in a southeastern state of the United States. The firm provides gas and electricity in a county that surrounds a large metropolitan area with a population of more than a million people. The firm is relatively small when compared to other utility companies. However, the firm's total assets stand around $400 million with an annual revenue of over $80 million. The firm's annual growth rate is about 11 percent, which is higher than the industry average of 8 percent. The total number of employees was about 2,500 as of January 1, 1970.

The firm is a publicly owned (by the municipal government) utility that is required by its charter to make payments to the city in lieu of taxes and to maintain certain "funds" that would not normally be required in a privately owned corporation. All revenues of the firm are deposited in a general account and then paid out according to certain priorities. Of primary importance to this firm is payment of current operating and maintenance expenses. After these payments are met, the firm must pay all the principal and interest that will come due on its bonds during the current year. Also, the firm must maintain a "reserve account" that is sufficient to pay the principal and interest that will come due in the fiscal year immediately succeeding the close of the current year.

The third set of payments that must be met each year are the payments to the city. The firm must render to the city a preliminary sum of 1.51 percent of its net assets and then make final payments that will bring the total remuneration to a level equal to 14 percent of the gross revenues in the current fiscal year. The fourth priority expense is the depositing of an annual sum equal to at least 12.5 percent of gross revenues into the improvements and contingencies fund (I&C fund). This payment is of a lower priority than the preliminary payment to the city, but is of a higher priority than the final payments to the city. After all obligations to the city are met, all remaining funds are placed in the I&C fund until such fund amounts to 25 percent of the value of fixed capital assets.

Any revenues in excess of all of the above payments are placed in a "surplus fund" that is to be used by the firm for various specified purposes. Also of importance is the fact that the firm does not desire to issue any more bonds in the high interest market that currently prevails. Also, since the issuing of bonds requires the approval of the city board and a vote of the population of the city, it is a very complicated procedure indeed.

Table 11.1 provides a summary of the various revenues and expenses of the firm and the variables that are to be used in the goal programming analysis for 1970. The goal programming model that was developed contained 7 choice variables, 26 deviational variables, and 13 constraints. The model constraints are presented below in both descriptive and algebraic forms.

## 1. OPERATING EXPENSES AND PAYROLL

(electricity revenue + gas revenue + bond fund + beginning balance + nonoperating revenue + depreciation + amortization + contribution from customers + salvage + Article VII funds + antitrust funds) − (operating expenses + payroll & employee benefits) ≥ 0

$(a_1x_1 + a_2x_2 + x_3 + c_2 + a_3 + a_4 + a_5 + a_8 + a_9 + a_{10} + a_{11}$
$- b_3 + b_4) \geqslant 0$
$(\$4,455,440.0x_1 + 28,292.0\,x_2 + 0\,x_3 + 30,509.0 + 2,275.0$
$+ 9,425.0 + 6.0 + 1,230.0 + 354.0 + 52.0 + 58.0) - (16,709.0$
$+ 18,200.0) \geqslant 0$, or

$$(11.1) \quad 4,445.440\,x_1 + 28,292.0\,x_2 + d_1^- - d_1^+ = 21,509.0$$

Table 11.1

Revenue and Expenditure Variables

| | Constants or Variables | Assumed Value* |
|---|---|---|
| **Revenue 1970** | | |
| Electricity revenue | $a_1x_1$ | ----- |
|   Rate per KWH ($1,000) | $x_1$ | ----- |
|   Estimated demand (1,000 KWH) | $a_1$ | $4,455,440.0 |
|   Minimum or maximum desired rate | $a_{12}$ | 0.0144/KWH |
| Gas revenue | $a_2x_2$ | ----- |
|   Rate per MCF ($1,000) | $x_2$ | ----- |
|   Estimated demand (MCF) | $a_2$ | 28292 |
|   Minimum or maximum desired rate | $a_{13}$ | 0.492/MCF |
| Nonoperating revenue ($1,000) | $a_3$ | $2,275.0 |
| Depreciation ($1,000) | $a_4$ | $9,425.0 |
| Amortization ($1,000) | $a_5$ | 6.0 |
| Construction funds | | |
|   Bond fund ($1,000) | $x_3$ | ----- |
|     Assumed Upper Limit | $a_6$ | 0.0 |
|   Improvement and contingency fund ($1,000) | $x_4$ | ----- |
|     Ratio to earned revenue | $a_7$ | 12.5% |
| Contribution from customers ($1,000) | $a_8$ | $1,230.0 |
| Salvage ($1,000) | $a_9$ | 354.0 |
| Article VII ($1,000)** | $a_{10}$ | 52.0 |
| Antitrust ($1,000) | $a_{11}$ | 58.2 |

**Expenses 1970**

| | Constants or Variables | Assumed Value[*] |
|---|---|---|
| Bond retirement ($1,000) | $b_1$ | $3,130.0 |
| Interest payment ($1,000) | $b_2$ | 2,601.0 |
| Operating expenses ($1,000) | $b_3$ | 16,709.0 |
| Payroll and employee benefits ($1,000) | $b_4$ | 18,200.0 |
| Technical improvement ($1,000) | $x_5$ | ----- |
| Desired ratio to earned revenue | $b_5$ | .005 |
| Payments to the city—preliminary ($1,000) | $b_6c_1$ | ----- |
| Ratio of payments to total assets | $b_6$ | .0151 |
| Payments to the city—final ($1,000) | $x_6$ | ----- |
| Ratio of payments to earned revenue | $b_7$ | .14 |
| Construction ($1,000) | $x_7$ | ----- |
| Desired construction ($1,000) | $b_8$ | $41,220.0 |
| Bond reserve fund ($1,000) | $b_9$ | 3,220.0 |
| Interest reserve fund ($1,000) | $b_{10}$ | 2,514.7 |

**Beginning and Ending Balances**

| | | |
|---|---|---|
| Total assets at the beginning of the year | $c_1$ | $463,000.0 |
| Beginning balance | $c_2$ | 30,509.0 |
| Ratio of surplus to total assets to lower rates | $c_3$ | .25 |

[*]For the dollar amounts, the unit of $1,000 is used.
**Revenue resulting from the sale of property.

## 2. PAYMENT OF PRINCIPAL, INTEREST, AND RESERVE FUNDS FUNDS

(electricity revenue + gas revenue + bond fund + beginning balance + nonoperating revenue + depreciation + amortization + contribution from customers + salvage + Article VII funds + antitrust funds) − (bond retirement + interest payment + operating expenses + payroll and employee benefits + bond reserve fund + interest reserve fund) $\geq 0$

$$(a_1 x_1 + a_2 x_2 + x_3 + c_2 + a_3 + a_4 + a_5 + a_8 + a_9 + a_{10} + a_{11})$$
$$- (b_1 + b_2 + b_3 + b_4 + b_9 + b_{10}) \geqslant 0,$$
$$(\$4,455,440.0\ x_1 + 28,292.0\ x_2 + 0\ x_3 + 30,509.0 + 2,275.0$$
$$+ 9,425.0 + 6.0 + 1,230.0 + 354.0 + 52.0 + 58.0) - (3,130.0$$
$$+ 2,601.2 + 16,709.0 + 18,200.0 + 3,220.0 + 2,514.7) \geqslant 0, \text{ or}$$

(11.2)    $4,455,440.0\ x_1 + 28,292.0\ x_2 + d_2^- - d_2^+ = 27,240.2$

## 3. PRELIMINARY PAYMENT TO I&C FUND

(electricity revenue + gas revenue + bond fund + beginning balance + nonoperating revenue + depreciation + amortization + contribution from customers + salvage + Article VII funds + antitrust funds) − (bond retirement + interest payment + operating expenses + payroll and employee benefits + bond reserve fund + interest revenue fund + preliminary payments to the city) $\geqslant 0$

$$(a_1 x_1 + a_2 x_2 + x_3 + c_2 + a_3 + a_4 + a_5 + a_8 + a_9 + a_{10} + a_{11})$$
$$- (b_1 + b_2 + b_3 + b_4 + b_9 + b_{10} + b_6 c_1) \geqslant 0,$$
$$(\$4,455,440.0\ x_1 + 28,292.0\ x_2 + 0\ x_3 + 30,509.0 + 2,275.0$$
$$+ 9,425.0 + 6.0 + 1,230.0 + 354.0 + 52.0 + 58.0) - (3,130.0$$
$$+ 2,601.2 + 16,709.0 + 18,200.0 + 3,220.0 + 2,514.7$$
$$+ 0.0151(463,000.0)) \geqslant 0, \text{ or}$$

(11.3)    $4,455,440.0\ x_1 + 28,292.0\ x_2 + d_3^- - d_3^+ = 34,231.5$

## 4. PRELIMINARY PAYMENTS TO THE CITY

(electricity revenue + gas revenue + nonoperating revenue) − (bond retirement + interest payment + operating expenses + payroll and employee benefits + bond reserve fund + interest reserve fund + preliminary payments to the city + ratio of I&C fund to earned revenue [electricity revenue + gas revenue + nonoperating revenue] ) $\geqslant 0$

$$(a_1 x_1 + a_2 x_2 + a_3) - (b_1 + b_2 + b_3 + b_4 + b_9 + b_{10} + b_6 c_1$$
$$+ a_7(a_1 x_1 + a_2 x_2 + a_3)) \geqslant 0,$$
$$(\$4,455,440.0\ x_1 + 28,292.0\ x_2 + 2,275.0) - (3,130.0 + 2601.2$$
$$+ 16,709.0 + 18,200.0 + 3,220.0 + 2,514.7 + 0.0151(463,000.0)$$
$$+ 0.125(\$4,455,440.0\ x_1 + 28,292.0\ x_2 + 2,275.0)) \geqslant 0, \text{ or}$$

(11.4)    $3,898,510.0\ x_1 + 24,755.5\ x_2 + d_4^- - d_4^+ = 51,375.1$

## 5. FINAL PAYMENTS TO THE CITY

(final payments to the city + preliminary payments to the city) $\geq$ (ratio of city payments to earned revenue) (electricity revenue + gas revenue + nonoperating revenue)

$$x_6 + b_6 c_1 \geq b_7 (a_1 x_1 + a_2 x_2 + a_3)$$
$$x_6 + 0.0151(463,000.0) \geq 0.14(4,455,440.0 \, x_1 + 28,202.0 \, x_2 + 2,275.0), \text{ or}$$

(11.5)  $623,761.6 \, x_1 + 3,960.8 \, x_2 - x_6 + d_5^- - d_5^+ = 6,672.8$

## 6. BREAKEVEN CONSTRAINT

(electricity revenue + gas revenue + bond fund + nonoperating revenue + depreciation + amortization + contribution from customers + salvage + Article VII fund + antitrust fund) $-$ (technical improvement + final payments to city + construction + bond retirement + interest payment + operating expenses + payroll and employee benefits + bond reserve fund + interest reserve fund + preliminary payments to the city) = 0

$$(a_1 x_1 + a_2 x_2 + x_3 + a_3 + a_4 + a_5 + a_8 + a_9 + a_{10} + a_{11}) - (x_5 + x_6 + x_7 + b_1 + b_2 + b_3 + b_4 + b_9 + b_{10} + b_6 c_1) = 0,$$
$$(4,455,440.0 \, x_1 + 28,292.0 \, x_2 + x_3 + 2,275.0 + 9,425.0 + 6.0 + 1,230.0 + 354.0 + 52.0 + 58.0) - (x_5 + x_6 + x_7 + 3,130.0 + 2,601.2 + 16,709.0 + 18,200.0 + 3,220.0 + 2,514.7 + 6,991.3) = 0, \text{ or}$$

(11.6)  $4,455,440.0 \, x_1 + 28,292.0 \, x_2 + x_3 + x_4 - x_5 - x_6 - x_7 - x_8 + d_6^- - d_6^+ = 40,052.5$

## 7. NEW BONDS

bond funds $\leq$ assumed bond upper limit

$$x_3 \leq a_6$$
$$x_3 \leq 0.0 \text{ or}$$

(11.7)  $x_3 + d_7^- + d_7^+ = 0.0$

## 8. MINIMUM ELECTRICITY RATE

electricity rate/KWH $\geqslant$ desired rate

$$x_1 \geqslant a_{12}$$
$$x_1 \geqslant 0.0144/\text{KWH, or}$$

(11.8)   $x_1 + d_8^- - d_8^+ = 0.0144$

## 9. MINIMUM GAS RATE

gas rate/MCF $\geqslant$ desired rate

$$x_2 \geqslant a_{13}$$
$$x_2 \geqslant 0.492/\text{MCF, or}$$

(11.9)   $x_2 + d_9^- - d_9^+ = 0.492$

## 10. TECHNOLOGICAL IMPROVEMENT

technological improvement fund $\geqslant$ desired ratio of technological improvement to earned revenue (electricity revenue + gas revenue + nonoperating revenue)

$$x_5 \geqslant b_5 (a_1 x_1 + a_2 x_2 + a_3),$$
$$x_5 \geqslant .005 (4{,}455{,}440.0 \, x_1 + 28{,}292.0 \, x_2 + 2{,}275.0), \text{ or}$$

(11.10)   $x_5 - 22{,}277.2 \, x_1 - 141.5 \, x_2 + d_{10}^- - d_{10}^+ = 11.4$

## 11. FINAL PAYMENT TO I&C FUND

I&C fund $\geqslant$ I&C ratio to earned income (electricity revenue + gas revenue + nonoperating revenue) + (depreciation + amortization + contribution from customers + salvage + Article VII funds + antitrust funds)

$$x_4 \geqslant a_7 (a_1 x_1 + a_2 x_2 + a_3) + (a_4 + a_5 + a_8 + a_9 + a_{10} + a_{11})$$
$$x_4 \geqslant .125(4{,}455{,}440.0 \, x_1 + 28{,}292.0 \, x_2 + 2{,}275.0) + (9{,}452.0$$
$$+ 6.0 + 1{,}230.0 + 354.0 + 52.0 + 58.2), \text{ or}$$

(11.11)   $x_4 - 556{,}930.0 \, x_1 - 3{,}536.5 \, x_2 + d_{11}^- - d_{11}^+ = 11{,}409.4$

## 12. CONSTRUCTION

construction $\geqslant$ desired construction

$$x_7 \geqslant b_8,$$
$$x_7 \geqslant 41{,}220.0, \text{ or}$$

(11.12)   $x_7 + d_{\overline{12}} - d_{12}^+ = 41{,}220.0$

The construction fund must be derived from the bond fund and I&C fund. However, there should be at least \$5 million in the I&C fund for contingencies.

$$x_7 = x_3 + (x_4 - 5{,}000.0), \text{ or}$$

(11.13)   $x_3 + x_4 - x_7 + d_{13}^+ = 5{,}000.0$

## 13. PRIORITY STRUCTURE FOR MANAGEMENT GOALS

$P_1$ = the first priority of the firm's top management, to avoid issuing any new bonds in 1970. The city issued bonds in 1969, and there is no mood among the citizens to consider new bonds. Therefore, $P_1$ is to be assigned to $d_7^-$ and $d_7^+$.

$P_2$ = the second goal of the management, to meet the current operating and payroll expenses, as well as payments of the bond principal, interest, and reserve fund expenses. However, management feels that the payment of operating and payroll expenses is twice as important as paying other expenses. Hence, $2P_2$ is assigned to $d_1^-$ and $P_2$ is assigned to $d_2^-$.

$P_3$ = the third priority of the management, to provide the payments to the city. $2P_3$ is assigned to $d_3^-$ (preliminary payments) and $P_3$ to $d_5^+$ (overachievement of the final payment to the city).

$P_4$ = the fourth priority factor, the avoidance of a shortage in the I&C fund. $P_4$, therefore, is assigned to $d_{11}^-$.

$P_5$ = the fifth priority of the management, to secure desired funds for technological improvements of the firm's operation. Hence, $P_5$ is assigned to $d_{10}^-$.

$P_6$ = the sixth goal, to secure a desirable amount of funds for continuous construction projects. $2P_6$ is assigned to $d_{12}^-$ to ensure that construction not go under the desired level, which is based on the bond and I&C funds. Then $P_6$ is assigned to $d_{13}^-$.

$P_7$ = the last goal of the management, at least to maintain the current electricity and gas service rates. Therefore, $P_7$ is assigned to $d_8^-$ and $d_9^-$.

Now the objective function can be formulated as:

(11.14) $\text{Min } Z = P_1(d_7^- + d_7^+) + 2P_2 d_1^- + P_2 d_2^- + 2P_3 d_3^- + P_3 d_5^+ + P_4 d_{11}^- + P_5 d_{10}^- + P_6 d_{13}^- + P_7(d_8^- + d_9^-)$

## RESULTS AND DISCUSSION

### A. The First Run

In the first run of the computer solution, the breakeven goal is treated as the lowest-priority goal in order to identify the revenue requirements (to achieve all the listed goals, $P_8 d_6^- + P_8 d_6^+$). The results of the first run are presented in Table 11.2. These results can be summarized as follows:

1. The electricity rate $(x_1)$ is increased to $0.01646 per KWH as compared to the projected goal of $0.0144 per KWH. This situation occurred because of the low priority that was assigned to maintaining the electricity rate $(P_7)$. In order to achieve the higher priority goals this rate had to be increased, according to this model.

2. The gas rate $(x_2)$ of $0.492 per MCF was achieved even though it was of a low-priority $(P_7)$ goal. This result is due to the fact that we did not assign any differential weights to the maintenance of existing rates between electricity and gas, and we listed the goal of electricity rate ahead of the gas rate.

Table 11.2

The First Computer Solution Results

| | |
|---|---|
| Revenue variables | |
| Electricity rate $(x_1)$ | $0.01646/KWH |
| Gas rate $(x_2)$ | 0.492/MCF |
| Bond fund $(x_3)$ | 0 |
| I & C fund $(x_4)$ | 46,220,000.00 |
| Surplus account | 23,902,560.00 |
| Expense variables | |
| Technological improvement $(x_5)$ | $447,741.00 |
| Final city payment $(x_6)$ | 5,544,180.00 |
| Construction $(x_7)$ | 41,220,000.00 |

3. The breakeven objective ($P_8$) was achieved.

4. The construction fund ($x_7$) of $41,222,000 was derived from the solution. This represents the achievement of desired construction level for the year ($P_6$).

5. The amount of funds that are to be placed in the technological improvement fund ($x_5$) is $447,741. This represents the achievement of the fifth priority goal ($P_5$).

6. The I&C fund ($x_4$) of $46,220,000 was derived from the solution. This represents the achievement of the fourth goal ($P_4$).

7. The final payment to the city is $5,544,175, which represents the complete achievement of the third goal. This amount, when added to the preliminary city payment of $6,991,300 (1.51 percent of the total assets of the firm), equals a total payment to the city of $12,535,475.

8. The goals to achieve the required payments to the operating and payroll expense and to principal, interest, and the reserve account ($P_2$) were all achieved.

9. Additional bonds did not result in the solution, as the highest priority factor was assigned to the avoidance of a new bond issue.

10. A surplus account of $23,902,560 was generated from the solution for contingencies.

Since the electricity and gas revenues constitute the major component of the firm's revenues, a change in either of these rates will have a large effect on the other financial decisions for management. According to the firm's charter, however, any change in the service rates must be voted on and passed by the city council. Therefore, it is important that management recognize the financial consequences of not raising rates in order to weigh the utility of going to the council for an increase and to present a valid justification for any increase. In the second run, the effects of the change in service rates will be analyzed.

## B. The Second Run

The first computer solution indicated that in order to achieve all the goals of the firm the electricity rate should be increased from the current $0.0144/KWH to $0.01646/KWH. In formulating corporate planning, it is extremely important for management to obtain the information concerning the degree of its goal achievements with the given revenue level. In other words, management seeks the information of its goal achievement levels when the current service rates are maintained. In the second run, therefore, maintaining the current rates is treated as the second goal. Also, the breakeven goal is assigned the third priority factor. The objective function of the second run is:

$$(11.15) \ \text{Min } Z = P_1(d_7^- + d_7^+) + P_2(d_8^- + d_8^+ + d_9^- + d_9^+) + P_3(d_6^- + d_6^+) +$$
$$2P_4 d_1^- + P_4 d_2^- + 2P_5 d_3^- + P_5 d_5^+ + P_6 d_{11}^- + P_7 d_{10}^- + P_8 d_{13}^-$$

The result of the second computer solution is presented in Table 11.3. The second run is utilized in an effort to determine the effects of not changing the current gas and electricity rates on the various other financial decisions. The results of the solution are summarized below.

1. The current electricity $(x_1)$ and gas $(x_2)$ rates were held constant at $0.0144/KWH and $0.492/MCF respectively.

2. The breakeven goal is exactly achieved.

3. The construction fund $(x_1)$ shows a balance of $33,365,590. This represents the underachievement of the goal by the amount of $7,854,410.

4. The technological improvement fund $(x_7)$ received $401,808, which achieved the goal.

5. The amount placed in the I&C Fund $(x_4)$ was $38,365,990, thus achieving the goal.

Table 11.3

The Second Computer Solution Results

| | |
|---|---|
| Revenue variables | |
| Electricity rate $(x_1)$ | $0.0144/KWH |
| Gas rate $(x_2)$ | 0.492/MCF |
| Bond fund $(x_3)$ | 0 |
| I & C fund $(x_4)$ | 38,365,590 |
| Surplus account | 17,196,460 |
| Expense variables | |
| Technological improvement $(x_5)$ | $  401,810 |
| Final city payment $(x_6)$ | 4,258,070 |
| Construction $(x_7)$ | 33,365,590 |

6. The final payment to the city ($x_6$) amounted to $4,258,070. This, when added to the $6,991,300 preliminary payment, amounted to the total of $11,249,370 paid to the city in lieu of taxes.

7. The solution indicates that all payments to the operating and payroll expenses and to principal, interest and the reserve accounts were made.

8. The bond fund ($x_3$) was held at zero. This represents, of course, the achievement of the first goal.

9. A surplus account of $17,196,460 was generated in the solution. This is the money over and above the amount that is required to fulfill the I&C fund requirements.

Maintaining the electricity rate of $0.144/KWH and changing the priority structure have caused some marked changes in the solution. The payment into the I&C fund was decreased by $7,854,410. This effect was caused by the fact that the payment to the I&C fund is based on 12.5 percent of the firms gross revenues. Therefore, maintaining the electricity service rate, when coupled with a stable demand for the service, leads to a lowering of gross revenues. Consequently, the I&C fund payment must also be decreased. This implies that less money is available to improve customer service, which is a major goal of a publicly owned utility firm.

Keeping the electricity rate also resulted in the loss of $45,932 in the technological improvements fund. Again, this fund is based on a desired ratio of 0.005 to the total revenues. With less total revenues, a smaller amount was allocated to this fund, and this result would decrease the firm's financial ability to make technological advances to improve its operational efficiency.

The total payments to the city are also significantly decreased with the same electricity rate. The difference amounts to $1,286,106 and is caused by the standard relationship between the final payment to the city and gross revenues (14 percent). The preliminary payment is of course stable because it is based on 1.51 percent of the total assets of the firm.

The amount of money that can be allocated to the construction fund is lowered by $7,854,410. A lower I&C fund means that construction level should also be lowered. It is of prime importance for the firm to realize that the underachievement of the desired construction level may have profound impact on the future services of the firm because the increasing electricity and gas requirements put a premium on capacity.

Finally, the surplus account is decreased by $6,706,099 in the second solution. This is caused by the relationship of surplus to gross revenues. Although this fund is not of primary importance to the firm's management, it does allow for more management latitude in making sound financial decisions.

# CONCLUSION

In this chapter, goal programming is applied to corporate planning through an empirical example. In order to make the example simpler, only the major items of the organization's operation were included in the model. For a complete corporate model, there may be several models for each facet of the operation, i.e. production scheduling models, capital budgeting models, etc., that are related to the overall goal programming model. Further more, many of these models for subsystems may involve stochastic features that may further complicate the overall corporate model. This study is intended to demonstrate that goal programming can be an effective vehicle to conglomerate various models of an organization so as to formulate the corporate planning model.

Although goal programming is not a complete substitute for sound management discretion in the long-range planning of an organization, it can serve as the systematic base for many planning problems. By incorporating various components of the organization and alternatives, a goal programming corporate model can effectively analyze the interrelationships and the effects of a decision on a variety of organizational goals. For example, the study presented in this chapter, the effects of changing the electricity rate on many management objectives were clearly demonstrated.

It is important to realize that the validity of a goal programming model and its resultant solutions is entirely dependent upon how the model represents the key features of the real-world problem. This becomes more evident when the model in question is for a corporate planning problem. It is imperative that key personnel in each functional area of the organization be involved in the model design. Furthermore, top management must also be directly involved so that the priority structure of management can be effectively reflected in the model.

# REFERENCES

Ansoff, H. I. "A Model for Diversification." *Management Science,* vol. 4 (July 1958), pp. 392-414.

Bellman, R. "On the Computational Solution of Linear Programming Problems Involving Almost-Block-Diagonal Matrices." *Management Science,* vol. 3 (July 1957), pp. 403-06.

Bowman, E. H. "Consistency and Optimality in Managerial Decision-Making." *Management Science,* vol. 9, no. 5 (January 1963), pp. 1-10.

Briskin, L. E. "A Method of Unifying Multiple Objective Functions." *Management Science,* vol. 12, no. 10 (June 1966), pp. 406-16.

Buffa, E. S. and Taubert, W. H. "Evaluation of Direct Computer Search Method for the Aggregate Planning Problem." *The Industrial Management Review,* vol. 8 (Fall 1967), pp. 19-36.

Charnes, A. and Cooper, W. W. "Management Models and Industrial Applications of Linear Programming." *Management Science,* vol. 4 (October 1957), pp. 38-91.
_____ and Fergusen, O. "Optimal Estimation of Executive Compensation by Linear Programming." *Management Science,* vol. 1, no. 2 (January 1955), pp. 138-51.
Cooper, W. W. *An Introduction to Linear Programming.* New York: John Wiley & Sons, Inc., 1954.
Crecine, J. P. "A Computer Simulation Model for Municipal Budgeting." *Management Science,* vol. 13, no. 11 (July 1967), pp. 786-815.
Dantzig, G. B. "Linear Programming under Uncertainty." *Management Science,* vol. 1 (April-July 1955), pp. 197-206.
_____. "On the Status of Multistage Linear Programming." *Management Science,* vol. 6 (October 1959), pp. 53-72.
_____. "Recent Advances in Linear Programming." *Management Science,* vol. 2 (January 1956), pp. 131-44.
_____. "Thoughts on Linear Programming and Automation." *Management Science,* vol. 3 (January 1957), pp. 131-39.
Dzielinski, B. P., *et al.* "Simulation Tests of Lot Size Programming." *Management Science,* vol. 9 (January 1963), pp. 229-58.
Eppen, G. D., and Gould, F. J. "A Lagrangian Application to Production Models." *Operations Research,* vol. 16 (July-August 1968), pp. 819-29.
Everett, H. "Generalized Lagrange Multiplier Method for Solving Problems of Optimum Allocation of Resources." *Operations Research,* vol. 11 (May-June 1963), pp. 399-417.
Farag, S. M. "A Planning Model for the Divisionalized Enterprise." *Accounting Review,* vol. 43 (April 1968), pp. 312-20.
Fetter, R. B. "A Linear Programming Model for Long-Range Capacity Planning." *Management Science,* vol. 7 (July 1961), pp. 507-17.
Garvin, W. W., *et al.* "Applications of Linear Programming in the Oil Industry." *Management Science,* vol. 3 (July 1957), pp. 407-30.
Henderson, A., and Schlaifer, R. "Mathematical Programming." *Harvard Business Review,* vol. 32 (May-June 1954), pp. 75-100.
Hetrick, J. C. "A Formal Model for Long-Range Planning." *Long Range Planning,* vol. 1 (March 1969), pp. 16-23.
Holt, C. C., Modigliani, F., and Simon, H. A. "A Linear Decision Rule for Production and Employment Scheduling." *Management Science,* vol. 2 (October 1955), pp. 1-30.
Jones, C. H. "Parametric Production Planning." *Management Science,* vol. 13, no. 10 (July 1967), pp. 843-66.
Magee, J. P. *Production Planning and Inventory Control.* New York: McGraw-Hill, 1967.
Manne, A. S. "Programming of Economic Lot Sizes." *Management Science,* vol. 4 (January 1958), pp. 115-35.
_____. "Linear Programming and Sequential Decisions." *Management Science,* vol. 6 (April 1960), pp. 259-67.
Masse, P., and Gilbrat, R. "Applications of Linear Programming to Inventories in the Electric Power Industry." *Management Science,* vol. 3 (January 1957), pp. 149-66.
McGarrah, R. E. *Production and Logistics Management.* New York: John Wiley & Sons, 1963.
McMillan, C. *Mathematical Programming.* New York: John Wiley & Sons, 1970.
Morris, W. T. "Diversification." *Management Science,* vol. 4 (July 1958), pp. 382-91.
Petrovic, R. "Optimization of Resource Allocation in Project Planning." *Operations Research,* vol. 16 (May-June 1968), pp. 559-68.
Rogers, J. "A Computational Approach to the Economic Lot Scheduling Problem." *Management Science,* vol. 4 (April 1958), pp. 264-91.

Sengupta, S. K., Tintner, G., and Millham, C. "On Some Theorems of Stochastic Linear Programming with Applications." *Management Science,* vol. 10 (October 1963), pp. 131-42.

Sharpe, W. F. "A Linear Programming Algorithm for Mutual Fund Portfolio Selection." *Management Science,* vol. 13, no. 7 (March 1967), pp. 499-510.

Smalter, D. J. "Analytical Techniques in Planning." *Long Range Planning,* vol. 1 (September 1968), pp. 25-33.

Steiner, G. A. *Top Management Planning.* London: The Macmillan Co., 1969.

Thompson, G. E. "On Varying the Constraints in a Linear Programming Model of the Firm." *American Economic Review,* vol. 58 (June 1968), pp. 485-95.

Wardle, P. A. "Forest Management and Operations Research: A Linear Programming Study." *Management Science,* vol. 11, no. 12 (August 1965), pp. 260-70.

Wilde, D. J. *Optimum Seeking Methods.* Englewood Cliffs, N. J.: Prentice-Hall, 1964.

Zangwill, W. I. "Non-Linear Programming Via Penalty Functions." *Management Science,* vol. 13, no. 5 (January 1967), pp. 344-58.

# Chapter 12

## Goal Programming for Academic Planning

In a business enterprise, the allocation of resources is based upon the expected return on investment. In the educational institution, however, there is no such clear decision criterion for resource allocation. In fact, the receipt of income has very little relationship to the use of university services. This weak tie between revenue and expenditure makes an efficient resource allocation in institutions of higher learning both extremely difficult and extremely important.[1]

There are several developments that have aggravated the difficulty in university management. These factors are:

1. Increase in enrollment. Expansion of enrollment has been observed in all aspects of higher education, and this trend is expected to continue in the future. According to the projection by the Department of Health, Education, and Welfare, in 1975 approximately 1,152,000 degrees will be granted compared to only 575,000 degrees in 1963.[2]

2. Expansion of programs and curriculums. Because of new development in technology and its applications, new courses and programs must be added. Also, the increasing demand for graduates with master's or doctoral degrees has forced many institutions to expand their existing programs.[3]

3. Expansion of facilities. To accommodate expanding enrollment and programs, new physical facilities are needed. Furthermore, requirements set by college associations, as well as facilities required for government research grants, have forced the expansion of facilities.

4. Increased research and service functions. Because of the increased interest of private industry and government in university research, universities play a significant role in the creation and dissemination of technology.[4] Institutions of higher learning have increased their service functions, such as regional economic research, training programs of all kinds, military research, state technical service, etc.

5. Faculty. The need for more faculty members and increasing salaries of professors has a multiplier effect in the rising costs of higher education.

6. Growth of public institutions. A significant shift of enrollment from private to public institutions has been observed in recent years. In the 10-year period preceding 1965, the proportion of students attending public schools increased from 56 percent to 66 percent. Inevitably, this trend has increased the government tax appropriation for institutions of higher learning. For example, in 1958 the total appropriations for the U.S. Office of Education for higher education amounted to about $5 million whereas in 1967 they reached almost to $1.5 billion.[5]

7. Teaching methods. There have been many innovational teaching techniques developed in the past decade, such as computer-applied courses, educational T.V., audiovisual aids, programmed education, etc.

The above factors have contributed, in varying degrees, to the growing pains of universities. The two most urgent problems in university management are obtainment of necessary funds and optimum allocation of resources based on analytical techniques.

This chapter explores the application of goal programming to planning and administration for institutions of higher education. The study presented is somewhat limited in scope and magnitude, because it simply attempts to point out the value of goal programming to academic resource allocation through a simple illustration.

## A GOAL PROGRAMMING MODEL FOR ACADEMIC RESOURCE ALLOCATION[6]

Sang M. Lee and Edward R. Clayton

The rapid rate of technological development and the growing complexity of society has brought about renewed awareness of the importance of higher education. Never in the history of our country has society directed such attention toward the broad area of higher education. Today, higher education is a $10-billion-a-year enterprise that includes 6 million students, faculty, and staff.[7]

In the early 1960's there were generous supports of higher education from many state legislatures, the federal government, and the public in general, but the period from the late 1960's has been characterized by severe financial stringency for many institutions. There may be many reasons for this trend, such as gradual reduction of federal research grants and a rate of expenditure increase in higher education that is faster than the rate of increase in state revenues. However, another important reason seems to be the switch of high priorities from higher education to more pressing social problems that require immediate attention of the government.

The rising expenditure of higher education has caused lawmakers and the public to develop a keener and more critical view of the operational efficiency of educational institutions. Institutions can no longer request prodigious sums of money from the legislature without clear justification in terms of viable goals, alternatives, and expected results. One of the most important functions of the university administrator is to acquire ever-increasing operational funds. The increasing financial pressure has greatly enhanced the importance of efficient resource allocation on the part of the institutions.

Although management science and mathematical models are developed and taught within the confines of academies, the application of these techniques for their own operation has been generally neglected.[8] Perhaps in the past academic planning and operational efficiency were of no significant importance; this is, of course, no longer the case. It is an urgent necessity for many institutions to develop a dynamic and systematic planning model for efficient resource allocation for their survival.

In recognition of the importance of planning and rational decision processes, many universities have been and are in the process of establishing formal long-range planning models as well as utilizing scientific decisions techniques. For example, the application of the Planning-Programming-Budgeting System (PPBS) to academic planning has become popular in the past several years. The purpose of the PPBS approach in the university is to achieve a more efficient resource allocation in achieving its objectives over a long-term planning horizon. Under this system, long-range planning is established for program elements such as colleges, departments, and curricula. With the limited available resources for pursuing university goals, the administrators have to eliminate some goals, postpone some, and reduce others in scale in order to fit desirable goals into practical and feasible objectives.

The PPBS is a managerial process. This system, however, does not automatically provide solutions to problems. In order to implement the system, or in addition to the formal PPB system, or even in absence of such a system, many universities have developed various analytical models.

The majority of models have focused upon a specific segment of the total institution in great depth.[9] With a number of specialized models such as prediction of student population,[10] faculty growth,[11] facility requirements,[12] expenditure analysis,[13] etc., the overall planning of the institution is attempted. The larger models have attempted to encompass functions of the total university system as well as to analyze interactions among major components.[14]

The various models introduced thus far vary considerably in their mathematical sophistication, functions, methods, purposes, subjects, and data. However, the majority of the models attempt to reduce some degree of uncertainty based primarily upon past trends or data.

The crucial issue in the administration of higher education does not end at operational efficiency, but it embodies the very purpose, function, and concept of each institution. In essence, then, the operational policy is based on the combined philosophy of many conflicting ideas of the university community, funders, administrators, faculty, students, and staff. The present situation in many universities is one of compromise and tense coexistence among all parties involved.[15] Any effective model, therefore, must be capable of reflecting the administrators' judgment about the priority of desired goals within the constraints of the existing situation. Most models introduced thus far fail to meet this requirement. The goal programming approach appears to be the most appropriate technique in developing a model to attain multiple, competitive, and often conflicting goals with varying priorities. It is the purpose of this study to present a goal programming model for an optimum allocation of resources in institutions of higher learning. Although it is possible to formulate a complex, multi-time-period model that serves the purpose of long-range planning for the entire university, the scope of this study is limited to the planning of one college within the university. In addition, the planning horizon under consideration is limited to one year. This limited scope allows a clear presentation of the methodology development of the model, and of the application potential of the study. Once the basic model is completed for a year, it can be extended— although this is not a simple task—for a longer planning horizon by forecasting parameter changes.[16] The aggregative university model can be designed when models for major components are established and their interactions are identified. However, this is more easily said than done.

## THE GENERAL MODEL

A general resource allocation planning model of a college will first be introduced. The application of the model is demonstrated in the numerical example section. For the development of a goal programming model, the following variables, constants, and constraints should be examined:

### Variables

$x_1$ = number of graduate research assistants
$x_2$ = number of graduate teaching assistants
$x_3$ = number of instructors
$x_4$ = number of assistant professors without terminal degree[17]
$x_5$ = number of associate professors without terminal degree
$x_6$ = number of full professors without terminal degree
$x_7$ = number of part-time faculty without terminal degree
$x_8$ = number of special professors without terminal degree
$x_9$ = number of staff
$y_1$ = number of assistant professors with terminal degree
$y_2$ = number of associate professors with terminal degree
$y_3$ = number of full professors with terminal degree
$y_5$ = number of part-time faculty with terminal degree
$y_6$ = number of special faculty with terminal degree
$w_2$ = total payroll increase from prior year, composed of faculty, staff, and graduate assistant salary increases

### Constants

$a_1$ = percentage of the academic staff that is classified as full-time faculty
$a_2$ = percentage of academic staff at the undergraduate level with terminal degree
$a_3$ = percentage of academic staff at the graduate level with terminal degree
$a_4$ = estimated number of undergraduate student credit hours required per session
$a_5$ = estimated number of graduate student credit hours required per session
$a_6$ = desired undergraduate faculty/student ratio
$a_7$ = desired graduate faculty/student ratio
$a_8$ = desired faculty/staff ratio
$a_9$ = desired faculty/graduate research assistant ratio
$b_{14}$ = projected undergraduate student enrollment for the coming academic year

$b_{15}$ = projected graduate student enrollment for the coming academic year
$b_{16}$ = desired percentage increase in salary for graduate assistants
$b_{17}$ = desired percentage increase in salary for faculty
$b_{18}$ = desired percentage increase in salary for staff

Maximum teaching loads, desired proportion of each faculty type, and average salary defined as:

| | | Teaching Loads | | |
| *Variable* | *Proportion* | *Undergraduate* | *Graduate* | *Salary* |
| --- | --- | --- | --- | --- |
| $x_1$ | $c_1$ | $b_1$ | $b'_1$ | $s_1$ |
| $x_2$ | $c_2$ | $b_2$ | $b'_2$ | $s_1$ |
| $x_3$ | $c_3$ | $b_3$ | $b'_3$ | $s_2$ |
| $x_4$ | $c_4$ | $b_4$ | $b'_4$ | $s_3$ |
| $x_5$ | $c_5$ | $b_5$ | $b'_5$ | $s_4$ |
| $x_6$ | $c_6$ | $b_6$ | $b'_6$ | $s_5$ |
| $x_7$ | $c_7$ | $b_7$ | $b'_7$ | $s_6$ |
| $x_8$ | $c_8$ | $b_8$ | $b'_8$ | $s_7$ |
| $x_9$ | — | — | — | $s_8$ |
| $y_1$ | $c_9$ | $b_9$ | $b'_9$ | $s_3$ |
| $y_2$ | $c_{10}$ | $b_{10}$ | $b'_{10}$ | $s_4$ |
| $y_3$ | $c_{11}$ | $b_{11}$ | $b'_{11}$ | $s_5$ |
| $y_4$ | $c_{12}$ | $b_{12}$ | $b'_{12}$ | $s_6$ |
| $y_5$ | $c_{13}$ | $b_{13}$ | $b'_{13}$ | $s_7$ |

## Constraints

### A. ACCREDITATION

1. A certain percentage of the academic staff must be full-time faculty.

$$(12.1) \quad \left(\sum_{i=3}^{6} x_i + x_8 + \sum_{i=1}^{3} y_i + y_5\right) / \left(\sum_{i=2}^{8} x_i + \sum_{i=1}^{5} y_i\right) \geqslant a_i$$

where it is assumed that the denominator is positive in all constraints.

2. A given percentage of the faculty available for undergraduate and graduate teaching duties are usually required to possess the terminal degree. If we assume for this model that $x_2$ through $x_7$ and $y_1$ through $y_3$ are available for

undergraduate teaching assignments, and $x_8$ and $y_1$ through $y_5$ are available for graduate teaching responsibilities, we may write

$$(12.2) \quad \sum_{i=1}^{3} y_i / (\sum_{i=2}^{7} x_i + \sum_{i=1}^{3} y_i) \geqslant a_2$$

$$\sum_{i=1}^{5} y_i / (x_8 + \sum_{i=1}^{5} y_i) \geqslant a_3.$$

3. There is usually a maximum number of student credit hours per session (for both graduate and undergraduate) that a faculty member may teach. It is not necessary to formulate a separate constraint for this requirement since it is easily incorporated into later constraints by selecting appropriate desired class sizes and teaching loads.

## B. TOTAL NUMBER OF ACADEMIC STAFF

One of the most important determinants of the number of academic staff requirements is the estimated number of student credit hours (both graduate and undergraduate) needed per session. With this information plus the maximum desired teaching loads of faculty members, the requirement of academic staff can be determined.

$$(12.3) \quad \sum_{i=2}^{7} b_i x_i + \sum_{i=1}^{5} b_{i+8} y_i \geqslant a_4 \text{ (undergraduate)}$$

$$\sum_{i=2}^{7} b_i' x_i + \sum_{i=1}^{5} b_{i+8}' y_i \geqslant a_5 \text{ (graduate)}.$$

Another aspect to be considered in the determination of academic staff requirements is the desired faculty/student ratio.

$$(12.4) \quad (\sum_{i=2}^{7} x_i + \sum_{i=1}^{3} y_i)/b_{14} \geqslant a_6 \text{ (undergraduate)}$$

$$(x_8 + \sum_{i=1}^{5} y_i)/b_{15} \geqslant a_7 \quad \text{ (graduate)}$$

## C.  DISTRIBUTION OF ACADEMIC STAFF

It is necessary to impose some constraints on the distribution of the academic faculty. If there were no constraints, the model would call for the most productive type of faculty in terms of teaching load, salary, and accreditation, i.e., the assistant professors with terminal degrees and instructors. In this model, we assume that the college desires to minimize the number of faculty without terminal coverage and to maximize those with terminal degrees.

$$(12.5) \quad \prod_{i=2}^{8} c_i T \leqslant \prod_{i=2}^{8} x_i$$

$$c_{12} T \leqslant y_4$$

$$\prod_{i=1}^{3} c_i T \geqslant \prod_{i=1}^{3} y_i$$

$$c_{13} T \geqslant y_5$$

where "$\pi$" represents "product" of the indicated terms and "T" represents

$$\sum_{i=2}^{8} x_i + \sum_{i=1}^{5} y_i$$

## D.  NUMBER OF STAFF

Due to the ever-increasing amount of stenographic services required by the academic staff, it is imperative, if backlogs and bottlenecks are to be avoided, that an adequate staff be provided. This objective may be incorporated into the model by designing a constraint that reflects a desired faculty/staff ratio.

$$(12.6) \quad \left( \sum_{i=2}^{8} x_i + \sum_{i=1}^{5} y_i \right) / x_9 \geqslant a_8$$

## E.  NUMBER OF GRADUATE RESEARCH ASSISTANTS

To provide adequate research support for the academic staff, it is desired to assign graduate research assistants to faculty members. This can be handled by introducing a constraint for desired faculty/graduate research assistant ratio.

$$(12.7) \quad \left( \sum_{i=3}^{8} x_i + \sum_{i=1}^{5} y_i \right) / x_1 \geqslant a_9$$

## F. SALARY INCREASE

To maintain an adequate staff, it is necessary to provide periodic salary increases. Any academic community must be cognizant of the fact that there exists a keen competition for members of its faculty. One' of the most viable means of meeting this competition is to offer salary increases according to the policy of the institution. The payroll increase constraint is:

$$(12.8) \quad b_{16}(s_1 \sum_{i=1}^{2} x_i) + b_{17}(s_2 x_3 + \sum_{i=3}^{7} s_i x_{i+1} + \sum_{i=3}^{7} s_i y_{i-2}) + b_{18}(s_8 x_9) \leqslant w$$

## G. THE TOTAL PAYROLL BUDGET

The increase in the salaries of the faculty, the staff, and graduate assistants represents only one facet of the entire budget. The total payroll budget is a major concern in a situation where limited resources are involved. The total payroll constraint can be expressed:

$$(12.9) \quad s_1 \sum_{i=1}^{2} x_i + s_2 x_3 + \sum_{i=3}^{7} s_i x_{i+1} + \sum_{i=3}^{7} s_i y_{i-2} + s_8 y_9 + w = p$$

where p represents the total payroll budget.

### Objective Function

The objective function is to minimize deviations, either negative or positive, from set goals with certain "preemptive" priority factors assigned by the dean of the college in accordance with the university policies, existing conditions, and his judgment.

## A NUMERICAL EXAMPLE

A simplified numerical example will be presented to demonstrate the application of the general model.[18] Let us assume that the Dean of the College of Business in a university provided the following priority structure for academic goals and information on constants:

### A. Priority Structures

$P_1$ = Maintain the necessary requirements for accreditation by AACSB.

$P_2$ = Assure adequate salary increases for the academic staff, graduate assistants, and general staff.

$P_3$ = Assure adequate number of faculty by meeting desired faculty/student ratios and by having instruction available for the needed student credit hours. The graduate faculty/student requirements are considered to be twice as important as the undergraduate requirement (note the weights at $P_3$ level in the objective function).

$P_4$ = Attain a desirable distribution of the academic staff with respect to rank.

$P_5$ = Maintain desired faculty/staff ratio.

$P_6$ = Maintain desired faculty/graduate research assistant ratio.

$P_7$ = Minimize cost.

Teaching Loads, Average Salaries, Desired Proportions of Total Staff

| | Teaching Load | | Desired Proportion | | |
| Variable | Undergraduate | Graduate | Maximum | Minimum | Salary |
|---|---|---|---|---|---|
| $x_1$ | 0 | 0 | — | — | $ 3,000 |
| $x_2$ | 6 | 0 | 7% | — | 3,000 |
| $x_3$ | 12 | 0 | 7 | — | 8,000 |
| $x_4$ | 9 | 0 | 15 | — | 13,000 |
| $x_5$ | 9 | 0 | 5 | — | 15,000 |
| $x_6$ | 6 | 0 | 2 | — | 17,000 |
| $x_7$ | 3 | 0 | 1 | — | 2,000 |
| $x_8$ | 0 | 3 | — | 1% | 30,000 |
| $x_9$ | — | — | — | — | 4,000 |
| $y_1$ | 6 | 3 | — | 21 | 13,000 |
| $y_1$ | 6 | 3 | — | 14 | 15,000 |
| $y_2$ | 3 | 3 | — | 23 | 17,000 |
| $y_3$ | 0 | 3 | 2 | — | 2,000 |
| $y_5$ | 0 | 3 | — | 2 | 30,000 |

## B. Constraints

The constraints in the model needed for accreditation were given the highest priority by the dean of the college. These goals will be considered first in the goal programming model, followed by the lower priority goals.

## 1. CONSTRAINTS FOR ACCREDITATION

It is required that 75 percent of the academic staff be full-time faculty according to AACSB. Since in our model $x_3$ to $x_6$, $x_8$, $y_1$ to $y_3$, and $y_5$ are considered full-time, we may write:

$$(12.10) \quad \sum_{i=3}^{6} x_i + x_8 + \sum_{i=1}^{3} y_i + y_5 - 0.75 \left( \sum_{i=1}^{8} x_i + \sum_{i=1}^{5} y_i \right) + d_1^- - d_1^+ = 0$$

It is also required that at least 40 percent of the academic teaching staff at the undergraduate level possess terminal coverage. This is expressed as:

$$(12.11) \quad \sum_{i=1}^{5} y_i - 0.40 \left[ \sum_{i=2}^{7} x_i + \sum_{i=1}^{3} y_i \right] + d_2^- - d_2^+ = 0$$

At least 75 percent of the academic staff teaching graduate studies are required to possess terminal coverage. This is expressed as:

$$(12.12) \quad \sum_{i=1}^{5} y_i - 0.75 \left[ x_8 + \sum_{i=1}^{6} y_i \right] + d_3^- - d_3^+ = 0$$

## 2. CONSTRAINTS FOR NUMBER OF ACADEMIC STAFF

To determine the faculty requirement, it is necessary to forecast the total number of student credit hours of instruction needed. In this example, the projected student enrollment is 1,820, the average number of credit hours/student taken at the college is 10, and the desired class size is set at 20. Therefore, 910 total student credit hours can be calculated by means of the following formula:

(projected enrollment) · (number of credit hours/student)/(desired class size)
$$(12.13) \quad 6x_2 + 12x_3 + 9x_4 + 9x_5 + 6x_6 + 3x_7 + 6y_1 + 6y_2 + 3y_3 + d_4^- - d_4^+ = 910$$

For the graduate student credit hours of instruction, we forecast 100 hours per session. The procedure is similar to the undergraduate forecast and the constraint becomes:

$$(12.14)\ 3x_8 + 3y_1 + 3y_2 + 3y_3 + 3y_4 + 3y_5 + d_5^- - d_5^+ = 100$$

The next aspect to be considered in the determination of the required academic staff is the desired faculty/student ratio at both the graduate and undergraduate level. The forecast enrollments in the next year at undergraduate and graduate levels are 1,820 and 100, respectively. The desired undergraduate faculty/student ratio is about 1/20 and the desired graduate faculty/student ratio is about 1/10.[19] These constraints then become, for the undergraduate requirement:

$$(12.15)\ \sum_{i=2}^{7} x_i + \sum_{i=1}^{3} y_i + d_6^- - d_6^+ = (0.05)(1{,}820) = 91$$

and for the graduate faculty:

$$(12.16)\ x_8 + \sum_{i=1}^{8} y_i + d_7^- - d_7^+ = (0.10)(100) = 10$$

## 3. CONSTRAINTS FOR THE DISTRIBUTION OF ACADEMIC STAFF

It is necessary to impose some constraints on the distribution of the academic faculty according to the desired proportion of the total faculty for each type of staff.

$$
\begin{aligned}
(12.17)\ & 0.07T - x_2 + d_8^- - d_8^+ = 0 \\
& 0.07T - x_3 + d_9^- - d_9^+ = 0 \\
& 0.15T - x_4 + d_{10}^- - d_{10}^+ = 0 \\
& 0.05T - x_5 + d_{11}^- - d_{11}^+ = 0 \\
& 0.02T - x_6 + d_{12}^- - d_{12}^+ = 0 \\
& 0.01T - x_7 + d_{13}^- - d_{13}^+ = 0 \\
& 0.01T - x_8 + d_{14}^- - d_{14}^+ = 0 \\
& 0.21T - y_1 + d_{15}^- - d_{15}^+ = 0 \\
& 0.14T - y_2 + d_{16}^- - d_{16}^+ = 0 \\
& 0.23T - y_3 + d_{17}^- - d_{17}^+ = 0 \\
& 0.02T - y_4 + d_{18}^- - d_{18}^+ = 0 \\
& 0.02T - y_5 + d_{19}^- - d_{19}^+ = 0
\end{aligned}
$$

where $T = \sum_{i=2}^{8} x_i + \sum_{i=1}^{5} y_i$.

In order to ensure adequate staff for clerical and administrative work, the desired faculty/staff ratio is set at 4 to 1 by the dean. The constraint is then:

$$(12.18) \quad T - 4x_9 + d_{20}^- - d_{20}^+ = 0$$

## 4. NUMBER OF GRADUATE RESEARCH ASSISTANTS

We set the desired faculty/graduate research assistant ratio at 5 to 1. Hence, the constraint is:

$$(12.19) \quad \sum_{i=3}^{8} x_i + \sum_{i=1}^{5} y_i - 5x_1 + d_{21}^- - d_{21}^+ = 0$$

## 5. COST OF ACADEMIC STAFF, GRADUATE ASSISTANTS, AND STAFF

The total salary increase constraint can be expressed as:

$$(12.20) \quad 0.06[3{,}000 \sum_{i=1}^{2} x_i] + 0.08(8{,}000x_3 + 13{,}000x_4 + 15{,}000x_5$$
$$+ 17{,}000x_6 + 2{,}000x_7 + 30{,}000x_8 + 13{,}000y_1 + 15{,}000y_2$$
$$+ 17{,}000y_3 + 2{,}000y_4 + 30{,}000y_5) + 0.06(4{,}000x_9) - w + d_{22}^-$$
$$- d_{22}^+ = 0$$

where there is a 6 percent increase for graduate students and staff and an 8 percent increase for faculty.

The total payroll constraint for the entire college will be:

$$(12.21) \quad 3{,}000x_1 + 3{,}000x_2 + 8{,}000x_3 + 13{,}000x_4 + 5{,}000x_5 + 17{,}000x_6$$
$$+ 2{,}000x_7 + 30{,}000x_8 + 13{,}000y_1 + 15{,}000y_2 + 17{,}000y_3$$
$$+ 2{,}000y_4 + 30{,}000y_5 + 4{,}000x_9 + w + d_{23}^- - d_{23}^+ = 0$$

## C. Objective Function

$$(12.22) \quad \text{Min } Z = P_1 \sum_{i=1}^{3} d_i^- + P_2 d_{22}^- + 2P_3 d_5^- + 2P_3 d_7^- + P_3 d_4^- + P_3 d_6^-$$

$$+ P_4 \sum_{i=8}^{13} d_i^- + P_4 d_{18}^- + P_4 \sum_{i=14}^{17} d_i^+ + P_4 d_{19}^+ + P_5 d_{20}^+ + P_6 d_{21}^+$$

$$+ P_7 d_{23}^+$$

## D. Solution

The goal programming model provides three types of solutions: (1) identification of the input (resource) requirements to attain all the desired goals; (2) the degree of goal attainments with the given inputs; and (3) the degree of goal attainments under various combinations of inputs and goal structures. This study presents three separate solutions in order to demonstrate the capability of the model. The solution is based on the GP computer program written by the first author.

### 1. THE FIRST RUN

In the first run, the above problem is solved to determine the input requirements necessary to achieve all the goals presented by the dean.[20] The results of the first run are presented below.

Goal Attainment

| | |
|---|---|
| Accreditation | Achieved |
| Salary increase | Achieved |
| Faculty/student ratios | Achieved |
| Faculty/staff ratio | Achieved |
| Faculty/distribution | Achieved |
| Faculty/graduate assistant ratio | Achieved |
| Minimize cost | $2,471,000 |

Variables

| | |
|---|---|
| $x_1 = 32$ | $x_8 = 1$ |
| $x_2 = 10$ | $x_9 = 38$ |
| $x_3 = 10$ | $y_1 = 42$ |
| $x_4 = 22$ | $y_2 = 20$ |
| $x_5 = 7$ | $y_3 = 34$ |
| $x_6 = 0$ | $y_4 = 0$ |
| $x_7 = 1$ | $y_5 = 3$ |
| $w = \$176,000$ | |

The solution of the first run indicates that all goals are achieved at the total cost of $2,471,000. Since the minimization of cost is treated as the goal with the lowest priority factor, the solution identifies the input requirements necessary to attain all the goals.

Although the result of the above solution provides valuable information to the administrator, there are a couple of points to be evaluated. First, as is usually the case, the desired faculty distribution may be impossible to obtain in reality. Second, the total cost derived by the first solution may far exceed the amount of funds the dean is able to obtain.[21]

Suppose, for example, that 1 percent of the academic staff are professors with no terminal coverage. The optimum solution called for zero. There is nothing that can be done about this situation so the constraint for this type of academic staff must be changed to read: $0.01T - x_6 - d_{12}^- + d_{12}^+ = 0$. Further, suppose the administrator believes his maximum allocation of funds will be $1,850,000. Or, in fact, suppose this is all the funds allocated to the college. This forces the right-hand side of equation (2) to become $1,850,000 instead of 0 and we are no longer considering the cost minimization as the lowest priority.

## 2. THE SECOND RUN

In the second run, the dean treats the avoidance of deficit operation as the second priority goal, after meeting accreditation requirements. Also, he adjusts the constraint concerning faculty distribution of full professors with no terminal degrees. The new objective function and the solution are presented below.

$$(12.23) \quad \text{Min } Z = P_1 \sum_{i=1}^{3} d_i^- + P_2 d_{23}^+ + P_3 d_{22}^- + 2P_4 d_5^- + 2P_4 d_7^- + P_4 d_4^- + P_4 d_6^-$$
$$+ P_5 \sum_{i=8}^{11} d_i^- + P_5 d_{13}^- + P_5 d_{18}^- + P_5 d_{12}^- + P_5 \sum_{i=14}^{17} d_i + P_5 d_{19}^+$$
$$+ P_6 d_{20}^+ + P_7 d_{21}^+$$

Goal attainment

| | |
|---|---|
| Accreditation | Achieved |
| Avoid deficit | Achieved |
| Salary increase | Achieved |
| Faculty/student ratio | Achieved |
| Faculty distribution | Not achieved—several ranks were not represented in this solution |

| Faculty/staff ratio | Not achieved—no staff |
| Faculty/graduate research assistant ratio | Not achieved |

Variables

| | |
|---|---|
| $x_1 = 0$ | $x_8 = 0$ |
| $x_2 = 9$ | $x_9 = 0$ |
| $x_3 = 20$ | $y_1 = 28$ |
| $x_4 = 20$ | $y_2 = 18$ |
| $x_5 = 7$ | $y_3 = 30$ |
| $x_6 = 1$ | $y_4 = 0$ |
| $x_7 = 1$ | $y_5 = 0$ |
| $w_1 = 135,000$ | |
| $\text{cost} = \$1,850,000$ | |

The result of the second run indicates that with $1,850,000 appropriated to the college the dean is unable to achieve all the desired goals. In fact, because of the priority structure of the goals, there is no fund available to hire any clerical staff after achieving the higher priority goals. Also, the desired faculty/graduate research assistant ratio was not attained.

Now, let us suppose that the dean of the college presented the result of the second run to the president of the university and that he was successful in obtaining an additional $120,000. Based on the result of the second computer run, the dean is aware of the fact the he should assign higher priorities to the faculty/staff and faculty/graduate research assistant ratios for an efficient operation of the college.

## 3.  THE THIRD RUN

In the third run, the dean again assigned the highest priority to the accreditation requirements, and the second priority factor on the cost minimization of $1,970,000. To ensure an adequate staff support he assigned the third priority to the faculty/staff ratio and the fourth priority to the faculty/graduate research assistant ratio. The faculty/student ratio was assigned the sixth priority, followed by the faculty distribution ratios, given the lowest priority factor.

The objective function for the third program is:

$$(12.24) \quad \text{Min } Z = P_1 \sum_{i=1}^{3} d_i^- + P_2 d_{23}^+ + P_3 d_{22}^- + P_4 d_{20}^+ + P_5 d_{21}^+ + 2P_6 d_5^-$$

$$+ 2P_6 d_7^- + P_6 d_4^- + P_6 d_6^- + P_7 \sum_{i=8}^{11} d_i^- + P_7 d_{13}^- + P_7 d_{18}^-$$

$$+ P_7 d_{12}^+ + P_7 \sum_{i=14}^{17} d_i^+ + P_7 d_{19}^+$$

The results of the program are shown below.

| | |
|---|---|
| Accreditation | Achieved |
| Salary increase | Achieved |
| Faculty/staff ratio | Achieved |
| Faculty/graduate research | |
|     assistant ratio | Achieved |
| Faculty/student ratios | Achieved |
| Faculty distribution | Not achieved—again several ranks were not presented in this solution |

Variables

| | |
|---|---|
| $x_1 = 26$ | $x_9 = 32$ |
| $x_2 = 9$ | $y_1 = 27$ |
| $x_3 = 22$ | $y_2 = 18$ |
| $x_4 = 19$ | $y_3 = 26$ |
| $x_5 = 6$ | $y_4 = 0$ |
| $x_6 = 1$ | $y_5 = 0$ |
| $x_7 = 0$ | $w = 144,000$ |
| $x_8 = 0$ | $\text{cost} = \$1,970,000$ |

As is apparent from the result above, the most important academic goals of the college are met by restructuring the priority levels and by acquiring an additional $120,000.[22]

## CONCLUSION

Virtually all models developed for university management have focused upon the analysis of input (resource) requirements. They have generally neglected or often ignored the system outputs, unique institutional values, and bureaucratic decision structures. However, these are important environmental factors that greatly influence the decision process. In this study the goal programming approach is utilized because it allows the optimization of goal attainments while permitting an explicit consideration of the existing decision environment.

Development and solving the goal programming model points out where some goals cannot be achieved under the desired policy and, hence, where trade-off must occur due to limited resources. Furthermore, the model allows the administrator to review critically the priority structure in view of the solution derived by the model.

The goal programming approach is not the ultimate solution for all budgeting and planning problems in an academy. It requires that administrators be capable of defining, quantifying, and ordering objectives. The goal programming model simply provides the best solution under the given constraints and priority structure. Therefore, some research questions concerning the identification, definition, and ranking of goals still remain. There is the need for future research to develop a systematic methodology to generate such information.

The purpose of this study is to demonstrate the application potential of goal programming to complex decision problems in university management. The model presented is a simple illustration. No doubt, each constraint requires an in-depth analysis, and it may well be a research area in itself. Furthermore, departmental interactions, boundary conditions, the administrator's own preferences, and the bureaucratic decision structure are important areas that require continuing research. It is hoped that this study will provide a guide for developing more complete models closer to reality that will perhaps encompass an entire university or a university system.

## REFERENCES

Abbey, D. and Jones, C. R. "On Modeling Education Institutions." *The Bulletin of the Institute of Management Science,* vol. 15 (August 1969), p. 67.

Bartholomew, D. J. "A Mathematical Analysis of Structural Control in a Graded Manpower System." Ford Foundation Research Program in University Administration, Paper P-3, December 1969.

Barton, R. F. "On Optimization in the American University." Texas Tech University, June 25, 1970.

Crandall, R. H. "A Constrained Choice Model for Student Housing." *Management Science,* vol. 16, no. 2 (October 1969), pp. 112-20.

Durstine, R. M. "Modeling the Allocation Process in Education." Center for Studies in Education and Development, Graduate School of Education, Harvard University, 1970.

Gani, J. "Formulae for Projecting Enrollments and Degrees Awarded in Universities." *Journal of the Royal Statistical Society*, A 126 (1963), pp. 400-09.

Halpern, J. "Bounds for New Faculty Positions in a Budget Plan." Ford Foundation Research Program in University Administration Paper P-10, University of California, Berkeley, 1970.

Judy, W., and Levine, J. B. *A New Tool for Educational Administrators: A Report to the Commission on the Financing of Higher Education.* Toronto: University of Toronto Press, 1965.

_____, and Center, S. I. Campus V Documentation, Vols. 1-6, Systems Research Group, 1970.

Koenig, H. E., Kenny, M. G. and Zemach, R. "A Systems Model for Management Planning and Resource Allocation in Institutions of Higher Education." Michigan State University, 1968.

Lee, S. M., Lerro, A. and McGinnis, B. "Optimization of Tax Switching for Commercial Banks." *Journal of Money, Credit, and Banking*, vol. 3, no. 2 (May 1971), pp. 293-303.

Lee, S. M., and Sevebeck, W. "An Aggregative Model for Municipal Economic Planning." *Policy Sciences*, vol. 1, no. 2 (June 1971), pp. 99-115.

Leimkuhler, F. F., and Cooper, M. D. "Analytical Planning for University Libraries." Ford Foundation Research Program in University Administration Paper P-1, University of California, 1970.

Marshall, K. T., Oliver, R. M., and Suslow, S. S. "Undergraduate Enrollments and Attendance Patterns." University of California Administrative Studies Project in Higher Education Report No. 4, 1970.

Menges, G., and Elstermann, G. "Capacity Models of University Management." 17th International Conference of TIMS, London, 1970.

Morse, P. M. *Library Effectiveness: A Systems Approach.* Cambridge, Mass.: M. I. T. Press, 1968.

Nordell, L. P. "A Dynamic Input-Output Model of the California Educational System" Office of Naval Research Technical Report No. 25. Berkeley: Center for Research in Management Science, University of California, 1967.

Office of Institutional Research. *OIR Series.* Toronto: University of Toronto, 1966.

Oliver, R. M. "An Equilibrium Model of Faculty Appointments, Promotions and Quota Restrictions." Ford Foundation Research Program in University Administration Report No. 68-3. Berkeley: University of California, 1968.

_____. "Models for Predicting Gross Enrollments at the University of California." Ford Foundation Research Program in University Administration, Report No. 68-3. Berkeley: University of California, 1968.

_____, Hopkins, D. S. and Armacost, R. "An Academic Productivity and Planning Model for a University Campus." University of California Administrative Studies Project in Higher Education Report No. 3, 1970.

Oliver, R. M., and Marshall, K. T. "A Constant Work Model for Student Attendance and Enrollment." Ford Foundation Research Program in University Administration, Report No. 69-1. Berkeley: University of California, 1969.

Palmour, V. E., and Wiederkehr, R. V. "A Decision Model for Library Policies on Serial Publications." 17th International Conference of TIMS, London, 1970.

Rowe, S. M., and Weathersby, G. B. "A Control Theory Solution to Optimal Faculty Staffing." Ford Foundation Research Project in University Administration Paper P-11, Berkeley: University of California, 1970.

Simon, K. A., and Lind, C. G. "Expenditure for Education." In *Projections of Educational Statistics*, pp. 41-51. Washington, D. C.: U. S. Government Printing Office, 1965.

Smith, Robert L. "Accommodating Student Demand for Courses by Varying the Classroom Size Mix." University of California Administrative Studies Project in Higher Education Report No. 4, 1970.

Spiegelman, Robert G. "A Benefit/Cost Model to Evaluate Educational Programs." *Socio-Economic Planning Sciences,* vol. 1, pp. 443-60.

Swanson, J. E. *Financial Analysis of Current Operations of Colleges and Universities.* Ann Arbor: Institute of Public Administration, University of Michigan, 1966.

Thompson, Robert K. "Higher Education Administration: An Operating System Study Utilizing a Dynamic Simulation Model." In *Corporate Simulation Models,* ed. Schrieber. Seattle: University of Washington, 1970.

Turksen, Ismail B., and Holzman, Albert B. "Micro Level Resource Allocation Models for Universities." 37th ORSA Meeting, Washington, 1970.

Weathersby, G. B. *The Development and Applications of a University Cost Simulation Model,* Berkeley: Graduate School of Business Administration and Office of Analytical Studies, University of California, 1967.

_____, and Weinstein, M. C. *A Structural Comparison of Analytical Models for University Planning.* Ford Foundation Research Project in University Administration Paper P-12. Berkeley: University of California, 1970.

Williams, Gordon, *et al.,* "Library Cost Models: Owning Versus Borrowing Serial Publications." Washington, D. C.: U. S. Department of Commerce, National Bureau of Standards, 1968.

Williams, H. *Planning for Effective Resource Allocations in Universities.* Washington, D. C.: American Council on Education, 1966.

Young, A., and Almond, G. "Predicting Distributions of Staff." *Computer Journal,* no. 4, pp. 246-50.

# Chapter 13

## Goal Programming for Decision Analysis in Government

In the same sense that a business corporation is an economic system that requires efficient planning, execution, and control by management aided by sound decision analysis, so may a government agency be regarded as such—indeed, should be. Basically, as an economic system a governmental agency has two primary functions:

1. Allocating the scarce resources in the most efficient manner to the production of goods and services that are to be consumed or invested.

2. Distributing the goods and services that it produces to the various groups that constitute the system in the most equitable manner.

As an economic system becomes more complex, the need for improved knowledge of the growing number of relationships and for an effective means of controlling these relationships becomes obvious. Government agencies have grown at a phenomenal rate during the past twenty years in their complexity and impact on the lives of the citizens. The widening scope and complexity of the government have been matched by the difficulties of coordination and decision analysis to provide needed services for the citizens. The basic decision

298

problem in government is selection of the optimum alternative among many competing programs because there is a scarcity of public resources in relation to overall demands and objectives.

There have been an abundance of studies and research on improving the efficiency of decision analysis in federal government agencies. However, only slight attention has been focused on decision analysis at municipal government level, especially for small to medium-sized municipalities.

Probably the most significant aspect of municipal government requiring effective decisions and controls is economic planning based on proper budgeting process. To date, relatively little has been done to apply an effective, fully integrated budget planning process to municipal economic planning with the aid of scientific decision techniques. This chapter presents a study that applies goal programming to economic planning in a small rural town. The model is specifically developed for the capital improvement requirements of a town for a three-year planning horizon. The author gratefully acknowledges Mayor John M. Barringer, Town Manager Mr. George Smith, and Treasurer Mrs. R. P. Brown of Blacksburg, Virginia, for their assistance in obtaining information and data for the study.

## AN AGGREGATIVE MODEL FOR MUNICIPAL ECONOMIC PLANNING[1]

### Sang M. Lee and William R. Sevebeck

## INTRODUCTION

Municipal governments have never been so complex, numerous, and expensive to operate as they are today. Thus, economic planning of the municipal government has become one of the most difficult policy decision problems for local government officials. Policy analysis on the part of the administrator takes on a greater significance as the growing pains of the municipality multiply while available resources remain relatively steady. To accomplish the most efficient resource allocation, the administrator must establish sound long-range goals and priorities among these goals. This process is the most important part of policy analysis. Therefore, policy analysis is the foundation of economic planning that establishes means of coping with growing problems and scarce resources. This paper presents a goal programming model as an aggregative model for municipal economic planning.[2]

There are at least two reasons why a study of municipal economic planning is important. First, economic systems analysis of municipal governments, especially for small local municipalities, has been generally neglected by economists.[3] The second reason for undertaking a study in the problem area is that tools and facilities exist that are effective in designing and utilizing an aggregative economic planning model.

With the increasing size and complexity of the municipal government, a systematic analysis of relationships among the growing number of factors and an effective control of these relationships become imperative. Specifically, with the limited staffs and resources of many municipal governments, the burden of complex policy analysis falls more and more heavily upon a few administrators. Probably the most important and difficult aspect requiring effective decisions and control is budget planning.

To date, there have been only limited applications of effective, fully integrated budget planning processes aided by scientific decision techniques for municipal governments. One application has been suggested by Crecine, in which a computer simulation model comprising three basic stages of the municipal budgeting process was designed.[4] This suggested process is basically one of estimating revenues and expenditures, determining whether a change in controllable revenue sources is required, and deciding on the proper course of action to achieve the desired revenue. However, this simulation model does not consider the priority structure of the municipal government in planning economic activities based on the estimated revenues.

Another popular systems-oriented procedure for the budgeting process is the planning-programming-budgeting system, or PPBS.[5] This application has received general acceptance by agencies of the federal government, as well as many state and local governments. The PPB system is an output-oriented administrative process. The underlying goal is to achieve a broad, common objective with the minimum resources in a long-range planning horizon. The general objective must be broken down into subobjectives or subgoals, which are further reduced until a set of specific program elements emerges. The unit cost of these program elements forms the basis for cost-effectiveness analysis,[6] which compares the quantity and quality of output per dollar of expenditure for alternative programs. This approach to governmental decisions parallels the "heuristic programming" procedure for ill-structured or highly complex problems. The PPB system involves systematic thinking about objectives and alternative courses of action with regard to resource constraints.

The PPBS seems to be particularly well suited to municipal economic planning programs.[7] The planning of municipal expenditures involves such problems as multiple subobjectives relating to overall goals, numerous alternatives to achieve these goals, interrelationships among subgoals, and a system of constraints, such as limited financial and temporal resources. The PPB system is a management process, yet it does not automatically provide an optimization model for economic planning. There is a need for an effective tool to design an aggregative economic model if PPBS implementation is to be truly effective.

The problem of multiple goals with varying degrees of priority levels in municipal governments quickly compounds the policy analysis process, and it prohibits the solution by a simple linear programming approach. With the application of goal programming, it is possible to render optimal solutions to complex economic planning problems for the municipal government.

## AN EMPIRICAL STUDY

### A. The Data

In order to illustrate the design of the model, empirical data acquired from the municipal government of Blacksburg, Virginia, will be used. Blacksburg has a town manager form of municipal government. It is responsible for building and maintaining public rights-of-way, municipal planning and zoning, enforcement of traffic regulations, control of public utilities, and related operations. Municipal tax and public utility rates are also set by the town. The Town Council provides funds for town beautification, recreational areas, and equipment, as well as public land and institutions.

Blacksburg is a rapidly growing university town.[8] The natural population increase, the growth of the university, and the expansion of existing industries clearly indicate the need for new business areas and additional services to be provided by the local government in the not too distant future. But, in many cases, the immediate satisfaction of these needs will not be possible by the time that they make themselves obvious. This type of municipal responsibility also entails long-range planning and the efficient allocation of financial resources to ensure that future needs are satisfactorily met on time.

A capital improvement budget for the town is prepared on an annual basis. There is no formal long-term program which includes those projects that cannot be financed during the fiscal year but can be undertaken sometime within the next five years. For the illustrative purposes of the model, a three-year planning period will be used. The town's budgeting process takes place through essentially two funds, the general fund and the water and sewer fund. In the event that bond financing is used for some series of projects, a bond fund is also maintained. However, in the model the activity pertaining to bond financing will take place in the general fund. This has been done since all capital improvements

Table 13.1

Revenue and Expenditure Variables

**A. Revenue**

| Fund | Current | Year 1 | Year 2 | Year 3 |
|---|---|---|---|---|
| General fund | | | | |
| Amount of 1-year bank loan ($1,000) | 0 | $x_1$ | $x_{12}$ | $x_{21}$ |
| Assumed credit limits ($1,000) | 0 | 100.00 | 100.00 | 100.00 |
| Amount of new bond issue ($1,000) | 0 | $x_2$ | $x_{13}$ | $x_{22}$ |
| Assumed upper limits of issue ($1,000) | | 0 | 200.00 | 250.00 |
| Property tax rate/$1,000 of assessed value | $20.00 | $b_1$ | $b_5$ | $b_9$ |
| Assumed upper limits | | 20.00 | 25.00 | 25.00 |
| Effective business license tax rate/$1,000 of receipts | $2.10 | $b_2$ | $b_6$ | $b_{10}$ |
| Assumed upper limits | | 2.30 | 2.50 | 2.50 |
| Average garbage collection charge/collection | $2.50 | $b_3$ | $b_7$ | $b_{11}$ |
| Assumed upper limits | | 2.50 | 2.75 | 3.00 |
| Real property valuation base ($1,000) | $8,503.05 | $a_1$ | $a_1$ | $a_1$ |
| Gross business receipts ($1,000) | $20,134.00 | $a_2$ | $a_2$ | $a_2$ |
| Number of garbage collections | $29,800.00 | $a_3$ | $a_3$ | $a_3$ |

| | Value | Year 1 | Year 2 | Year 3 |
|---|---|---|---|---|
| Water and sewer fund | | | | |
| Transfer from general fund | 0 | $x_5$ | $x_{16}$ | $x_{25}$ |
| Assumed lower limits | | 0 | 0 | 0 |
| Water and sewer service rate/10,000 gal. | $7.00 | $b_4$ | $b_8$ | $b_{12}$ |
| Assumed upper limits | | 7.00 | 7.50 | 8.00 |
| Water and sewer service charge base | $47,616.00 | $a_6$ | $a_6$ | $a_6$ |
| Fixed Revenues | | | | |
| General fund | | | | |
| Value used | — | $a_{12}$  240,493.00 | $a_{14}$  240,493.00 | $a_{16}$  240,493.00 |
| Water and sewer fund | | | | |
| Value Used | — | $a_{19}$  16,459.00 | $a_{20}$  16,459.00 | $a_{21}$  16,459.00 |

**B. Expenditure**

| Fund | Value | Year 1 | Year 2 | Year 3 |
|---|---|---|---|---|
| General fund | | | | |
| Percent of completion to Main St. (Clay to Roanoke) | — | $x_3$ | $x_{14}$ | $x_{23}$ |
| Percent of completion of Main St. (Faculty to north corporate limits) | — | $x_4$ | $x_{15}$ | $x_{24}$ |

(Continued)

Table 13.1 (Continued)

**B. Expenditure** (Continued)

| Fund | Value | Year 1 | Year 2 | Year 3 |
|---|---|---|---|---|
| Transfer to water and sewer fund | —— | $x_5$ | $x_{16}$ | $x_{25}$ |
| Repayment to prior year's loan ($1,000) | —— | | $x_1$ | $x_{12}$ |
| Amount of principle due on first-year bonds | —— | | $a_9 x_2$ | $a_9 x_2$ |
| Amount of principle due of second-year bonds | —— | | | $a_9 x_{13}$ |
| Prevailing interest rate on new bond issue | 0.05 | $a_{18}$ | $a_{18}$ | $a_{18}$ |
| Prevailing interest rate on bank loan | 0.06 | $a_{11}$ | $a_{11}$ | $a_{11}$ |
| Main St. improvement cost (Clay to Roanoke) ($1,000) | $87.4 | $a_4$ | $a_4$ | $a_4$ |
| Main St. improvement cost (Faculty to north corporate limits) | $1,018.9 | $a_5$ | $a_5$ | $a_5$ |
| Percent of new bond issue to be returned each year ($1,000) | 10.0 | $a_9$ | $a_9$ | $a_9$ |
| Water and sewer fund | | | | |
| Percent of completion of water tank A | —— | $x_9$ | $x_{18}$ | $x_{27}$ |
| Percent of completion of water tank B | —— | $x_{10}$ | $x_{19}$ | $x_{28}$ |
| Cost of water tank A ($1,000) | 70.0 | $a_7$ | $a_7$ | $a_7$ |
| Cost of water tank B ($1,000) | 70.0 | $a_8$ | $a_8$ | $a_8$ |

Fixed expenditures

| | | $a_{13}$ | $a_{15}$ | $a_{17}$ |
|---|---|---|---|---|
| General fund | | | | |
| Value used | — | $637,516 | 655,816 | 653,586 |
| | | $a_{22}$ | $a_{23}$ | $a_{24}$ |
| Water and sewer fund | | | | |
| Value used | — | $374,161 | 374,161 | 374,161 |

## C. Beginning and Ending Balances

| | | | |
|---|---|---|---|
| Beginning balances | | | |
| General fund | $x_6$ | | |
| Water and sewer fund | $x_8$ | | |
| Ending balances | | | |
| General fund | $x_7$ | $x_{17}$ | $x_{26}$ |
| Water and sewer fund | $x_{11}$ | $x_{20}$ | $x_{29}$ |

will be made out of either the general fund or the water and sewer fund, and provisions for transfers to the water and sewer fund will be made.

If the town is to provide an environment attractive to new industry and business, and also capable of meeting the needs of a growing population, it must consider the type of capital improvements that will contribute to achieving this goal and the relative priorities of their completion. It must also take into consideration what costs and revenue requirements will be involved and whether or not additional financing will be required.

The general fund and the water and sewer fund were broken down into variable revenues and expenditures and fixed revenues and expenditures. Included in the variable expenditures were terms that represented the amount of money to be spent in a year for each capital improvement project. Each set of variables (general fund and water and sewer fund) was given a different designation for each year. There were six basic subgoals for the three-year planning period, these subgoals being to remain solvent in each fund at the end of each fiscal year. In addition to the subgoals, each variable was constrained according to what were believed to be the objectives of the town. The list of revenue variables, revenue variable coefficients, expenditure variables, expenditure variable coefficients, fixed revenues, fixed expenditures, beginning balances, and ending balances is given in Table 13.1.

## B. Subgoals, Constraints, and Objective Functions

To summarize the long-range goals of the town, there is a serious need to increase the number of businesses and industries in the town in order to create a broader economic base, so that the town may become economically less dependent on population as such. At the same time, however, the physical needs of a rapidly expanding population must be adequately met. The financial resources of the town are limited and must be efficiently allocated to meet the present needs, and in addition provisions must be made for future needs. According to the Blacksburg Planning Commission, the most viable area for new business development is along Main St. from Faculty St. north. In addition, Main Street must be widened and improved from Clay Street to Roanoke Street. It is desirable that both of these stretches should be completed within the next three years.[9] Also, it is essential to improve water storage capacity within this period, which will require two new water tanks. These may be constructed concurrently.[10]

The goal programming model thus developed contained 29 real variables, 70 deviational and/or slack variables, and 35 constraints. The general subgoal equations, in both descriptive and algebraic forms, are presented below. It should be noted that since there are no deviational variables in the subgoal equation, slack variables $(S_i)$ will have to be inserted.

## GENERAL FUND, YEAR 1:

(short-term bank loan + amount of new bond issue + property tax revenue + business license taxes + garbage collection charges) + (fixed revenue) − [(amount spent on Main St., Clay to Roanoke) + (amount spent on Main Street, Faculty to north corporate limits) + (interest due on new bonds)] − (fixed expenditures) − (transfer to water and sewer fund) + (beginning balance) − (ending balance) = 0

$$(13.1) \quad 1000x_1 + 1000x_2 + a_1 b_1 + a_2 b_2 + a_3 b_3 + a_{12} - a_4 x_3 - a_5 x_4 - a_{18} x_2 - x_5 + x_6 - x_7 + S_1 = 0$$

## WATER AND SEWER FUND, YEAR 1:

(water and sewer service charges) − (amount spent on water tank A) − (amount spent on water tank B) + (transfer from general fund) − (ending balance) = 0

$$(13.2) \quad a_6 b_4 + a_{19} - a_7 x_9 - a_8 x_{10} - a_{22} + x_5 + x_8 - x_{11} + S_2 = 0$$

## GENERAL FUND, YEAR 2:

(short-term bank loan + amount of new bond issue + property taxes + business license taxes + garbage collection charges) + (fixed revenue) − [(amount spent on Main Street, Clay to Roanoke, year 2) + (amount spent on Main Street, North, year 2) + (interest due on year 1 bonds) + (principle due on year 1 bonds) + (interest due on year 2 bonds) + (repayment of Year 1 bank loan) + (interest due on year 1 bank loan)] − (fixed expenditures) − (transfer to water and sewer fund) + (beginning balance) − (ending balance) = 0

(13.3)   $1000x_{12} + 1000x_{13} + a_1 b_5 + a_2 b_6 + a_3 b_7 + a_{14} - a_4(x_{14} - x_3)$
$- a_5(x_{15} - x_4) - a_{18}(x_2 - a_9 x_2/1000) - a_9 x_2 - a_{18}x_{13} - 1000x_1$
$- a_{11}x_1 - a_{15}x_{16} + x_7 - x_{17} + S_3 = 0$

## WATER AND SEWER FUND, YEAR 2:

(water and sewer service charges) − (amount spent on water tank A, year 2) − (amount spent on water tank B, year 2) + (transfer from general fund) + (beginning balance) − (ending balance) = 0

(13.4)   $a_6 b_8 + a_{20} - a_7(x_{18} - x_9) - a_8(x_{19} - x_{10}) - a_{23} + x_{16} + x_{11}$
$- x_{20} + S_4 = 0$

## GENERAL FUND, YEAR 3:

(short-term bank loan + amount of new bond issue + property taxes + business license taxes + garbage collection charges) + (fixed revenue) − [(amount spent on Main Street, North, year 3) + (amount spent on Main Street, Clay to Roanoke, year 3) + (interest due on year 1 bonds) + (principle due on year 1 bonds) + (interest due on year 2 bonds) + (principle due on year 2 bonds) + (interest due on year 3 bonds) + (principle due on year 3 bonds) + (repayment of year 2 bank loan) + (interest due on year 2 bank loan)] − (fixed expenditures) − (transfer to water and sewer fund) + (beginning balance) − (ending balance) = 0

(13.5)   $1000x_{21} + 1000x_{22} + a_1 b_9 + a_2 b_{10} + a_3 b_{11} + a_{16} - a_4(x_{23} - x_{14})$
$+ a_5(x_{24} - x_{15}) - a_{18}(x_2 - 2a_9 x_2/1000) - a_9 x_2 - a_{18}(x_{13} -$
$a_9 x_{13}/1000) - a_9 x_{13} - a_{18}x_{22} - 1000x_{12} - a_{11}x_{12} - a_{17} - x_{25} + x_{17}$
$- x_{26} + S_5 = 0$

## WATER AND SEWER FUND, YEAR 3:

(water and sewer service charges) − (amount spent on water tank A, year 3) − (amount spent on water tank B, year 3) + (transfer from general fund) + (beginning balance) − (ending balance) = 0

(13.6)   $a_6 b_{12} + a_{21} - a_7(x_{17} - x_{18}) - a_8(x_{28} - x_{19}) - a_{24} + x_{25} + x_{20}$
$- x_{29} + S_6 = 0$

The category "variable revenue" (bank loans, bonds, property taxes, business license taxes, and garbage collection charges) provides for two common forms of external financing, as well as three of the most significant sources of revenue in the town budget. In addition to the above revenue sources the water and sewer fund utilizes the combined water and sewer service charges. The bank loans and bonds were left as unknown in the model, since it was desired to solve for only the amount of financing needed and no more. However, these external sources of funds were constrained by establishing upper limits of borrowing for each year. Thus, if external funds were required, they could only be appropriated up to the amounts set by the credit limits. As for the internal sources of revenue, these were precalculated before insertion into the model; therefore, they were treated as constraints in the development of the final model. The values used for each of these rates were assumed to be the permissible upper limits for each year as determined by the existing financial policies of the town. These rates multiplied by their respective computation bases yielded the maximum amounts of revenue expected in each year from these sources. This is a type of procedure that would normally be used in budget planning regardless of goal programming. The "fixed revenue" terms in the subgoals consisted of budget items that, although variable to some extent, were not considered to contribute significantly to revenue individually, nor were they related to the planned capital expenditures.

As it may be noted from the subgoal relationships, the primary financial activity takes place through the general fund, which is common to many municipalities. With respect to major capital expenditure, the general fund has been utilized for all new road construction or street improvements. The water and sewer fund has been used for construction of new water tanks. Although it is possible for needed extra funds to come into the water and sewer fund from the general fund if they are available, it was felt that any revenues accruing from water or sewer sources should be applied only to water and sewer expenditures. Also, the general fund has access to external financing, but the water and sewer fund does not. Whereas the subgoal equations allow for transfers into the water and sewer fund, there were no transfers permitted out of it into the general fund. It was also necessary to assume for purposes of the model that all capital expenditures would be on a "pay-as-you-go" basis since construction can conceivably take place in all years of the planning period. Furthermore, many of the variables that were developed had to be simplified in order to include them successfully in the model. For instance, the business license tax actually consists of a schedule rather than one flat rate. However, this does not detract from the usefulness of the model for budget programming. The model will pinpoint the expected degree of completion of projects based on the expected values of various revenues, not rates in particular.

Although the general model can be adapted to changing environmental conditions by forecasting the expected values of the revenue and expenditure coefficients for each year in the planning period, for the purpose of clarity, this was not done in this paper. Instead, it was assumed that for the three-year period under consideration, factors such as gross business receipts, water consumption, etc., remained constant.

By substituting the various coefficients, constants, and deviational variables into the subgoals and rearranging, the appropriate constraints may be written as shown in Table 13.2.

### Table 13.2

### Model Constraints

General fund, year 1:
(13.7) $1000x_1 + 950x_2 - 87.4x_3 - 1018.9x_4 - x_5 + x_6 - x_7 + S_1 = 47,305$

Water and sewer fund, year 1:
(13.8) $-x_5 - x_8 + 70x_9 + 70x_{10} + x_{11} + S_2 = 3,129$

General fund, year 2:
(13.9) $-1060x_1 - 145x_2 + 87.4x_3 + 1018.9x_4 + x_7 + 1000x_{12} + 950x_{13} - 87.4x_{14}$
$$-1018.9x_{15} - x_{16} - x_{17} + S_3 = 1,614$$

Water and sewer fund, year 2:
(13.10) $70x_9 + 70x_{10} + x_{11} + x_{16} - 70x_{18} - 70x_{19} - x_{20} + S_4 = 582$

General fund, year 3:
(13.11) $-140x_2 - 1060x_{12} - 145x_{13} + 87.4x_{14} + 1018.9x_{15} + x_{17} + 1000x_{21}$
$$+940x_{22} - 87.4x_{23} - 1018.9x_{24} - x_{25} - x_{26} + S_5 = 1,934$$

Water and sewer fund, year 3:
(13.12) $-70x_{18} - 70x_{19} - x_{20} - x_{25} + 70x_{27} + 70x_{28} + x_{29} + S_6 = 23,226$

Other constraints:

(13.13)    $x_1+d_1^-=100.0$                          $x_{16}-d_{13}^++S_{13}=0$

$x_2+S_7=0$                                $x_{17}-d_{14}^++S_{14}=0$

$x_3+d_2^-=1000.00$                        $x_{18}+d_{15}^-=1000.00$

$x_4+d_3^-=1000.00$                        $x_{19}+d_{16}^-=1000.00$

$x_5-d_4^++S_8=0$                          $x_{20}-d_{17}^++S_{15}=0$

$x_6+S_9=0$                                $x_{21}+d_{18}^-=100.00$

$x_7-d_5^++S_{10}=0$                       $x_{22}+d_{19}^-=250.00$

$x_8+S_{11}=0$                             $x_{23}+d_{20}^-=1000.00$

$x_9+d_6^-=1000.00$                        $x_{24}+d_{21}^-=1000.00$

$x_{10}+d_7^-=1000.00$                     $x_{25}-d_{22}^++S_{16}=0$

$x_{11}-d_8^++S_{12}=0$                    $x_{26}-d_{23}^++S_{17}=0$

$x_{12}+d_9^-=100$                         $x_{27}+d_{24}^-=1000.00$

$x_{13}+d_{10}^-=200.00$                   $x_{28}+d_{25}^-=1000.00$

$x_{14}+d_{11}^-=1000.00$                  $x_{29}-d_{26}^++S_{18}=0$

$x_{15}+d_{12}^-=1000.00$

## C. Solutions

The goal programming solution is primarily based upon the priority structure of the established goals. In other words, the model dictates the solution according to the policy of the administration. In this study three separate solutions are presented according to the priority structure of goals.

Because of the nature of the simplex algorithm for solving goal programming problems, if there is no evident first solution by inspection, which is the case in this model, an artificial slack variable must be provided whenever no such real variables $d_i$ exist. Because it is necessary to eliminate these slack variables from the program first, they must be assigned the uppermost priority, $P_1$, to be minimized to zero. These slack variables include the variable that represents no new bonds in year 1. This is necessary because the town issued $850,000 bonds in 1968. In all three solutions, the first goal remained the same.

## Solution 1

The administration has no definite priority structure of goals. However, the road improvement projects are thought to be the most immediate problems of the town.

### 1. PRIORITIES OF GOALS[11]

$P_2$: Since it is desirable to obtain external financing before attempting to start any projects, the second priority factor, $P_2$, is assigned to minimizing underachievement of the credit limits for bank loans and bonds.

$P_3$: The third goal is to complete the improvement of two Main Street sections. The completion of a short section between Clay Street and Roanoke Street is assumed to be twice as urgent as one between Faculty Street and the corporate limits.

$P_4$: The fourth goal is to complete the construction of two water tanks. It is desired, however, to complete tank A before tank B is considered. Therefore, twice the weight is assigned to the completion of tank A.

$P_5$: It is desirable to be able to utilize any surpluses resulting in the general fund for the water and sewer fund. Therefore, the fifth goal is the minimization of any overachievement of the transfer constraint from zero.

$P_6$: The last goal is the minimization of ending balances in the general fund and water and sewer fund.

### 2. OBJECTIVE FUNCTION

$$(13.14) \quad \text{Min } Z = P_1 \sum_{i=1}^{18} S_i + P_2 d_1^- + P_2 d_9^- + P_2 d_{10}^- + P_2 d_{18}^- + P_2 d_{19}^- + 2P_3 d_2^-$$
$$+ 2P_3 d_{11}^- + 2P_3 d_{20}^- + P_3 d_3^- + P_3 d_{12}^- + P_3 d_{21}^- + 2P_4 d_6^- + 2P_4 d_{15}^-$$
$$+ 2P_4 d_{25}^- + P_4 d_7^- + P_4 d_{16}^- + P_4 d_{15}^- + P_5 d_4^+ + P_5 d_{13}^- + P_5 d_{22}^-$$
$$+ P_6 d_5^+ + P_6 d_8^+ + P_6 d_{14}^- + P_6 d_{17}^- + P_6 d_{23}^- + P_6 d_{26}^-$$

subject to the constraints set forth above.

### 3. THE RESULTS

The results of the computer solution are presented in Table 13.3. With the output variables shown in the table, the following achievement of goals resulted.

$P_1$: Achieved—no bonds were issued in year 1

$P_2$: Achieved—the full borrowing limits were utilized for bank loans and bonds.

$P_3$: Not achieved—the Clay Street to Roanoke Street section is completed in year 2, but the Faculty Street to north corporate limits section is completed only 24.36% in year 3.

$P_4$: Not achieved—water tank A is completed only 36,82% at the end of year 3, and tank B is not even considered.

$P_5$: Achieved—road improvements exhausted funds in the general fund and there was none to be transferred to the water and sewer fund.

$P_6$: Achieved—all funds were exhausted and there were no ending balances.

Table 13.3

Solution 1:  Model Results

### A.  Revenue Variables

| | Computer Value | | |
| Fund | Year 1 | Year 2 | Year 3 |
|---|---|---|---|
| General Fund | | | |
| Amount of 1-year bank loan ($1,000's) | 100.00 | 100.00 | 100.00 |
| Amount of new bond issue ($1,000's) | 0 | 200.00 | 250.00 |
| Water and sewer fund | | | |
| Transfers from general fund | 0 | 0 | 0 |

### B.  Expenditure Variables

| | Year 1 | Year 2 | Year 3 |
|---|---|---|---|
| General fund | | | |
| Percent of completion of Main St. (Clay St. to Roanoke St.) | 60.29 | 100.00 | 100.00 |
| Percent of completion of Main St. (Faculty St. to north corporate limits) | —— | 14.49 | 24.36 |
| Ending balances | 0 | 0 | 0 |
| Water and sewer fund | | | |
| Percent of completion of water tank A | 4.47 | 8.11 | 36.82 |
| Percent of completion of water tank B | —— | —— | —— |
| Ending balances | 0 | 0 | 0 |

## Solution 2

When the outcome of the first solution was presented, the town administration realized the fallacies in its priority structure of goals. The most urgent immediate need is an additional water tank. It must be completed by the end of year 2. It is also desirable to start the second water tank as soon as possible. The second solution is based upon this modification of priorities.

## *1. PRIORITIES OF GOALS*

$P_2$: Same as solution 1.

$P_3$: The third goal is to complete the needed water tanks. Since it is desirable to complete tank A in year 2 before tank B is considered, twice the weight is assigned to the completion of tank A in comparison to tank B.

$P_4$: The fourth goal is to complete the improvement of the road sections. The shorter of the two sections, Clay Street to Roanoke Street, is assumed to be twice as urgent as Faculty Street to north corporate limits.

$P_5$: Same as solution 1.

$P_6$: Same as solution 1.

## *2. THE RESULTS*

The objective function will not be repeated here, since the only change in the function will be the transposition of $P_3$ and $P_4$. The results of the computer solution are presented in Table 13.4. With the output variables shown in the table, the following goal attainments are possible.

$P_1$: Achieved.

$P_2$: Achieved.

$P_3$: Achieved—water tank A is completed 79.75% at year 1 and 100% in year 2; water tank B is also 100% completed in year 2.

$P_4$: Not achieved—the improvement of Clay Street to Roanoke Street is completed in year 2; however, the Faculty Street to north corporate limits section is completed only 1% in year 2 and 20.54% in year 3.

$P_5$: Achieved—appropriate transfers are made.

$P_6$: Achieved.

## Table 13-4

## Solution 2:  Model Results

### A.  Revenue Variables

| Fund | Year 1 | Computer Value Year 2 | Year 3 |
|---|---|---|---|
| General fund | | | |
| Amount of 1-year bank loan ($1,000's) | 100.00 | 100.00 | 100.00 |
| Amount of new bond issue ($1,000's) | 0 | 200.00 | 250.00 |
| Water and sewer fund | | | |
| Transfers from general fund | $52,695.00 | $84,757.75 | 0 |

### B.  Expenditure Variables

| | Year 1 | Year 2 | Year 3 |
|---|---|---|---|
| General fund | | | |
| Percent of completion of Main St. (Clay St. to Roanoke St.) | — | 100.00 | 100.00 |
| Percent of completion of Main St. (Faculty St. to north corporate limits) | — | 1.00 | 20.54 |
| Ending balances | 0 | 0 | 0 |
| Water and sewer fund | | | |
| Percent of completion of water tank A | 79.75 | 100.00 | 100.00 |
| Percent of completion of water tank B | — | 100.00 | 100.00 |
| Ending balances | 0 | 0 | 0 |

## Solution 3

The town administration was not completely happy with the outcome of the model. The completion of water tank A in year 2 is necessary, but the completion of tank B in the same year is not the most urgent project. The administration desires to start tank B with the water and sewer fund alone and use all other available funds for the completion of road improvements. The third solution is based upon this modification of goals.

## *1. PRIORITIES OF GOALS*

$P_2$ : Same as solution 1.

$P_3$ : The third goal is the completion of water tank A and the road improvement between Clay Street and Roanoke Street. However, completion of tank A is given twice the weight.

$P_4$ : The fourth goal is the completion of the road improvement between Faculty Street and north corporate limits and water tank B. The road improvement is assumed to be twice as urgent as the completion of water tank B.

$P_5$ : Same as solution 1.

$P_6$ : Same as solution 1.

## *2. THE RESULTS*

The results of the computer solution of the third run are presented in Table 13.5. The degrees of goal attainments are:

$P_1$ : Achieved.

$P_2$ : Achieved.

$P_3$ : Achieved—water tank A is 79.74% completed in year 1 and 100% completed in year 2. The improvement of the Clay Street to Roanoke Street section is completed in year 2.

$P_4$ : Not achieved—the Faculty Street to north corporate limits sections is 7.87% completed in year 2 and 17.74% complete in year 3; water tank B is completed only 33.18% in year 3.

$P_5$ : Achieved—appropriate transfers are made.

$P_6$ : Achieved.

Table 13.5

Solution 3:  Model Results

### A.  Revenue Variables

| Fund | Computer Value | | |
| --- | --- | --- | --- |
| | Year 1 | Year 2 | Year 3 |
| General fund | | | |
| Amount of 1-year bank loan ($1,000) | 100.00 | 100.00 | 100.00 |
| Amount of new bond issue ($1,000) | 0 | 200.00 | 250.00 |
| Water and sewer fund | | | |
| Transfers from general fund | $52,695.00 | $14,757.98 | 0 |

### B.  Expenditure Variables

| | | | |
| --- | --- | --- | --- |
| General fund | | | |
| Percent of completion of Main St. | | | |
| (Clay to Roanoke St.) | 0 | 100.00 | 100.00 |
| Percent of completion of Main St. | | | |
| (Faculty St. to north corporate limits) | 0 | 7.87 | 17.74 |
| Ending balances | 0 | 0 | 0 |
| Water and sewer fund | | | |
| Percent of completion of water tank A | 79.74 | 100.00 | 100.00 |
| Percent of completion of water tank B | 0 | 0 | 33.18 |
| Ending balances | 0 | 0 | 0 |

The solution outcome provided by solution 3 indicates that the administration could achieve the most urgent goals and get a start toward other important goals. This solution provides the best outcome to be expected under the given decision environment. The model has shown that, with the existing limitations of financing, tax schedules, and service charges, and the priority structure, it was not possible to achieve full completion of all the projects. However, the most important goals were met to the greatest extent, consistent with the assigned priorities. This type of situation is quite common in municipal planning.

## CONCLUSION

A goal programming model has been developed specifically for the town of Blacksburg for illustrative purposes in this study. This approach can be applied, with some variations according to the characteristics of the municipal government, to many municipal planning problems. The model can be expanded to comprise a much larger scope of consideration, a finer treatment of goals, and a longer planning horizon.

The goal programming approach requires administration to identify and establish critical interrelationships, and define the relative importance of various objectives. The structure thus derived represents a given policy or set of policies. Goal programming modeling helps to determine where some incompatibility exists among goals under a given policy and where the policy must be reviewed and modified in view of the most desirable objectives.

The simulation capability of the goal programming model provides the following four important advantages:

1. It can save considerable time in planning economic activities of the municipal government.

2. The results of the model may be used for further planning, realignment of goals, and reevaluation of constraints. In this case, the upper limits of property tax, business license tax, water service, or garbage collection rates may have to be reconsidered.

3. Further, if the absolute upper limit of credit has not been used, the information derived from the model can be useful in planning for additional external financing.

4. This approach to economic planning stresses the organized and integrated understanding of objectives, the interrelationships among variables in the system, and a careful consideration of alternatives and constraints. It is believed that this type of systematic thinking by management will lead to long-range effectiveness in planning, operations, and control.

# REFERENCES

Blacksburg Planning Commission. *Business Zoning Study of Blacksburg Virginia.* Blacksburg, Virginia: 1967.

Charnes A., and Cooper, W. W. *Management Models and Industrial Applications of Linear Programming.* New York: John Wiley & Sons, 1961.

____, *et al.,* "A Goal Programming Model for Media Planning." *Management Science,* vol. 14, no. 8 (April 1968), pp. 423-30.

____. "An Extended Goal Programming Model for Manpower Planning." Management Science Research Paper No. 188. Carnegie-Mellon University, June 1969.

Committee for Economic Development. *Budgeting for National Objectives.* New York: Committee for Economic Development, 1966.

Crecine, J. P. "A Computer Simulation Model of Municipal Budgeting." *Management Science,* vol. 13, no. 11 (July 1967), pp. 786-815.

Hatry, H. P. and Cotton, J. F. *Program Planning for State, County, and City.* Washington, D. C.: George Washington University, 1967.

Ijiri, Y. *Management Goals and Accounting for Control.* Chicago: Rand McNally, 1965.

Jaaskelainen, V. "A Goal Programming Model of Aggregate Production Planning." *Swedish Journal of Economics,* vol. 71, no. 2 (1969), pp. 14-29.

Lee, S. M. and Clayton, E. "A Goal Programming Model for Academic Resource Allocation." *Management Science,* vol. 18, no. 8 (April 1972), pp. 395-408.

McGregor, R. N. "Capital Budgeting in a Small City." *Municipal Finance* (November 1961), pp. 96-100.

McKean, R. N. *Efficiency in Government Through Systems Analysis.* New York: John Wiley & Sons, 1958.

Smithies, A. "Conceptual Framework for the Program Budget." In *Program Budgeting,* ed. D. Novick. Cambridge, Mass.: Harvard University Press, 1965.

Virginia Polytechnic Institute. *Central Business District Study, Blacksburg, Virginia.* Department of Urban and Regional Planning, 1966.

Younger and Associates, Inc. "*Administrative Survey for Town of Blacksburg, Virginia– Draft.*" Blacksburg, Virginia: 1968.

# Chapter 14

## Goal Programming for Medical Care Planning

The decade of the sixties was a time of radical changes in the health care area in the United States. There were tremendous increases in the number of company-sponsored programs of health and major medical coverage, as well as significant increases in privately supported medical insurance plans. During this period, government also introduced Medicare and Medicaid programs for the aged. Congress has been studying a national health scheme that would cover everyone in the country. It appears certain that there will be more dramatic developments in the medical system in the years to come.

There are two basic factors underlying the dramatic change: inefficient delivery of health services to the population as a whole and skyrocketing medical costs.[1] These two factors are interrelated and they deserve a close investigation. The nation's total health expenditure in 1955 was $17 billion, and it reached $40 billion in 1966. It now stands at $61 billion, which is approximately 7 percent of the gross national product.[2] The national expenditure per person for medical goods and services is up to $300 a year, more than double what it was a decade ago, and is still rising rapidly.[3] Since 1966 the cost of health care has been increasing at an average annual rate of 7 percent, faster than the rate of increase in consumer prices. The fastest rising cost in health care is hospital cost, which has increased over 160 percent in the past 10 years.

In view of the rising health care expenditures and the fact that the United States spends a greater proportion of national resources for health care services than any other nation, we might expect the quality and delivery of health care services to improve. Yet, the United States has been falling rapidly behind other nations in the key indexes of national health; its rank is 14th in infant mortality rate and 18th in average life expectancy. There are many causes, such as the shortage of trained medical personnel, ineffective health insurance plans, inefficient allocation of resources, lack of division of labor, etc. Some of these are social and political factors and are hard to change. However, an efficient delivery of health services through a more scientific management is well within the powers of the medical profession.

The alarming fact remains that, although steps are being taken to find solutions, we are still far behind in establishing efficiency in health care services. It must be recognized that health care clinics are business organizations, and they are faced with many of the same problems experienced by profit-oriented firms. It is now essential to apply modern management techniques to assure the efficient utilization of medical facilities.

We have already seen that management is facing the problem of spiraling medical costs with a much less rapidly growing increase of personal income. This, however, is not the only problem that must be faced. Industrial systems have finite inputs, finite products, and a manageable balance between the use of resources and the marketability of the product. But health care systems are not so easily controllable. Their basic raw material is a morbidity statistic, and the marketing of the finished product depends not only on the uncertain level of input and the skill of processing, but on the willingness of a market to commit itself to paying for the product.[4] Furthermore, the causes of the spiraling costs indicate not only inefficient operation, but also obsolescence of facilities and functions and improper use of manpower and resources. Phillip R. Lee goes on to say:

Our national health service system is disorganized, self-duplicating, and haphazard, and it does not fulfill its mission in the case of all the people of the U.S. It seems to be obvious that the existing system requires analysis and restructuring, in the practical and businesslike manner with which any given business would approach an organizational problem . . . I suggest that the approach in health care might well follow the example of industry. We must consider the cost of operating inefficient facilities.[5]

In recognizing such management problems, the assistance of systems analysis or the management science process has been suggested to determine the most efficient way to operate the medical facility, to schedule its patients, to schedule surgery, and to free physicians and others from tasks not directly related to the care of the patient.[6] The need for a systematic management approach is also echoed by Carl Anderson, who states that "the operating expenses of the modern medical office make it imperative that efficient administrative methods be used to take care of the expenses and provide reasonable renumeration for the doctor himself."[7]

This chapter is devoted to the application of goal programming to medical care planning. More specifically, the chapter presents two separate studies. First, the goal programming approach is applied to the budget planning of relatively small health care clinics. The second study presents a goal programming resource allocation model for hospital administration.

## A BUDGET PLANNING MODEL
## FOR HEALTH CARE CLINICS[8]

### HEALTH CARE CLINICS

The development of group practice has been, in most cases, an attempt by physicians at economy, convenience, and efficiency. Group practice is a system for a cooperative practice of medicine among physicians for the purpose of pooling experience and specialities, facilities and equipment, and technical and other supporting staff, and sharing operating expenses. In short, the purpose of group practice is the improvement of the quality, quantity, and effectiveness of medical care and the reduction of operating costs. Studies have noted that group doctors see more patients (up to 25 percent more), have more free time to keep up with the latest research in their fields, have better equipment, and treat patients at a lower cost than the individual practitioner.[9]

There have been four basic origins for today's group practices (1) simple partnerships forming into complex varieties of partnerships, then moving into an integrated group practice; (2) the hospital staff where specialists practice within the hospital in an integrated manner (pressures of practice force the doctors to create their own office facilities close to the hospital for efficiency and convenience for themselves and their patients); (3) a sponsored group originated by a nonmedical third party; and (4) the organization of medical school facilities into group practice.[10] No matter what the origin, however, they all end up looking more and more alike, each occupying and staffing a medical complex or medical center and each available to the general public on the basis of service.

The matter of economics, as earlier stated, has been a profound proponent for the trend toward group practice. These clinics benefit the patient, the health professions, and the community, and they also provide extended and better service in a convenient way and help attain consistency and continuity of treatment. They also give the physicians the opportunity for consultation, research and postgraduate study, a satisfactory income, alternating working days, nights and weekends, and paid vacations.[11]

## THE MODEL

The characteristics of the budget planning model for the clinic are based upon many factors, such as the type of clinic, the medical specialty, the location and size of the clinic, etc. Hence, it is difficult to design a general model that can be applied to all types of clinics. However, once a budget planning model is developed, it can be modified to fit many other types of clinics.

In this study an orthopedic clinic is selected for the model design. It is possible to formulate a complex multi-time-period model that serves the purpose of long-range planning for the clinic. The scope of this study is limited, however, to the planning horizon of one year. It is felt that this limited scope will allow a clearer presentation of the model development. Once it is completed for a year, the basic model can be expanded for a longer planning horizon by forecasting parameter changes.

The orthopedic clinic under study is located in a medium-sized city in Virginia. It is solely concerned with the treatment of patients in need of orthopedic care on an outpatient basis.

Personnel employed by the clinic are:

> 6 orthopedic surgeons
> 1 full-time and 2 part-time nurses
> 1 full-time and 2 part-time x-ray technicians
> 1 business manager
> 6 secretaries
> 2 receptionists
> 4 office personnal
> 2 maintenance personnel

The doctors schedule their services in such a manner that they can see the majority of their patients at the clinic. However, they are responsible for filling the orthopedic needs of two hospitals in this city, as well as conducting a clinic for the treatment of those unable to pay for private treatment. The doctors' billing is handled through the clinic for all their services, and provide the sole income to the business itself.

Tables 14.1 and 14.2 outline the pertinent information needed for this study. The salaries given are an average of the salaries earned by each person in the individual category. The number of hours stated as being the physicians' weekly hours is necessarily an average; however, the physicians are salaried, so the figure given by multiplying the (hours/week) by the (salary/hour) will be an accurate average for the six doctors' income. Figures given for the x-ray, medical, and administrative and miscellaneous expenses are accurate. The total

Table 14.1

Clinic Personnel, Working Hours, and Wages[*]

| Position | Number Employed | Hours/ Week | Total Hours/ Position/Year | Salary Hour | Salary After 7% Increase | Priority for Wage Increase |
|---|---|---|---|---|---|---|
| Orthopedic surgeon | 6 | 65ea | 20,280 | $14.25 | $15.25 | 10th |
| Full-time nurse | 1 | 40 | 2080 | 2.50 | 2.68 | 5th |
| Part-time nurse | 2 | 20ea | 2080 | 2.48 | 2.65 | 6th |
| Full-time x-ray tech | 1 | 40 | 2080 | 2.19 | 2.34 | 1st |
| Part-time x-ray tech | 2 | 20ea | 2080 | 2.20 | 2.35 | 7th |
| Business Manager | 1 | 40 | 2080 | 5.31 | 5.68 | 9th |
| Secretary | 6 | 40ea | 12,480 | 2.08 | 2.23 | 2nd |
| Business office personnel | 4 | 40ea | 8320 | 1.97 | 2.11 | 4th |
| Receptionist | 2 | 40ea | 4160 | 1.94 | 2.08 | 3rd |
| Maintenance | 2 | 14ea | 1456 | 1.24 | 1.33 | 8th |

[*]Figures are based on total for the year ending December 31, 1969.
Salaries are averages of all personnel in each position category.

number of patients seen at the clinic is not a measure of individual patients, since it would be virtually impossible to determine this figure because of the number of patients who have more than one visit per year. This does not affect the accuracy of the model, however, since there is no contract or group plan billing system.

The information found in the tables is a compilation of operating revenues and expenses for the past year. In order to provide for the rising costs of the

Table 14.2

Patients, Expenses, and Equipment Replacement[*]

**Patients**

| | |
|---|---|
| Total patients last year = | 27,850 |
| Expected increase for coming year (5%) = | 1,393 |
| Total expected patients for planning year = | 29,243 |
| Average charge per patient = | $19.89 |

**Expenses**

| | Total for the Past Year | Average per Patient | Average/Patient After 5% Increase |
|---|---|---|---|
| X-ray | $11,784.64 | $0.42 | $0.44 |
| Medical supplies | 10,538.08 | 0.38 | 0.39 |
| Administrative and miscellaneous | 95,113.64 | 3.42 | 3.59 |

**Reserves for other expenses**

| | |
|---|---|
| X-ray replacement | $20,000 |
| Typewriter | 2,163.20 |
| Dictaphone | 1,359 |
| Retirement fund | 15% of total salaries |
| Continuing education of doctors | $10,000 ($8,000 last year) |

[*]Figures are based on totals for the year ending December 31, 1969.

coming year, all of the figures for the categories that will be affected by this steady rise are multiplied by 1.05. This, of course, assumes a 5% increase in costs, which has been determined to be an accurate approximation. However, the average salary increase for the clinic's personnel is set at 7 percent.

For the model design, the following variables, constants, and constraints are to be defined:[12]

# VARIABLES

$x_1$ = new hourly pay rate for physicians
$x_2$ = new hourly pay rate for full-time nurse
$x_3$ = new hourly pay rate for part-time nurse
$x_4$ = new hourly pay rate for full-time x-ray technician
$x_5$ = new hourly pay rate for part-time x-ray technician
$x_6$ = new hourly pay rate for business manager
$x_7$ = new hourly pay rate for secretaries
$x_8$ = new hourly pay rate for business office personnel
$x_9$ = new hourly pay rate for receptionists
$x_{10}$ = new hourly pay rate for maintenance personnel
$x_{11}$ = retirement fund
$x_{12}$ = fund for continuing education of physicians
$x_{13}$ = expense for new x-ray machine
$x_{14}$ = expense for new typewriters
$x_{15}$ = expense for new dictaphones
$y_1$ = required number of physicians' hours/year
$y_2$ = required number of full-time nurse hours/year
$y_3$ = required number of part-time nurse hours/year
$y_4$ = required number of full-time x-ray technician hours/year
$y_5$ = required number of part-time x-ray technician hours/year
$y_6$ = required number of business manager's hours/year
$y_7$ = required number of secretaries' hours/year
$y_8$ = required number of business office personnel hours/year
$y_9$ = required number of receptionists' hours/year
$y_{10}$ = required number of maintenance personnel hours/year
$z_1$ = x-ray expenses per patient
$z_2$ = medical expenses per patient
$z_3$ = administrative and miscellaneous expenses per patient
$z_4$ = average charge per patient

## CONSTRAINTS AND/OR GOALS

### A. Wages

It is desired that all personnel receive a 7 percent increase over the past year.

$$
\begin{aligned}
(14.1) \quad x_1 + d_1^- - d_1^+ &= \$15.27 \\
x_2 + d_2^- - d_2^+ &= 2.68 \\
x_3 + d_3^- - d_3^+ &= 2.65 \\
x_4 + d_4^- - d_4^+ &= 2.34 \\
x_5 + d_5^- - d_5^+ &= 2.35 \\
x_6 + d_6^- - d_6^+ &= 5.68 \\
x_7 + d_7^- - d_7^+ &= 2.23 \\
x_8 + d_8^- - d_8^+ &= 2.11 \\
x_9 + d_9^- - d_9^+ &= 2.08 \\
x_{10} + d_{10}^- - d_{10}^+ &= 1.33
\end{aligned}
$$

### B. Expenses

### 1. RETIREMENT FUND

The retirement fund = 15 percent of the total yearly salaries.

$$
\begin{aligned}
(14.2) \quad x_{11} - .15 \, [20,&280x_1 + 2,080x_2 + 2,080x_3 + 2,080x_4 + 2,080x_5 + \\
&2,080x_6 + 12,480x_7 + 8,320x_8 + 4,160x_9 + 1,456x_{10}] + d_{11}^- - \\
&d_{11}^+ = 0
\end{aligned}
$$

### 2. CONTINUING EDUCATION FUND

$$
(14.3) \quad x_{12} + d_{12}^- - d_{12}^+ = \$10,000
$$

### 3. X-RAY REPLACEMENT FUND

The estimated cost for a new machine is $22,000. The estimated salvage on old equipment is $2,000.

(14.4)  $x_{13} + d_{13}^- - d_{13}^+ = \$20,000$

## 4. TYPEWRITER REPLACEMENT FUND

Four new typewriters are needed at the cost of $540.80 each. The estimated total salvage on old equipment is $200.

(14.5)  $x_{14} + d_{14}^- - d_{14}^+ = \$2,163.20$

## 5. DICTAPHONE FUND

Three new dictaphones are required at the cost of $453 each.

(14.6)  $x_{15} + d_{15}^- - d_{15}^+ = \$1,359$

## C. Personnel Requirement

It is determined that the present personnel manpower level will be adequate to provide satisfactory service to the patients.

$$
\begin{aligned}
(14.7) \quad y_1 &+ d_{16}^- - d_{16}^+ = 20,280 \\
y_2 &+ d_{17}^- - d_{17}^+ = 2,080 \\
y_3 &+ d_{18}^- - d_{18}^+ = 2,080 \\
y_4 &+ d_{19}^- - d_{19}^+ = 2,080 \\
y_5 &+ d_{20}^- - d_{20}^+ = 2,080 \\
y_6 &+ d_{21}^- - d_{21}^+ = 2,080 \\
y_7 &+ d_{22}^- - d_{22}^+ = 12,480 \\
y_8 &+ d_{23}^- - d_{23}^+ = 8,320 \\
y_9 &+ d_{24}^- - d_{24}^+ = 4,160 \\
y_{10} &+ d_{25}^- - d_{25}^+ = 1,456
\end{aligned}
$$

## D. Expenses per Patient

The expenses per patient are broken down into three classifications: x-ray expenses, medical expenses, and administrative and miscellaneous expenses.

## 1. X-RAY EXPENSES PER PATIENT

(14.8)   $z_1 + d_{26}^- - d_{26}^+ = \$0.44$

## 2. MEDICAL EXPENSES PER PATIENT

(14.9)   $z_2 + d_{27}^- - d_{27}^+ = \$0.39$

## 3. ADMINISTRATIVE AND MISCELLANEOUS EXPENSES

(14.10) $z_3 + d_{28}^- - d_{28}^+ = \$3.59$

### E. Break-Even Constraint

In order to determine the reasonable charge per patient ($z_4$) that will provide enough resources to achieve desired goals, a breakeven constraint must be introduced. This constraint can be used in two different ways: (1) to determine the required charge per patient to achieve all the goals; and (2) to determine the degree of goal achievement with a given charge per patient.

$$(14.11)\ 29{,}243z_4 - [(20{,}280x_1 + 2{,}080x_2 + 2{,}080x_3 + 2{,}080x_4 + 2{,}080x_5 \\ + 2{,}080x_6 + 12{,}480x_7 + 8{,}320x_8 + 4{,}160x_9 + 1{,}456x_{10}) + (x_{11} + \\ x_{12} + x_{13} + x_{14} + x_{15}) + (29{,}243z_1 + 29{,}243z_2 + 29{,}243z_3)] + \\ d_{29}^- - d_{29}^+ = 0$$

## MODEL RESULTS

### A. The First Run

The business manager must determine the economic goals of the clinic for the coming year in order to establish the budget planning model. This process usually involves a group decision by the business manager and physicians. The business manager lists the following goals in descending order of importance:

1. Provide job security to all personnel by avoiding underutilization of their regular working hours.
2. Provide an adequate (7 percent) wage increase to all personnel in keeping with the economic trend (see Table 14.1 for priority weights).
3. Provide funds for expenses per patient.
4. Provide funds for equipment replacements.
5. Provide reserve for the retirement fund.
6. Provide funds for continuing education fund.
7. Achieve the breakeven goal in the operation.

The objective function for the first run is:

$$(14.12) \quad \text{Min } Z = P_1 \sum_{i=16}^{25} d_i^- + (10P_2 d_4^- + 9P_2 d_7^- + 8P_2 d_9^- + 7P_2 d_8^- + 6P_2 d_2^-$$
$$+ 5P_2 d_3^- + 4P_2 d_5^- + 3P_2 d_{10}^- + 2P_2 d_6^- + P_2 d_1^-) + P_3(d_{26}^- +$$
$$d_{27}^- + d_{28}^-) + P_4(d_{13}^- + d_{14}^- + d_{15}^-) + P_5 d_{11}^- + P_6 d_{12}^- +$$
$$P_7(d_{29}^- + d_{29}^+)$$

In the first run, the above model is solved to determine the input requirements necessary to achieve all the goals presented by the business manager. Consequently, the breakeven goal is rated as the least important. The results of the first run are presented below.

*GOAL ATTAINMENT*

      Job security: Achieved
      Wage increase: Achieved
      Patient expenses: Achieved
      Equipment replacement: Achieved
      Retirement fund: Achieved
      Continuing education: Achieved
      Breakeven: Achieved

*VARIABLES*

| | | |
|---|---|---|
| $x_1 = 15.27$ | $x_5 = 2.35$ | $x_9 = 2.08$ |
| $x_2 = 2.68$ | $x_6 = 5.68$ | $x_{10} = 1.33$ |
| $x_3 = 2.65$ | $x_7 = 2.23$ | $x_{11} = 59{,}684.98$ |
| $x_4 = 2.34$ | $x_8 = 2.11$ | $x_{12} = 10{,}000$ |

$$x_{13} = 20,000 \qquad y_4 = 2,080 \qquad y_{10} = 1,456$$
$$x_{14} = 2,163.20 \qquad y_5 = 2,080 \qquad z_1 = 0.44$$
$$x_{15} = 1.359 \qquad y_6 = 2,080 \qquad z_2 = 0.39$$
$$y_1 = 20,280 \qquad y_7 = 12,480 \qquad z_3 = 3.59$$
$$y_2 = 2,080 \qquad y_8 = 8,320 \qquad z_4 = 21.21$$
$$y_3 = 2,080 \qquad y_9 = 4,160$$

The solution of the first run indicates that all goals are achieved at the total cost of $620,341.12. The charge per patient $(z_4)$ required to break even is $21.21, which is a 6.6 percent increase from the last year's figure of $19.89. Since the breakeven in the operation is treated as the goal with the lowest priority factor, the solution identifies the input requirements necessary to attain all the goals. It is clear that the set of goals defined by the business manager are quite realistic as they can be completely attained with a charge per patient that is only 6.6 percent above the last year's figure.

Although the result of the above solution provides some valuable information concerning the operation of the clinic, there is a point to be evaluated further. The charge per patient of $21.21 required to break even is a little more than what the business manager and physicians want to charge. They feel that the increase should not exceed 5 percent above the last year's charge. Hence, the charge per patient should be $20.88 or less. It is evident that this desired charge rate will not generate enough resources to attain all the goals listed by the business manager. Now, the problem becomes achievement of most desirable goals with the fixed resources rather than identifying the resource requirements to attain all goals.

## B. The Second Run

Based on the results of the first run, the business manager and physicians have decided to readjust the hierarchy of goals. Since it is decided that the charge per patient should not exceed $20.88, this should be treated as the first goal. The following additional goal constraint must be introduced.

$$(14.13) \quad z_4 + d_{30}^- - d_{30}^+ = 20.88$$

Furthermore, the breakeven goal is no longer the least important as in the first run. This goal will be treated as the second goal. Knowing the fact that some goals cannot be fully attained with the above fixed resources, the business manager and physicians have decided to treat their own salary increases as the lowest priority goal. In addition, they decided to limit the increase to only 5 percent. The following two more additional constraints must be introduced in the model:

$$(14.14) \quad x_1 + d_{31}^- - d_{31}^+ = 16.03$$
$$x_6 + d_{32}^- - d_{32}^+ = 5.96$$

The hierarchy of goals for the second run is:

1. Limit the increase of the charge per patient to 5 percent.
2. Achieve breakeven in the operation.
3. Provide job security to all personnel by avoiding underutilization of their regular working hours.
4. Provide funds for expenses per patient.
5. Provide an adequate (7 percent) wage increase to all personnel except the business manager and physicians (see Table 14.1 for priority weights) and provide funds for expenses per patient.
6. Provide funds for equipment replacement.
7. Provide reserve for the retirement fund.
8. Provide funds for continuing education fund.
9. Provide pay increase of 5 percent to the business manager and physicians.

The objective function for the second run is:

$$(14.15) \quad \text{Min } Z = P_1 d_{30}^+ + P_2 (d_{29}^- + d_{29}^+) + P_3 \sum_{i=16}^{25} d_i^- + P_4 (d_{26}^- + d_{27}^- + d_{28}^-) +$$
$$(8P_5 d_4^- + 7P_5 d_7^- + 6P_5 d_9^- + 5P_5 d_8^- + 4P_5 d_2^- + 3P_5 d_3^- +$$
$$2P_5 d_5^- + P_5 d_{10}^-) + P_5 (d_{26}^- + d_{24}^- + d_{28}^-) + P_6 (d_{13}^- + d_{14}^- + d_{15}^-)$$
$$+ P_7 d_{11}^- + P_8 d_{12}^- + (2P_9 d_{32}^- + P_9 d_{31}^-)$$

## GOAL ATTAINMENT

Charge per patient: Achieved
Break even: Achieved
Job security: Achieved
Expenses per patient: Achieved
Wage increase and expenses per patient: Achieved

Equipment replacement: Achieved
Retirement fund: Achieved
Continuing education: Achieved
Pay increase for business manager and doctors: Not achieved

## VARIABLES

| | | |
|---|---|---|
| $x_1$ = 14.72 | $x_{11}$ = 59,684.98 | $y_6$ = 2,080 |
| $x_2$ = 2.68 | $x_{12}$ = 10,000 | $y_7$ = 12,480 |
| $x_3$ = 2.65 | $x_{13}$ = 20,000 | $y_8$ = 8,320 |
| $x_4$ = 2.34 | $x_{14}$ = 2,163.20 | $y_9$ = 4,160 |
| $x_5$ = 2.35 | $x_{15}$ = 1,359 | $y_{10}$ = 1,456 |
| $x_6$ = 5.58 | $y_1$ = 20,280 | $z_1$ = 0.44 |
| $x_7$ = 2.23 | $y_2$ = 2,080 | $z_2$ = 0.39 |
| $x_8$ = 2.11 | $y_3$ = 2,080 | $z_3$ = 3.59 |
| $x_9$ = 2.08 | $y_4$ = 2,080 | $z_4$ = 20.88 |
| $x_{10}$ = 1.33 | $y_5$ = 2,080 | |

The solution of the second run indicates that all goals are achieved except the 5 percent pay increase for the business manager and doctors. In fact, the business manager receives the 5 percent increase, but doctors receive only a 3.3 percent increase. As is apparent from the results above, the most important goals of the clinic are met by restructuring the priority levels while holding the increase of the charge per patient to only 5 percent above the last year's figure.

## CONCLUSION

This study presents the goal programming approach to aggregative budget planning model for health care clinics. The study indicates that the model results provide a sound planning basis for the health care clinic. Through the analysis of resource requirements, the desired charge per patient, and the trade-offs among the set of goals, the business manager can establish an aggregative budget planning on a more sound basis.

## A RESOURCE ALLOCATION MODEL
## FOR HOSPITAL ADMINISTRATION[13]

In recent years, hospital administration has become a very complex management process. The great demand for hospital care is understandable in view of the increased concern for health care on the part of the American population as a whole, increased institutional protection for health and accident, and of course increasing population. Rapidly rising salaries of medical personnel, coupled with these factors, have accelerated the increase of hospital costs. However, another important contributor to the cost increase is inefficient resource allocation and ineffective utilization of existing facilities, a result of the increased complexity of hospital operations.

The administration of virtually every hospital is a unique management problem. It would be difficult to find two hospitals that offer identical services to the same type of patients through identical management processes. Hence, it is difficult to design a general model that can be applied to all hospitals. However, the basic functions of the hospital are more or less universal among all types of medical facilities. Therefore, once an aggregative resource allocation model is designed for a hospital, it can be easily modified to fit the unique characteristics of the hospital for application.

In this study, a community hospital located in a small city (population approximately 25,000) in southeastern Virginia is selected for the model design. The community hospital under study serves a predominantly rural community within a radius of twenty miles. With no resident physicians, patients are generally admitted by their personal physicians or through the emergency ward. The hospital's emergency room is staffed by local doctors on a rotation basis according to an agreement with the hospital. The hospital has 200 beds and employs 184 employees, excluding local physicians.

## A SIMPLE MODEL

It is possible to formulate a complex multiyear resource allocation model that serves the purpose of long-range planning for the hospital. The scope of this study is limited, however, to the planning horizon of one year. It is felt that this limited scope will allow a clearer presentation of the model development. Once it is completed for one year, the basic model can be expanded for a longer planning horizon by forecasting parameter changes.

Tables 14.3 and 14.4 outline the model variables and other pertinent information needed for this study. The salaries given are an average of the salaries earned by each person in the individual personnel category. The figures for each category are arbitrarily determined upon the request of the hospital administrator. The personnel classifications were made in relation to the assignment of personnel expenses within the various accounting designations utilized by the hospital. Although a number of split assignments are possible and often practiced, an attempt is made here to minimize these for the model design.

For the model formulation, the following goals and constraints are to be examined:

## A. Goals and Their Priorities

The administrator must determine the goals of the hospital and their priorities in order to accomplish the optimum allocation of resources. This process usually involves a group decision by the hospital administrator and the board of directors. The administrator lists the following goals in order of importance.

1. Secure the necessary manpower to provide adequate services to the patient. The administrator feels that the existing personnel will be sufficient to provide adequate services for the coming year.
2. Replace and/or acquire new equipment that is required to provide the services of the hospital (this figure should be in addition to funds provided by depreciation).
3. Provide adequate pay increases to all personnel in keeping with the economy and the community labor market (see Table 14.3 for the administrator's desired pay increases).
4. Achieve the desired personnel/patient ratio so as to provide satisfactory service (see the constraints).
5. Achieve the desired distribution of each personnel category (see Table 14.3).
6. Achieve the desired nurse/staff, nurse/professional[14] and professional/staff ratios (see the constraints).
7. Minimize costs and break even in the operation.

## B. Constraints

## 1. PERSONNEL REQUIREMENT

The hospital presently employs 184 persons, and the administrator feels that the existing personnel must be retained in order to provide satisfactory services to the patient.

$$(14.16) \quad \sum_{i=1}^{27} x_i + d_1^- - d_1^+ = 184$$

Table 14.3

Hospital Personnel, Desired Personnel Proportions,

Average Salaries, and Desired Pay Increases

| Variable[*] | Position | Desired % of Total Employees | Average Salary | Desired % Pay Increase |
|---|---|---|---|---|
| $x_1$ | Nursing service administration | 2.61 | $ 9,850 | 7 |
| $x_2$ | Medical & surgical nurse | 24.98 | 8,850 | 8 |
| $x_3$ | Medical & surgical orderly | 4.89 | 6,000 | 6 |
| $x_4$ | Pediatric nurse | 1.63 | 9,525 | 8 |
| $x_5$ | Obstetric nurse | 2.72 | 8,925 | 8 |
| $x_6$ | Newborn nursery nurse | 4.34 | 7,350 | 8 |
| $x_7$ | Operating & recovery room nurse | 2.72 | 11,175 | 9 |
| $x_8$ | Delivery room nurse | 2.18 | 8,100 | 8 |
| $x_9$ | Service & supply room nurse | 1.09 | 6,675 | 6 |
| $x_{10}$ | Sterile supply nurse | 1.63 | 7,525 | 6 |
| $x_{11}$ | Emergency room nurse | 1.63 | 9,900 | 9 |
| $x_{12}$ | Intensive care nurse | 1.09 | 13,450 | 10 |
| $x_{13}$ | Nurse in practical nurse program | 0.11 | 6,000 | 7 |
| $x_{14}$ | Laboratory technician | 6.78 | 6,300 | 6 |
| $x_{15}$ | Pathologist | 3.26 | 11,500 | 10 |
| $x_{16}$ | Electrocardiology technician | 0.27 | 6,400 | 7 |
| $x_{17}$ | Cardiologist | 0.54 | 10,750 | 10 |
| $x_{18}$ | Radiologist technician | 2.18 | 9,350 | 7 |
| $x_{19}$ | Radiologist | 3.26 | 22,400 | 10 |
| $x_{20}$ | Pharmacist | 0.54 | 2,400 | 0 |
| $x_{21}$ | Anesthesiologist | 1.63 | 15,800 | 10 |
| $x_{22}$ | Medical recorder | 2.72 | 8,550 | 7 |
| $x_{23}$ | Dietician | 12.49 | 5,525 | 6 |
| $x_{24}$ | Plant operation & maintenance | 2.18 | 7,025 | 6 |
| $x_{25}$ | Housekeeping | 5.97 | 5,125 | 6 |
| $x_{26}$ | Laundry & linen | 0.54 | 5,150 | 6 |
| $x_{27}$ | Administrative service | 5.97 | 10,200 | 5 |

[*]The variables indicate the number of employees in each classification.

## 2. NEW EQUIPMENT

A new x-ray equipment is required if the x-ray service is to be continued for the coming year. The new equipment is estimated to cost $8,000. Also, it is desired to reserve $200,000 in the contingency fund for emergencies.

$$(14.17) \quad z_1 + d_2^- - d_2^+ = \phantom{00}8,000$$
$$z_2 + d_3^- - d_3^+ = 200,000$$

Table 14.4

Patient Days, Expenses, and Reserves

**Patient days**

| | |
|---|---|
| Last year's total | 54,619 |
| Expected increase for coming year (5%) | 2,731 |
| Expected total for the planning year | 57,350 |

**Expenses**

| Variable | Category | Total for past year | Total for coming year[*] (5% increase) |
|---|---|---|---|
| $y_1$ | Nursing division | $130,071 | $136,575 |
| $y_2$ | Physicians' fee (emergency ward) | 231,762 | 243,350 |
| $y_3$ | General services (x-ray, medical supplies, etc.) | 245,167 | 257,425 |
| $y_4$ | Administration | 36,667 | 38,500 |
| $y_5$ | Miscellaneous | 182,524 | 191,650 |

**Reserves for coming year**

| Variable | Category | Amount |
|---|---|---|
| $z_1$ | Radiology equipment | $ 8,000 |
| $z_2$ | Contingency reserve | 200,000 |

[*]It is assumed that all expenses will increase by the amount of 5 percent in the planning year. This figure appears to be quite a reasonable estimate according to the past records.

## 3. EMPLOYEE PAY INCREASE

The administrator feels that the minimum pay increase should be 5 percent and the maximum should be 10 percent for any given personnel category. The figure before each group of variables (also see Table 14.3) is the proposed pay increase.

(14.18) $.05(10,200x_{27}) + .06(6,000x_3 + 6,675x_9 + 7,525x_{10} + 6,300x_{14} + 5,523x_{23} + 7,025x_{24} + 5,125x_{25} + 5,150x_{26}) + .07(9,850x_1 + 6,000x_{13} + 6,450x_{16} + 9,350x_{18} + 8,550x_{22}) + .08(8,850x_2 + 9,525x_4 + 8,925x_5 + 7,350x_6 + 8,100x_8) + .09(11,175x_7 + 9,900x_{11}) + .10(13,450x_{12} + 11,500x_{15} + 10,750x_{17} + 22,400x_{19} + 15,800x_{21}) + d_4^- - d_4^+ = z_3$

In the above constraint, $z_3$ represents the total amount of salary increase for the hospital personnel.

## 4. PERSONNEL/PATIENT RATIO

The administrator feels that the desired personnel/patient ratio is 0.8/1, that is, to have four employees for every five patients. The hospital forecasts the total of 57,350 patient days for the coming year, including the baby days. Each employee is assumed to receive an average of two weeks' vacation for the year. No overtime is considered, although this is often inevitable to meet unexpected service loads. The desired personnel/patient ratio can be expressed by:

(14.19) $(5 \cdot 50) \sum_{i=1}^{27} x_i / 57,350 + d_5^- - d_5^+ = 0.80$

## 5. PERSONNEL DISTRIBUTION

According to the trend of demand for hospital services, the administrator has established the desired number of employees in each personnel classification as a proportion of the total employees as shown in Table 14.3. If we denote $a_i$ as the desired number of employees in the ith category as a proportion of the total number of employees, 27 separate equations can be expressed by a general equation as:

(14.20) $x_i - a_i + d_{i+5}^- + d_{i+5}^+ = 0$ $(i = 1, 2, 3, \ldots, 27)$

For example, for the desired number of nurses in the nursing service administration, the constraint will be $x_1 - 4 + d_6^- - d_6^+ = 0$.

## 6. NURSE/STAFF, NURSE/PROFESSIONAL, AND PROFESSIONAL/STAFF RATIOS

In addition to the adequate distribution of each personnel classification, it is considered important that there be one staff member for every four nurses, 10 nurses for every professional employee, and two staff members for every professional employee.

$$(14.21) \quad x_1 + x_2 + \sum_{i=4}^{13} x_i - 4 \left(x_3 + \sum_{i=22}^{27} x_i\right) + d_{33}^- - d_{33}^+ = 0$$

$$x_1 + x_2 + \sum_{i=4}^{13} x_i - 10 \sum_{i=14}^{21} x_i + d_{34}^- - d_{34}^+ = 0$$

$$x_3 + \sum_{i=22}^{27} x_i - 2 \sum_{i=14}^{21} x_i + d_{35}^- - d_{35}^+ = 0$$

## 7. COST MINIMIZATION

The total cost of the hospital operation is calculated in this constraint. Hence, this constraint identifies the resource requirements to achieve the set of goals presented by the administrator. If a certain maximum resource is previously determined, it could be used so as to identify the degree of goal achievements with the given resources. In order to simplify the constraint, let $b_i$ represent the average salary figure for the ith personnel category as shown in Table 14.3 (i.e., \$9,850 for the nursing service administration, etc.). Then, the cost minimization constraint will be:

$$(14.22) \quad \sum_{i=1}^{27} b_i x_i + \sum_{i=1}^{5} y_i + \sum_{i=1}^{2} z_i + d_{36}^- - d_{36}^+ = 0$$

## 8. THE OBJECTIVE FUNCTION

The objective function of the model is to minimize deviations for the goal constraints with certain priorities assigned to them.

$$(14.23) \quad \text{Min } Z = p_1 d_1^- + p_2(d_2^- + d_3^-) + p_3 d_4^- + p_4 d_5^- + p_5 \sum_{i=6}^{32} d_i^- + p_6 \sum_{i=33}^{35} d_i^+ + p_7 d_{36}^-$$

## C. The Solution

The above goal programming problem is solved by the computer program written by the author. The results of the run are presented below:

### 1. GOAL ATTAINMENT

Manpower for service ($p_1$): Achieved
Equipment acquisition ($p_2$): Achieved
Employee pay increases ($p_3$): Achieved
Personnel to patient ratio ($p_4$): Achieved
Distribution of personnel ($p_5$): Achieved (not precisely)
Personnel ratios ($p_6$): Achieved (not precisely)
Minimize cost: Not possible

### 2. VARIABLES

| | | |
|---|---|---|
| $x_1$ = 4.8 | $x_{13}$ = .2 | $x_{25}$ = 11 |
| $x_2$ = 46 | $x_{14}$ = 12.5 | $x_{26}$ = 1 |
| $x_3$ = 9 | $x_{15}$ = 6 | $x_{27}$ = 11 |
| $x_4$ = 3 | $x_{16}$ = .5 | $y_1$ = \$136,575 |
| $x_5$ = 5 | $x_{17}$ = 1 | $y_2$ = 243,350 |
| $x_6$ = 8 | $x_{18}$ = 4 | $y_3$ = 257,425 |
| $x_7$ = 5 | $x_{19}$ = 6 | $y_4$ = 38,500 |
| $x_8$ = 4 | $x_{20}$ = 1 | $y_5$ = 191,650 |
| $x_9$ = 2 | $x_{21}$ = 3 | $z_1$ = 8,000 |
| $x_{10}$ = 3 | $x_{22}$ = 5 | $z_2$ = 200,000 |
| $x_{11}$ = 3 | $x_{23}$ = 23 | $z_3$ = 120,250 |
| $x_{12}$ = 2 | $x_{24}$ = 4 | $d_{35}^+$ = 2,307,085 |

The solution of the above model indicates that all the goals can be achieved at the total cost of \$2,307,085. Since cost minimization is treated as the goal with the lowest priority factor, it is impossible to minimize the cost to zero. Some of the personnel ratios are not completely compatible. Therefore, there were slight deviations from the desired figures, but they were very minor indeed.

Over 75 percent of the hospital's expenses are billed on a "cost plus" basis. As long as the hospital can justify these expenses as real and essential for the care of the patient and these expenses do not exceed the limits set by such agencies as Medicare or Blue Cross and Blue Shield, the existing "essential" services need not be curtailed as a cost-saving measure. The services included in this model were all considered essential to the operation of the hospital, thus the model results would be most useful in planning the budget and resource allocation for the planning period.

Although the results of the above solution provide some valuable information concerning the operation of the hospital, there are some points to be examined further. First, the above model ignored the sources of resource, i.e., the expected number of patient days and revenue per patient day. Secondly, the administrator's plan for service improvements was very limited in order to simplify the model. For example, the simple model is strictly for the resource allocation to maintain the *status quo* of hospital services. However, the hospital is to open a new wing at the beginning of the new fiscal year. The new facilities include 65 new beds, a new emergency ward, new x-ray rooms, new delivery rooms, and new offices. With such new facilities, the hospital is planning to expand its services and functions. This will of course increase the operating costs. Since the hospital has been running at near capacity and often overcrowded, a significant increase in patient days is expected, which in turn will result in increased revenues. The expanded model will be presented to encompass all the realistic features of the hospital operation with the expanded facilities.

## THE EXPANDED MODEL

The example of the community hospital discussed above can be expanded by using another approach to goal programming. A number of additions are made to the simple model to demonstrate the multitude of uses and applications of goal programming. Some of the goals are changed, others have been shifted in the priority structure, and still others are new to the example.

In addition to the three types of variables presented in the simple model, a group of revenue variables are added in this model. These revenue variables are generated to estimate resources at the various operating conditions possible for the hospital. Any number of revenue variables may be employed depending upon the particular accounting system of the hospital and according to the types of patients with different insurance plans and other sources of funds. The total resources of the hospital will be determined by the sum of the product of revenue variables multiplied by the forecast number of patients. This total expected revenue actually determines the degree of goal attainment for the planning year.

Hospitals operate on a continuous basis. There are few major changes in personnel possible or desirable. Hence, the average salary of each personnel classification has no significance in the planning of resource allocation. In this example, therefore, the total payroll figures for each personnel classification are used for the analysis. Tables 14.5 and 14.6 provide the variables and other pertinent information for the model formulation. As is apparent in the tables, the expanded model is entirely based on the ambitious plans for a dramatic expansion of the hospital services as a result of the opening of the new wing at the hospital. The increased services will require additional personnel. It is estimated that there should be approximately 5 percent increase in the number of nurses; 4 percent increase in professionals, medical recorder, and dietician;

Table 14.5

Hospital Personnel, Total Salaries, Desired Salary Increases,
and Expected Personnel Increases

| Variable[*] | Position | Total Salary | Desired % Pay Increase | Required % Personnel Increase |
|---|---|---|---|---|
| $x_1$ | Nursing service administration | $ 47,275 | 5 | 5 |
| $x_2$ | Medical & surgical nurse | 407,050 | 5 | 5 |
| $x_3$ | Medical & surgical orderly | 54,025 | 5 | 5 |
| $x_4$ | Pediatric nurse | 28,600 | 5 | 5 |
| $x_5$ | Obstetric nurse | 44,600 | 5 | 5 |
| $x_6$ | Newborn nursery nurse | 58,850 | 5 | 5 |
| $x_7$ | Operating & recovery room nurse | 55,900 | 5 | 5 |
| $x_8$ | Delivery room nurse | 32,425 | 5 | 5 |
| $x_9$ | Service & supply room nurse | 13,325 | 5 | 5 |
| $x_{10}$ | Sterile supply nurse | 22,600 | 5 | 5 |
| $x_{11}$ | Emergency room nurse | 29,600 | 5 | 5 |
| $x_{12}$ | Intensive care nurse | 26,900 | 5 | 5 |
| $x_{13}$ | Nurse in practical nurse program | 6,000 | 5 | 5 |
| $x_{14}$ | Laboratory technician | 78,975 | 4 | 4 |
| $x_{15}$ | Pathologist | 69,075 | 4 | 4 |
| $x_{16}$ | Electrocardiology technician | 3,225 | 4 | 4 |
| $x_{17}$ | Cardiologist | 10,775 | 4 | 4 |
| $x_{18}$ | Radiology technician | 37,450 | 4 | 4 |
| $x_{19}$ | Radiologist | 134,450 | 4 | 4 |
| $x_{20}$ | Pharmacy technician | 2,400 | 4 | 4 |
| $x_{21}$ | Anesthesiologist | 47,425 | 4 | 4 |
| $x_{22}$ | Medical recorder | 42,725 | 3 | 4 |
| $x_{23}$ | Dietician | 126,900 | 3 | 4 |
| $x_{24}$ | Plant operation & maintenance | 28,075 | 5 | 10 |
| $x_{25}$ | Housekeeping | 56,500 | 5 | 10 |
| $x_{26}$ | Laundry & linen | 5,150 | 5 | 10 |
| $x_{27}$ | Administrative service | 112,125 | 5 | 10 |

[*]The variables represent the total amount of payroll in each personnel classification.

and 10 percent increase in the maintenance, housekeeping and administrative staff.

It is also expected that the number of patient days will increase approximately 20 percent in the coming year. The increased patient days will

## Table 14.6

### Patient Days, Expenses, and New Equipments and Facilities

**Forecast patient days**

| | |
|---|---:|
| Patients generating $44 per day (20% increase) | $ 17,400 |
| Patients generating $48 per day (20% increase) | 30,624 |
| Patients generating $51 per day (20% increase) | 21,576 |
| Outpatients generating average of $10 (30% increase) | 2,500 |

**Expenses**

| Variable | Category | |
|---|---|---:|
| $y_1$ | Nursing administration (25% increase) | $170,719 |
| $y_2$ | Physicians' fee (emergency room) – (25% increase) | 304,188 |
| $y_3$ | General service (x-ray, medical supplies, etc.) (30% increase) | 334,653 |
| $y_4$ | Administration (30% increase) | 50,050 |
| $y_5$ | Miscellaneous (20% increase) | 229,980 |

**New equipment and facilities**

| Variable | Category | |
|---|---|---:|
| $z_1$ | Radiology equipment | $ 10,000 |
| $z_2$ | Staff physician | 80,000 |
| $z_3$ | Research–children's disease & maladjustment | 100,000 |
| $z_4$ | Improvement of emergency room facilities | 118,000 |
| $z_5$ | Mobile clinic facilities | 160,000 |
| $z_6$ | Research on birth defects | 200,000 |
| $z_7$ | Contingency reserve | 200,000 |
| $z_8$ | Foundation grant for research on birth defects | 60,000 |

generate additional revenues. However, the revenue source will be more diverse in this model. Patients with different insurance coverage generate different revenue per patient day, i.e. those with Medicare or Medicaid contribute $44 per patient day, those with group insurance plans (Blue-Cross, Blue-Shield) contribute $48, people with private insurance plans contribute $51, etc. The average patient who requires emergency-room facilities generates $10. Table 14.6 also indicates the increases in the hospital expenses, as well as costs of new equipment and facilities required of the expanded hospital services.

## A. Goals and Their Priorities

With the limited resources, the cost minimization or breakeven goal is no longer the least important as was the case in the simple model. The administrator lists the following goals in the ordinal ranking of importance:

1. Secure adequate personnel and funds for the increased expenses of maintaining the existing level of hospital services.
2. Break even in the operation of the hospital.
3. Employ additional personnel for the expanded hospital services.
4. Purchase the new radiology equipment.
5. Reserve $200,000 in the contingency reserve for emergencies.
6. Set up the research facilities for birth defects.
7. Grant the desired salary increases.
8. Employ two staff physicians.
9. Improve the facilities of the emergency room.
10. Acquire the mobile clinic facilities.
11. Set up the research facilities for children's diseases and maladjustments.

## B. Constraints

## 1. PERSONNEL REQUIREMENTS AND FUNDS FOR EXPENSES

The first goal of the administrator is to maintain the present level of hospital services by securing the existing personnel and funds to meet the inflated expenses. To formulate this constraint, payroll and other expenses must be analyzed while estimating the total revenue. If we denote $c_1$ as the total payroll for all personnel classifications and $c_2$ as the total expenses (see Table 14.5 for actual value), the constraint will be:

$$(14.24) \quad \sum_{i=1}^{27} x_i + d_1^- - d_1^+ = c_1$$

$$\sum_{i=1}^{5} y_i + d_2^- - d_2^+ = c_2$$

$$TR - (\sum_{i=1}^{27} x_i + \sum_{i=1}^{5} y_i) + d_3^- - d_3^+ = 0$$

where TR = 17,400(44) + 30,624(48) + 21,576(51) + 2,500(10).

## 2. EMPLOYMENT OF ADDITIONAL PERSONNEL

To accommodate the expanded and improved services resulting from the opening of the new wing, employment of additional personnel is necessary. The greatest need will be in the maintenance, housekeeping, and administrative positions (10 percent), followed by the nurses (5 percent), professionals and supporting staff (4 percent). The increase in personnel will be expressed by the increase of payroll in the above mentioned four basic personnel groups. The actual employment and assignment of new employees will not be discussed here.

$$(14.25) \quad d_3^+ - (.05 \sum_{i=1}^{13} x_i + .04 \sum_{i=14}^{23} x_i + .10 \sum_{i=24}^{27} x_i) + d_4^- - d_4^+ = 0$$

## 3. NEW RADIOLOGY EQUIPMENT

$$(14.26) \quad z_1 + d_5^- - d_5^+ = 10,000$$

## 4. CONTINGENCY RESERVE FOR EMERGENCIES

$$(14.27) \quad z_7 + d_6^- - d_6^+ = 200,000$$

## 5. RESEARCH FACILITIES FOR BIRTH DEFECTS

The hospital has received an assurance from a foundation that if the hospital engages in birth-defects research it will be granted $60,000 support. The administrator would like to establish this research facility (including the research staff), although it will take a sizable amount of funds initially on the part of the hospital, because it may receive future research grants from government and/or foundations.

$$(14.28) \quad z_6 + d_7^- - d_7^+ = 200,000 - 60,000 = 140,000$$

## 6.  SALARY INCREASES

The administrator has determined that salary increases should be relatively small because of the major expenses the hospital has to absorb in the coming year. He assigns the greatest weight to salary increases for the nurses, followed by increases for the maintenance-housekeeping-administrative group, medical recorder and dietician, and the professionals (see Table 14.5 for the desired percentage increases).

$$(14.29) \quad d_4^+ - .05 \sum_{i=1}^{13} x_i + d_8^- - d_8^+ = 0$$

$$d_8^+ - .04 \sum_{i=14}^{21} x_i + d_9^- - d_9^+ = 0$$

$$d_9^+ - .03 \sum_{i=22}^{23} x_i + d_{10}^- - d_{10}^+ = 0$$

$$d_{10}^+ - .05 \sum_{i=24}^{27} x_i + d_{11}^- - d_{11}^+ = 0$$

## 7.  STAFF PHYSICIANS

One of the chief complaints at the hospital is that there is no resident physician available. To eliminate this criticism, the administrator would like to employ two staff physicians.

$$(14.30) \quad z_2 + d_{12}^- - d_{12}^+ = 80,000$$

## 8.  IMPROVEMENT OF THE EMERGENCY ROOM FACILITIES

With the addition of new wing, it is expected that an increased number of emergency cases will be handled. Accordingly, it is desirable to improve the facilities with new and additional equipment.

$$(14.31) \quad z_4 + d_{13}^- - d_{13}^+ = 118,000$$

## 9. MOBILE CLINIC FACILITIES

The hospital is located in the heart of Appalachia. Numerous pockets of poverty exist in the area served by the hospital. A service that could be a great help for these areas would be a mobile clinic facility. Once the project is on the way, there are strong possibilities of receiving government grants to continue the service. The total project is expected to cost $160,000 to start.

$$(14.32) \quad z_5 + d_{14}^- - d_{14}^+ = 160,000$$

## 10. RESEARCH FACILITIES FOR CHILDREN'S DISEASES AND MALADJUSTMENTS

The hospital serves a community with a large number of young children. This is partially because of the two large colleges within its area of service. A realistic addition to the hospital's services would be expanded research facilities and personnel for the testing of children's diseases and maladjustments.

$$(14.33) \quad z_3 + d_{15}^- - d_{15}^+ = 100,000$$

## 11. BREAKEVEN

The hospital must at least break even in its operation in order to continue providing health care services in the community. In this constraint, employment of additional personnel and salary increases are combined for the purpose of simplicity.

$$(14.34) \quad TR - (\sum_{i=1}^{27} x_i + \sum_{i=1}^{5} + .10 \sum_{i=1}^{13} x_i + .08 \sum_{i=14}^{21} x_i + .07 \sum_{i=22}^{23} x_i +$$
$$.15 \sum_{i=24}^{27} x_i) - (z_1 + z_2 + z_3 + z_4 + z_5 + z_6 + z_7 + z_8) + d_{16}^- - d_{16}^+$$
$$= 0$$

## 12. OBJECTIVE FUNCTION

When the model treats the breakeven as the goal with the lowest priority, the solution identifies the resource requirements to achieve the stated goals. In this model, however, the resource is already determined as the result of the forecast patient days and expected research grants. Therefore, we are no longer permitted to assign the lowest priority to breakeven, and consequently the second priority factor is assigned to this constraint. This procedure will identify the degree of goal achievements with the predetermined total resources.

$$(14.35) \quad \text{Min } Z = p_1 d_3^- + p_2(d_{16}^- + d_{16}^+) + p_3 d_4^- + p_4 d_5^- + p_5 d_6^- + p_6 d_7^- + p_7(4d_8^- + 3d_9^- + 2d_{10}^- + d_{11}^-) + p_8 d_{12}^- + p_8 d_{13}^- + p_{10}d_{14}^- + p_{11}d_{15}^-$$

## C. Solution

The solution of the model indicates the following goal attainments and variables.

### 1. GOAL ATTAINMENT

Personnel and expenses for service: Achieved
Employment of additional personnel: Achieved
Radiology equipment: Achieved
Contingency fund for emergencies: Achieved
Research facilities for birth defects: Achieved
Salary increases: Achieved
Staff physician: Achieved
Improvement of the emergency room facilities: Achieved
Mobile clinic facilities: Not completely achieved
Research of children's disease and maladjustment: Not achieved

### 2. VARIABLES

The new payroll figures and expenses are as follows:

| Personnel Group | Salary |
|---|---|
| Nursing service administration | $ 52,003 |
| Medical and surgical nurse | 447,755 |
| Medical and surgical orderly | 59,428 |
| Pediatric nurse | 31,460 |
| Obstetric nurse | 49,060 |
| Newborn nursery nurse | 64,735 |
| Operating and recovery room nurse | 61,490 |

| | |
|---|---:|
| Delivery room nurse | 35,667 |
| Service and supply room nurse | 14,657 |
| Sterile supply nurse | 24,860 |
| Emergency room nurse | 32,560 |
| Intensive care nurse | 29,590 |
| Nurse in practical nurse program | 6,600 |
| Laboratory technician | 85,293 |
| Pathologist | 74,601 |
| Electrocardiology technician | 3,483 |
| Cardiologist | 11,637 |
| Radiology technician | 40,446 |
| Radiologist | 145,206 |
| Pharmacy technician | 2,592 |
| Anesthesiologist | 51,219 |
| Medical recorder | 45,716 |
| Dietician | 135,783 |
| Plant operation and maintenance | 32,286 |
| Housekeeping | 64,975 |
| Laundry and linen | 5,923 |
| Administrative service | 119,974 |

*Expenses*

| | |
|---|---:|
| Nursing division | $170,719 |
| Physicians fee | 304,188 |
| General service | 334,653 |
| Administration | 50,050 |
| Miscellaneous | 229,980 |

*New Equipment and Facilities*

| | |
|---|---:|
| Radiology equipment | $ 10,000 |
| Staff physician | 80,000 |
| Research on children's disease and maladjustment | 0 |
| Improvement of emergency room facilities | 118,000 |
| Mobile clinic facilities | 148,639 |
| Research on birth defects | 200,000 |
| Reserve for emergency | 200,000 |
| Research grant from government or foundation for birth defects | 60,000 |

The above result of the model indicates that most of the important goals are achieved with the total revenue of $3,435,228. The mobile clinic is short of the target of $160,000. However, the research on children's disease and maladjustment has to be canceled because of the lack of funds.

The goal programming analysis allows management to evaluate the hospital's current rates, cost structures, personnel distribution, and other desired programs in an effort to identify the optimum resource allocation mix. This approach can be used as a general framework for continuous management planning in order to ensure an efficient operation of the hospital.

## CONCLUSION

This study attempts to apply the approach of goal programming to design an aggregative resource allocation model of a hospital. From the literature available it appears that thus far no such application has ever been made.

Most of the management science models developed for health care delivery systems have focused upon the analysis of input (resource) requirements and demand forecasts. They have often ignored the system outputs, unique organizational characteristics, and behavioral aspects of the decision process. However, these are important decision variables. The goal programming approach allows the optimization of goal attainments while permitting an explicit consideration of the resource requirements, sources of resources, and the administrator's priority among multiple conflicting goals.

## REFERENCES

Anderson, Carl W. "Administration of Group Practice–Business Manager's Point of View." *Group Practice*, vol. 18 (July 1968), pp. 43-49.

Burney, Leroy E. "Impact Forces on Rising Costs of Health Care and Health Care Facilities." In *Conference on Costs of Health Care Facilities.* Washington, D. C.: National Academy of Sciences, 1968.

Davis, Joseph B. "Increasing Productivity of Physicians." *Group Practice*, vol. 18 (July 1968), pp. 19-23.

Degen, Joseph W. "The Conference." In *Conference on Costs of Health Care Facilities,* Washington, D. C.: National Academy of Sciences, 1968.

Lee, Phillip R. "Charge to the Conference." In *Conference on Costs of Health Care Facilities.* Washington, D. C.: National Academy of Sciences, 1968.

*Physicians' Visits.* Vital and Health Statistics, Series No. 10, No. 49.

Souder, James J. "The Planning Process in Health Care." In *Conference on Costs of Health Care Facilities.* Washington, D. C.: National Academy of Sciences, 1968.

"Special Report: The $60-Billion Crisis over Medical Care." *Business Week*, January 17, 1970, pp. 50-64.

Strickey, J. M. "Administration of Group Practice–A Physician's Point of View." *Group Practice,* vol. 18 (July 1968), pp. 36-42.

# PART 4

## FINAL REMARKS

# Chapter 15

## Implementation of Goal Programming

During the past twenty years, there has been significant progress in the field of management science. Many new techniques have been developed through technical breakthroughs, and new applications of existing techniques have been explored in various decision problems. The greatest advance in management science, however, has occurred in the implementation of management science for decision analysis.

### THE SELLING OF GOAL PROGRAMMING

In the early stages of management science development, a practicing management scientist had to be a supersalesman as well as a competent scientist. Indeed, management scientists were lonely people; they were regarded by executives as whiz-kids who talked only to each other and computers. As time has passed, however, the function of management science has changed dramatically. Most executives and administrators have seen the value of management science. They recognize that it is inevitably a factor in survival and success in today's technological society. In fact, managers not only recognize the contribution of management science but also take pride in the fact that their decision analysis is based on a scientific approach. The basic question managers raise today concerning management science is not whether to use it but how they can use it to maximum benefit. The selling of goal programming, therefore, should not be as hard a job as it would have been 20 years ago.

In order to apply goal programming to decision analysis, a management scientist should be able to present the concept and benefits of goal programming to management. Furthermore, he must obtain the full cooperation and confidence of top management. The application of goal programming requires information such as management philosophy, policies, management goals and their priorities, and the relationships among pertinent decision variables. Much of such vital information can be obtained only from top management. The management scientist must have the full confidence of the manager, so that some of his philosophy and confidential data can be reflected in the model. Unless the model contains such information, the solution derived by the model may simply be an expensive exercise that would never be implemented by the decision maker.

In order to put goal programming to managerial use, we must be able to present some advantages of goal programming. Discussion of this subject necessarily involves the changing nature of problems that make many conventional techniques less effective for decision analysis.

Many of the formal solution techniques we have used for decision analysis have been primarily concerned with making the "right" choice among a set of alternatives. In other words, the basic problem has been selecting the "superior" over "inferior" alternatives. However, the nature of major decision problems has changed drastically in recent years, and a serious doubt has been raised as to the adequacy of many solution techniques for contemporary problems. Many of today's decision problems that require scientific analysis do not lend themselves to a clear-cut solution by certain cardinal criteria. For example, many social problems involve multiple incompatible goals in the jungle of conflicting interests. The question is not selecting the "social good" over the "social evil," but often the choice of evil vs. evil or social good vs. social good.

Let us consider the example of strip mining. In this day of environmental concern, any sensible person, without mentioning the naturalists and conservationists, can easily tell us all the problems involved in strip mining. However, we should not forget the increasing energy demand in the United States. It is estimated that the national energy demand will double by 1980. With the rapidly diminishing natural gas reserves and the remote possibility of sufficient nuclear energy production for the total energy demand, increased coal production is an important national concern. We simply cannot afford the luxury of deciding whether we should or shouldn't mine coal. Rather, the decision problem is that of determining the most efficient way to mine coal, i.e., by conventional underground mining, strip mining, or some combination of both methods. Both methods of mining contribute to social good. Also, both contribute to social problems. Strip mining, although it is in most cases the more economical method of producing coal, involves many ecological problems. On the other hand, conventional underground mining creates health problems for workers—black-lung disease, the constant danger, working in a cramped space, etc.

Could we say, then, that strip mining ought to be banned altogether to preserve the environment? That implies that we would rather allow greater danger to miners working underground to satisfy the greater demand for coal. As it stands now, our society simply cannot afford to abandon coal production from strip mining. At least one thing is certain. Unless a new technological breakthrough dictates otherwise, the national demand for coal has to be met in some way. In achieving this goal, there are multiple conflicting subgoals and interests.

One of the basic questions may be whether we should meet the coal requirement at the expense of ecology or at the expense of workers and increased taxes. It may be possible to achieve the goal with minimal damage to both the ecology and the workers. For example, it may be possible to restrict strip mining to certain terrains with limited degree of slope so that the reclaimation of land can be relatively easily achieved. Furthermore, as the mining firm may be required to post a bond, which is sufficient to redevelop the land, in order to receive a strip mining permit; some states are already enforcing such regulations. On the other hand, efficiency of the conventional underground mining operation can be further improved with the application of new technology.

The point of this discussion is that problems such as this cannot easily be solved by the traditional numerical analysis. In fact, many contemporary decision problems faced by industry, government, and other institutions will increasingly require more elusive and abstract objective functions. The objective function no longer will be restricted to a cardinal criterion; rather it will involve general criteria related to public value. Certainly, costs will remain an important decision variable, because they determine the resource requirements. However, their function will be shifted from that of the objective criterion to a decision constraint.[1]

The question still remains as to whether we can use the conventional numerical objective function approach (e.g., linear programming) for today's complex decision problems. It is apparent that solution methods based on numerical criteria are not capable of analyzing problems that involve highly abstract objective criteria such as welfare to the taxpayer, public health, academic standards, consumer protection and satisfaction, community image, etc. It should be pointed out that numerical solution techniques are still being used for contemproary social decision problems by estimating the numerical measures of abstract objective criteria in terms of a convenient numerical value, such as utilities, profits, etc. However, the process often results in a considerable degree of fabrication and distortion of information in order to express abstract criteria in numerical values. Hence, the model solution is of very little value to the decision maker. The only alternate method to the numerical approach for problems involving multiple conflicting objective criteria is the ordinal solution approach. Goal programming appears to be the most appropriate technique, at least at the present stage of management science development, for such complex decision problems. This is the primary advantage of goal programming.

In discussing the weaknesses of the numerical approach, Bartee states:

> A central issue as to the adequacy of our solution techniques seems to lie within the assumption that to be scientific requires one to confine his measurements to numerical entities. Although numerical measurement has considerable merit, it is neither a necessary nor a sufficient condition for science. A numerical attribute is an attribute that has a number assigned to it in accordance with certain rules. If the concepts within a science are not numerical, then its laws cannot be expressed in the form of equations, yet, they are laws all the same. A discipline that formulates and tests such laws and theories is a science without numerical measurement being a necessary condition.[2]

Although the author does not necessarily agree with Bartee's statement that non-numerical concepts cannot be expressed in the form of equations, his basic idea is clear. Bartee continues:

> The numerical methods, when confronted with an abstract objective function must either (1) distort and fabricate knowledge of the utilities and the relations between courses of action or (2) declare that these problems which involve abstract utility are outside the scope of the numerical approach. To pursue the first alternative can never be justified. The second alternative would not be too serious if we could be confident that the excluded problems would not be of any particular consequence or importance. As already discussed, this does not seem to be true. Therefore, the numerical approach is clearly faced with either fabrication and distorting

the measurement of utility or ignoring some of the more significant problems that are to confront man in the future.[3]

Now it should be obvious that goal programming on the basis of ordinal solution is an effective approach for decision problems involving abstract objective function. A successful selling of the goal programming approach also requires recognition of its limitations. Goal programming is not the answer to all decision problems. In fact there are a great number of problems that cannot be solved by goal programming. Furthermore, goal programming will not replace the subjective aspects of decision making for the decision maker. It is indeed quite possible that goal programming may impose additional burdens on the decision maker by forcing him to analyze many aspects of the problem that have been completely neglected before. This process is indeed an important characteristic of goal programming that distinguishes it from other techniques.

It is important to return to some of the limitations of goal programming presented in Chapter 1. The inherent limitation of a mathematical model is that it is an approximate representation of reality. Thus, it should be obvious that there is no guarantee that the optimal solution of a model will also be the optimal solution for the real problem under study. Nevertheless, if the model solution provides significant improvement over other solutions, it would warrant the use of goal programming.

## SOME THOUGHTS ON THE ANALYSIS OF MULTIPLE GOALS

Even if we accept the notion that an ordinal solution is superior to the numerical approach for problems involving multiple conflicting goals, there still remains the problem of establishing priorities for the goals. It is usually not possible for a management scientist to be fully aware of the goals of the organization, since these are usually only implicit in the philosophy of management, rather than explicitly stated. Also, some outcomes of a decision contain aspects that are relevant to the objectives of the organization but cannot be easily evaluated by the management scientists in terms of certain objective criteria. In order to realize the benefits of the goal programming approach, explicitness and consistency in model formulation are imperative.

One possible method of obtaining a priority scale for multiple goals may be to present a list of goals to the decision maker and ask him to rank them in the order of importance or preference. A device that makes this process simple and provides some check on the consistency in the value judgment of the decision maker is the paired comparison method.[4] In this method, the decision maker is simply asked to compare the goals two at a time and indicate which is more important in the pair. This procedure is carried on until all possible pairs of goals are analyzed for comparative value judgment. From this analysis a complete ordinal ranking may be obtained for the goals in terms of their importance to the organization.

Suppose there are five goals, $G_1$ (market share), $G_2$ (profit), $G_3$ (customer satisfaction), $G_4$ (employee relations), and $G_5$ (community reputation). The paired comparison method suggests that a pair of two goals should be evaluated at a time by the decision maker. If there are n goals, the total combination of paired comparisons will be n(n-1)/2. For example, for the above stated five goals there are 10 possible pairs. Let us assume that the decision maker's judgments for the paired comparisons are as follows:

| | |
|---|---|
| $G_1 > G_2$ | $G_2 > G_4$ |
| $G_1 > G_3$ | $G_2 < G_5$ |
| $G_1 > G_4$ | $G_3 > G_4$ |
| $G_1 > G_5$ | $G_3 < G_5$ |
| $G_2 < G_3$ | $G_4 < G_5$ |

To derive the complete ordinal ranking of goals, we should rearrange the preference judgments so that all the "more important than" signs will point in the same direction, as follows:

| | |
|---|---|
| $G_1 > G_2$ | $G_2 > G_4$ |
| $G_1 > G_3$ | $G_5 > G_2$ |
| $G_1 > G_4$ | $G_3 > G_4$ |
| $G_1 > G_5$ | $G_5 > G_3$ |
| $G_3 > G_2$ | $G_5 > G_4$ |

The most important goal must be more important than four other goals. In other words, the most important goal should appear on the "more important than" side of the list four times. Goal $G_1$ indeed appears four times on that side. The second goal in importance similarly should appear three times on the "more important than" side. Goal $G_5$ qualifies as the second priority goal. In general, if there are n conflicting goals, the goal whose ranking is r in the ordinal priority scale should appear n-r times on the "more important than" side of the preference judgment list. In this example, the ordinal ranking of the five goals considered will be:

| Priority | Goal | n-r |
|----------|------|-----|
| 1 | $G_1$ | 4 |
| 2 | $G_5$ | 3 |
| 3 | $G_3$ | 2 |
| 4 | $G_2$ | 1 |
| 5 | $G_4$ | 0 |

The above-described procedure should work in constructing the priority structure of multiple goals provided that the decision maker's judgment is consistent. However, sometimes the decision maker may not be consistent in value judgments, and consequently the paired comparison procedure may not produce the desired ordinal ranking of goals. For example, if the decision maker rated $G_2 > G_5$ instead of the original $G_2 < G_5$, then goals $G_2$, $G_3$ and $G_5$ will appear twice on the left-hand side of the "more important than" list. Does that mean these three goals are equally important? Let us review the reasoning of the decision maker:

$G_2$ is more important than $G_5$

$G_5$ is more important than $G_3$

$G_3$ is more important than $G_2$

Obviously, there exists logical inconsistency in the decision maker's judgment. Hopefully, a review of the situation by the decision maker will clear up the problem. The paired comparison method can be effectively used not only to establish the priority structure of the multiple conflicting goals but also as a checking device for the consistency of the decision maker's judgment.

It may appear that the paired comparison method is an impossible technique for decision problems that involve a large number of conflicting goals. Fortunately, however, for many decision problems there are only a limited number of conflicting goals, usually less than 10. Hence, the paired comparison would be a simple enough task. Furthermore, often several goals are commensurable, thereby allowing them to be analyzed on the same priority level with varying differential weights according to the opportunity cost as discussed in Chapter 2.

## IMPLEMENTATION OF GOAL PROGRAMMING ANALYSIS

The most important phase of management science application is the implementation of the model. Now that we have equipped ourselves with the concept, advantages, and limitations of goal programming, it appears appropriate to outline several phases we should follow in order to launch a successful application.[5]

## 1. Management Participation

Most management science applications cut across the formal organization structure. It is imperative, therefore, that top management be part of the project. Furthermore, for many decision problems it is only top management that has a broad enough knowledge as to whether the project is being directed toward the overall organizational objectives rather than the interests of individual departments.

Because goal programming is based on organizational goals and their priorities, active participation of top management is of paramount importance for a successful implementation of the model. Also, participation of top management is often the only way to receive full cooperation from the operating manager, who is directly involved with the actual application of the model. Usually, the detailed information concerning variables, relationships, constraints, and work procedures that are important for the model formulation are obtained from the operating manager. His participation and interest in the study often determine the creditibility of the model, as well as an objective analysis of the model results.

## 2. Analysis of Objectives, Goals, and Policies

Effective decision analysis obviously requires that organizational objectives should be analyzed as operational goals, so that the degree of goal attainment can be evaluated. It does not have any significant meaning if a firm simply presents such broad objectives as "customer satisfaction," "leadership in industry," "social responsibility," etc. Based on a qualitative description of organizational objectives, operational goals must be established. Goals must be defined in such a way that they allow a means for quantitative self-evaluation. Moreover, relationships of the decision system under consideration must be quantitatively related to management goals.

Another important factor to be considered in goal programming analysis is the analysis of organizational policies. It should be apparent that management goals cannot be completely specified until operating policies of the organization have been clearly determined. As we have seen in various application examples presented in the preceding chapters, management policies often are important constraints in goal attainment. Hence, an incomplete analysis of operating policies often results in an unrealistic objective function and consequently misleading model outputs. Management policies not only encompass the operational aspects of the organization, but also reflect the philosophy of management concerning various exogenous and endogenous factors. For example, such considerations as social responsibilities, good citizenship, public image of the firm, stable employment level, etc. have become important policy matters in recent years. Organizations also have formal policies concerning those who are directly affected by their operation—the owner, the employees, the customers, the dealers, and the government. Management policies often impose restrictions not only on the determination of goals for a decision problem but also on the very nature of the decision to be made. Indeed, management policies must be clearly reflected in the model in order to derive a valid solution to the problem.

## 3. Formulating a Goal Programming Model

After the decision problem is defined in terms of its goals, management policies concerning the problem, and the recognition of possible benefits of the analysis, the next step is to formulate the problem as a goal programming model. Mathematical models are abstract symbolic representations of reality. It is out of place to emphasize all the benefits of mathematical models. But it must be pointed out that there are many pitfalls to be avoided in formulating the model. It should be remembered that a mathematical model is a simple representation of the problem. Therefore, the model should be a great deal simpler than reality, otherwise there would be no reason to construct the model in the first place. Management scientists often take pride in formulating a model that is almost as complex and elaborate as the real-world problem. Once we heard a top executive of a petroleum firm announce at a professional meeting that the management science group in his firm designed such a fantastic inventory model that there is only one computer in the world that has a large enough capacity to test the model. He also added that he could hardly wait to get a new computer to run the model. Now we all wonder whether the executive and the management science group are still with the firm.

It should be clear that models must be simpler to understand and manipulate than reality. On the other hand, they should be capable of representing reality with an acceptable degree of accuracy and reliability. Fortunately, this is often possible because, although an analysis of a large number of components may be required to predict a phenomenon with perfect accuracy, a small number of components tend to account for most of the phenomenon.[6] It is very important, therefore, to identify the right components and the correct interrelationships among them, as well as their relationships to the objective function.

## 4. Testing the Model and Solution

Goal programming is an optimization technique. The objective of the model is search for an "optimal" solution that allows the maximum degree of goal attainments. The optimal solution is optimal only under the given model constraints and the objective function. As stated previously, since the model is only an abstract representation of the real problem, the model solution may not be the optimum solution for the real problem. However, as long as the model is carefully formulated to represent the real problem, the solution would provide an effective guide to decision making. In order to determine the accuracy and reliability as to whether the model functions with sufficient accuracy, the model and its solutions must be thoroughly tested. One of the advantages of goal programming is that it provides the decision maker an opportunity to review his judgment of goal priorities in view of the resulted solution. This procedure provides a good testing device not only for the consistency in goal analysis but also for the completeness of the model with regards to whether it contains all the important factors and relationships of the problem. It is also important to test the model on a continuous time interval basis so that significant changes in decision environment can be reflected in the model.

## 5. Final Implementation of the Solution

The final stage of goal programming application is the actual implementation of the final solution to the problem. This is, of course, the most critical phase, since this is where the benefits of the model are to be realized. In this stage, top management and operating managers play vital roles. Top management must approve the final solution after careful review, and operating managers take the responsibility of the model implementation. A continuous evaluation of the model to keep it current should also be the responsibility of the operating manager.

The primary function of management scientists in this phase is to develop a general procedure to be followed in the model implementation. Also they should assist operating personnel in an advisory capacity in analyzing the model results as to how they are related to the real operation of problems.

## EPILOGUE

The Epilogue provides time for a quiet and humble reflection upon the work presented in this book. It has certainly been an interesting and challenging project. It also has been a frustrating experience because of the limited number of references that could help the author expand the discussion by exploring more aspects of goal programming in depth.

It appears certain that goal programming will find a wide application for decision analysis of contemporary problems in industry, government, and nonprofit organizations. It is hoped that this book will help introduce the technique to students and practitioners of management science. It is the belief of the author that one of the reasons for the slow diffusion of the goal programming concept is the fact that many research studies that have explored it were oriented toward mathematical analysis rather than toward applications. In general, the mathematical sophistication of these studies was beyond the understanding of many management scientists. Consequently, their interest in goal programming has not been sufficiently stirred. For that reason, the author has attempted to avoid sophisticated mathematical notations whenever and wherever possible so as to make the presentation simple to understand.

The materials presented in this book represent the author's work for the past three years. Much of the contents was developed by the author. It is always possible that some of the logic presented may contain some errors. It is hoped that continuous research by others may improve the author's ideas. It should be mentioned again that the author is deeply indebted to Professors Yuji Ijiri and Veikko Jaaskelainen; many ideas in this book are based on their research, which has been inspirational in undertaking this project.

This book deals very little with the organizational and human aspects of decision analysis. Decisions are made by people in the organization according to certain accepted patterns. Decision results are also implemented by people. Clearly, the socio-psychological aspects of decision analysis are not only real but are also of paramount importance in the study of decision analysis. It is assumed that the readers are familiar with the basic process of decision making. Furthermore, the author assumes that the reader recognizes the importance of organizational and human aspects of decision analysis.

Finally, we should review the scope of decision analysis. Decision analysis assists the decision maker in taking effective action as long as the process recognizes: (1) organizational objectives, goals, and policies; (2) relationships among relevant variables; (3) an explicit and verifiable recommended course of action; (4) recommended course of action based on an adaptation of the scientific method; and (5) management participation. Goal programming is an effective technique for decision analysis within the realm of such a process.

# REFERENCES

Ackoff, Russell L., and Sasieni, Maurice W. *Fundamentals of Operations Research.* New York: John Wiley & Sons, Inc., 1968.

Bartee, Edwin M. "Problem Solving with Ordinal Measurement." *Management Science,* vol. 17, no. 10 (June 1971), pp. 622-33.

Churchman, C. West, Ackoff, Russell L., and Arnoff, E. Leonard. *Introduction to Operations Research.* New York: John Wiley & Sons, Inc., 1957.

Hillier, Frederick S. and Leiberman, Gerald J. *Introduction to Operations Research.* San Francisco: Holden-Day, Inc., 1967.

Morris, William T. *The Analysis of Management Decisions.* Homewood, Ill.: Richard D. Irwin, Inc., 1964.

Wagner, Harvey M. *Principles of Management Science.* Englewood Cliffs, New Jersey: Prentice-Hall, Inc., 1970.

# Notes

## CHAPTER 1

[1]Vincent Astrom, "Culture, Science, and Politics," in *The Making of Decisions: A Reader in Administrative Behavior,* ed. W. J. Gore and J. W. Dyson (London: The Free Press of Glencoe, 1964), pp. 85-92.

[2]Kenneth Boulding, "The Specialist with a Universal Mind," *Management Science,* vol. 14, no. 12 (August 1968), pp. 647-53.

[3]Astrom, "Culture, Science, and Politics."

[4]*Ibid.*

[5]Herbert A. Simon, "A Behavioral Model of Rational Choice," *Quarterly Journal of Economics,* vol. 69, no. 1 (February 1955), pp. 99-118.

[6]*Ibid.*

[7]*Ibid.*

[8]Astrom, "Culture, Science, and Politics."

[9]Simon, "Behavioral Model."

[10]Herbert A. Simon, *The New Science of Management Decision* (New York: Harper & Bros., 1960).

[11]Robert E. Schellenberger, *Managerial Analysis* (Homewood, Ill.: Richard D. Irwin, Inc., 1969).

[12]Melvin Anshen, "The Manager and the Black Box," *Harvard Business Review* (November-December 1960), p. 60.

[13]Martin Schubik, "Approaches to the Study of Decision-Making Relevant to the Firm," in *The Making of Decisions: A Reader in Administrative Behavior,* ed. Gore and Dyson, pp. 31-50.

[14]Philip Selznik, *Leadership in Administration* (Evanston, Ill.: Row, Peterson & Co., 1957).

[15]Joseph W. McQuire, *Theories of Business Behavior* (Englewood Cliffs, N.J.: Prentice-Hall, 1964).

## CHAPTER 2

[1]A. Charnes, W. W. Cooper, and R. Ferguson, "Optimal Estimation of Executive Compensation by Linear Programming," *Management Science,* vol. 1, no. 2 (January 1955), pp. 138-51.

[2]A. Charnes and W. W. Cooper, *Management Models and Industrial Applications of Linear Programming* (New York: John Wiley & Sons, Inc., 1961).

[3]*Ibid.,* pp. 215-16.

[4]Y. Ijiri, *Management Goals and Accounting for Control* (Chicago: Rand-McNally, 1965).

[5]A. Charnes, W. W. Cooper, *et al.,* "A Goal Programming Model for Media Planning," *Management Science,* vol. 14, no. 8 (April 1968), pp. 423-30; and "Note on an Application of a Goal Programming Model for Media Planning," *Management Science,* vol. 14, no. 8 (April 1968), pp. 431-36.

[6]A. Charnes, W. W. Cooper, and R. J. Nilhaus, "A Goal Programming Model for Manpower Planning," Management Science Research Report No. 115, Carnegie-Mellon University (August 1968). Also see Management Science Research Report No. 188.

[7]B. Contini, "A Stochastic Approach to Goal Programming," *Operations Research* (May-June 1968), pp. 576-86.

[8]Veikko Jaaskelainen, "A Goal Programming Model of Aggregate Production Planning," *Ekonomisk Tidskrift* (Swedish Journal of Economics), no. 2 (1969), pp. 14-19.

[9]The author's contributions include: with E. Clayton, "A Goal Programming Model for Academic Resource Allocation,"*Management Science,* vol. 18, no. 8 (April 1972) pp. 395-408; with A. Lerro and B. McGinnis, "Optimization of Tax Switching for Commercial Banks," *Journal of Money, Credit, and Banking,* vol. 3, no. 2 (May 1971), pp. 293-303; with V. Jaaskelainen, "Goal Programming for Financial Planning," *Liiketaloudellinen Aikakauskirja* (The Finnish Journal of Business Economics), vol. 3 (1971), pp. 291-303; with William Sevebeck, "An Aggregative Model for Municipal Economic Planning," *Policy Sciences,* vol. 2, no. 2 (June 1971), pp. 99-115; "An Aggregative Resource Allocation Model for Hospital Administration," a paper presented at the Third Annual Meeting of the American Institute for Decision Science, October 1971; and "An Aggregative Budget Planning Model for Hospital Administration," a paper presented at the 12th American Meeting of the Institute of Management Science, September 1971.

[10]Veikko Jaaskelainen, *Accounting and Mathematical Programming* (Helsinki: 1969).

[11]Sang M. Lee, "Decision Analysis Through Goal Programming,"*Decision Sciences,* vol. 2, no. 2 (April 1971), pp. 172-80.

[12]George B. Dantzig, *Linear Programming and Extensions* (Princeton, N.J.: Princeton University Press, 1963).

[13]The mathematical formulation of goal programming presented here is more clearly presented in Ijiri, *Management Goals and Accounting for Control.*

[14]*Ibid.,* p. 49.

[15]Lee, "Aggregative Resource Allocation Model for Hospital Administration"; Lee and Clayton, "Goal Programming Model for Academic Resource Allocation."

[16]Frederick S. Hillier and Gerald J. Lieberman, *Introduction to Operations Research* (San Francisco: Holden-Day, Inc., 1967), pp. 135-38.

[17]T. Ruefli, "A Generalized Goal Decomposition Model," *Management Science,* vol. 17, no. 8 (April 1971), pp. 505-18.

## CHAPTER 3

[1]Yuji Ijiri, *Management Goals and Accounting for Control* (Chicago: Rand-McNally, 1965), pp. 40-42.

[2]It is possible to assign a preemptive priority factor or regular weights to the choice variables and include them in the objective function. However, the same effect can be achieved by using deviational variables. For example, a goal programming model includes a cost minimization as one of the set goals. This can be achieved by minimizing $d^+$ when we formulate the cost function such as $c_j x_j + d^- - d^+ = 0$.

[3]This example is taken from: Eero Pitkanen, "Goal Programming and Operational Objectives in Public Administration," *Swedish Journal of Economics,* vol. 72, no. 3 (1970), pp. 207-14.

[4]In practical problems, differential weights are usually based on net profit ratios. The example presented here is not realistic in this sense. If the cost of running the plant (per hour) is known, net profit per unit for each product can be determined by subtracting the cost from the gross profit. It should be noted that if the profitability per hour of operation is used as the basis of differential weights as in the example, the simplex solution may be different from the graphical solution. This example is presented here only to illustrate the goal programming model formulation.

## CHAPTER 5

[1]Y. Ijiri, *Management Goals and Accounting for Control* (Chicago: Rand-McNally, 1965).

[2]For the exact steps of the simplex method of linear programming, consult one of the following: A. Charnes and W. W. Cooper, *Management Models and Industrial Applications of Linear Programming* (New York: John Wiley & Sons, Inc., 1961); George B. Dantzig, *Linear Programming and Extensions* (Princeton, N. J.: Princeton University Press, 1963); W. W. Garvin, *Introduction to Linear Programming* (New York: McGraw-Hill, Inc., 1962); G. Hadley, *Linear Programming* (Reading, Mass.: Addison-Wesley, Inc., 1962), F. Hillier and G. J. Lieberman, *Introduction to Operations Research* (San Francisco: Holden-Day, Inc., 1967); D. Teichroew, *An Introduction to Management Science: Deterministic Models* (New York: John Wiley & Sons, Inc., 1964); H. Wagner, *Principles of Operations Research with Applications to Managerial Decisions* (Englewood Cliffs, N. J.: Prentice-Hall, Inc., 1969).

[3]It is assumed that the reader understands not only the mechanics of simplex procedure but also the mathematical theorem of linear programming. For the fundamental theorem of linear programming, consult one of the following: Charnes and Cooper, *Management*

*Models;* Dantzig, *Linear Programming and Extensions;* Hadley, *Linear Programming;* Hillier and Lieberman, *Operations Research.*

[4]For a detailed discussion of the complications of linear programming, consult one of the following: Charnes and Cooper, *Management Models;* Dantzig, *Linear Programming and Extensions;* Hillier and Lieberman, *Operations Research.*

[5]Ijiri, *Management Goals and Accounting for Control.*

## CHAPTER 6

[1]A goal programming computer program was developed by Veikko Jaaskelainen in 1968, as explained in Chapter 2. Although the program is effective in solving various goal programming problems, it usually requires a voluminous data deck. It becomes a cumbersome problem if the model under analysis is a complex one. Jaaskelainen's program has not been widely circulated among management scientists.

[2]For a more detailed discussion of this point of view, see Chapter 15.

## CHAPTER 7

[1]For a detailed explanation of this procedure, see the simplex theorem of linear programming in F. S. Hillier and G. J. Lieberman, *Introduction to Operations Research* (San Francisco: Holden-Day, Inc., 1967), pp. 493-94.

[2]S. E. A. Elmaghraby, "An Approach to Linear Programming Under Uncertainty," *Operations Research,* vol. 7, no. 2 (March-April 1959), p. 208.

[3]A. Madansky, "Methods of Solution of Linear Programs Under Uncertainty," *Operations Research,* vol. 10, no. 4 (July-August 1962), pp. 463-64.

[4]R. J. Freund, "The Introduction of Risk into a Programming Model," *Econometrica,* vol. 24 (July 1956), pp. 253-63; G. Tintner, "A Note on Stochastic Linear Programming," *Econometrica,* vol. 28 (April 1960), pp. 490-95.

[5]A. Charnes and W. W. Cooper, "Chance Constrained Programming," *Management Science,* vol. 6, no. 2 (October 1959), pp. 73-79; *Management Models and Industrial Applications of Linear Programming* (New York: John Wiley & Sons, Inc., 1961).

[6]G. Dantzig, "Linear Programming Under Uncertainty," *Management Science,* vol. 1 (April 1955), pp. 196-207; "Recent Advances in Linear Programming," *Management Science,* vol. 2 (July 1956), pp. 131-44; *Linear Programming and Extensions* (Princeton, N. J.: Princeton University Press, 1963). S. E. A. Elmaghraby, "Programming Under Uncertainty" (Ph.D. diss., Cornell University, 1958), pp. 115-26; "Approach to Linear Programming Under Uncertainty"; "Allocation Under Uncertainty When the Demand Has Continuous D.F.," *Management Science,* vol. 6, no. 8 (April 1960), pp. 208-24.

[7]A. Charnes, W. W. Cooper, R. J. Niehaus, and D. Scholtz, "An Extended Goal Programming Model for Manpower Planning," Management Science Research Report No. 156 (Pittsburgh: Carnegie-Mellon University, Graduate School of Industrial Administration, December 1968).

[8]Dantzig, *Linear Programming and Extensions.*

[9]*Ibid.*

[10]The problem presented here is drawn in part from Yuji Ijiri, *Management Goals and Accounting for Control* (Chicago: Rand-McNally, 1965), pp. 45-49.

[11]Harry M. Markowitz, *Portfolio Selection: Efficient Diversification of Investments,* Cowles Foundation Monograph 16 (New York: John Wiley & Sons, Inc., 1959).

[12]See Chapter 15 for a detailed analysis.

## CHAPTER 8

[1]The transportation method is presented in E. H. Bowman, "Production Scheduling by the Transportation Method of Linear Programming," *Operations Research,* vol. 4, no. 1 (1956), pp. 100-03; the simplex method in F. Hanssman and S. W. Hess, "A Linear Programming Approach to Production and Employment Scheduling," *Management Technology,* 1960, pp. 46-51, and R. E. McGarrah, *Production and Logistics Management* (New York: John Wiley & Sons, Inc., 1963); and the linear decision rule model in C. Holt, F. Modigliani, and H. A. Simon, "A Linear Decision Rule for Production and Employment Scheduling," *Management Science,* vol. 2, no. 1 (1955), pp. 1-30.

[2]Elwood S. Buffa, *Production-Inventory Systems: Planning and Control* (Homewood, Ill.: Richard D. Irwin, 1968).

[3]E. A. Silver, "A Tutorial on Production Smoothing and Work Force Balancing," *Operations Research,* vol. 15, no. 5 (1967), pp. 985-1010.

[4]W. H. Taubert, "A Search Decision Rule for the Aggregate Scheduling Program," *Management Science,* vol. 14, no. 3 (1968), pp. 343-59.

[5]Buffa, *Production-Inventory Systems.*

[6]Taubert, "Search Decision Rule."

[7]B. Dzielinski, C. Baker, and A. Manne, "Simulation Test of Lot-Size Programming," *Management Science,* vol. 9, no. 3 (1963), pp. 229-53.

[8]Jaaskelainen, "A Goal Programming Model of Aggregate Production Planning;" Lee and Jaaskelainen, "Goal Programming Management's Math Model."

[9]This example is adapted from Jaaskelainen, "Goal Programming Model of Aggregate Production Planning," by permission of the author.

## CHAPTER 9

[1]The textbook case appears in Pearson Hunt and Victor L. Andrews, *Financial Management: Cases and Readings* (Homewood, Ill.: Richard D. Irwin, Inc., 1968), pp. 575-94. The study of the case as it appears in this chapter is based on the working paper "A Goal Programming Model for Financial Planning," on which the author collaborated with Veikko Jaaskelainen.

[2]Harry M. Markowitz, *Portfolio Selection: Efficient Diversification of Investments,* Cowles Foundation Monograph 16 (New York: John Wiley & Sons, Inc., 1959).

[3]Figure 9.1 is adapted from F. B. Renwick, "Asset Management and Investment Portfolio Behavior," *Journal of Finance,* vol. 24, no. 2 (May 1969), p. 183.

[4]B. A. Wallingford, "A Survey and Comparison of Portfolio Selection Models," *Journal of Financial and Quantitative Analysis,* vol. 3, no. 2 (June 1967), pp. 85-106.

[5]Markowitz, *Portfolio Selection.*

[6]R. M. Soldofsky, "Yield-Risk Performance Measurements," *Financial Analyst Journal,* vol. 24 (September-October 1968), pp. 93-100.

[7]For a detailed explanation see E. F. Renshaw, "Portfolio Balance Models in Perspective: Some Generalizations That Can Be Derived from the Two-Asset Case," *Journal of Financial and Quantitative Analysis,* vol. 3, no. 2 (June 1967), pp. 123-49.

[8]The variance, $s_i^2$, used in this paper is derived from the geometric mean. Hence, the portfolio model based on $s_i^2$ will be somewhat different from the traditional Markowitz-Sharpe models. For a random variable X,

$$VAR\ X = E(X - EX)^2 = E(X - a)^2 - (EX - a)^2$$

where $a$ is the geometric mean and $E(X - a)^2 = s_i^2$, the variance about the geometric mean.

If $R_p$ is the return from the portfolio, $R_i$ is the return from each stock, and $X_i$ is the $i$th stock, then the standard Markowitz model is:

$$R_p = \Sigma X_i R_i$$
$$ER_p = \Sigma X_i ER_i$$
$$VAR\ R_p = E(R_p - ER_p)^2$$
$$= \underset{i\ j}{\Sigma\Sigma} X_i X_j \sigma_i \sigma_i r_{ij}$$

where $\sigma_i$ is the standard deviation of $i$th stock and $r_{ij}$ is the correlation coefficient between i and j. Thus,

$$\Sigma\Sigma X_i X_j s_i s_j r_{ij} \neq \Sigma\Sigma X_i X_j \sigma_i \sigma_j r_{ij}$$

The difference shown above is small if $EX \approx a$. The use of $s_i$ has some advantages, as discussed in R. A. Levy, "Measurement of Investment Performance," *Journal of Financial and Quantitative Analysis,* vol. 3, no. 1 (1968), pp. 35-57.

[9]B. A. Wallingford, "Portfolio Selection Models."

[10]J. F. Weston and E. F. Brigham, *Managerial Finance,* 3rd ed. (New York: Holt, Rinehart & Winston, 1969).

[11]B. A. Wallingford, "Portfolio Selection Models."

[12]*Ibid.*

[13]D. E. Farrar, *The Investment Decision Under Uncertainty* (Englewood Cliffs, N.J.: Prentice-Hall, Inc., 1965).

[14]W. F. Sharpe, "Simplified Model for Portfolio Analysis," *Management Science,* vol. 9, no. 5 (January 1963), pp. 277-93.

[15]W. F. Sharpe, "A Linear Programming Algorithm for Mutual Fund Portfolio Selection," *Management Science,* vol. 13, no. 7 (March 1967), pp. 499-510.

[16]J. L. Treynor *et al.,* "Using Portfolio Composition to Estimate Risk," *Financial Analysts' Journal,* vol. 24 (September-October 1968), pp. 93-100.

[17]Securities Research Co., *3-Trend Cycli-Graphs,* 109th Ed. (Boston: Securities Research Co., January 1970).

[18]For a more comprehensive study of goal programming for mutual fund selection, see S. M. Lee and A. J. Lerro, "Optimizing the Portfolio Selection for Mutual Funds," working paper, Virginia Polytechnic Institute and State University, 1971.

[19]Sharpe, "Linear Programming Algorithm for Mutual Fund Portfolio Selection."

[20]As reported in *Business Week,* January 10, 1970, pp. 86-87.

## CHAPTER 10

[1]R. W. Bell and R. W. Paul, "Quantitative Determination of Manpower Requirements in Variable Activities," *Operational Research Quarterly,* vol. 17 (1966), pp. 381-412; Delbert J. Duncan and Charles F. Phillips, *Retailing: Principles and Methods* (Homewood, Ill.: Richard D. Irwin, Inc., 1963).

[2]There have been some recent studies dealing with various specific problems of sales effort allocation. For example, see O. A. Davis and J. U. Farley, "Allocating Sales Force Effort with Commissions and Quotas," *Management Science,* vol. 18, no. 4, III (1971), pp. 55-63; and D. B. Montgomery, A. J. Silk, and C. E. Zaragoza, "A Multiple-Product Sales Force Allocation Model," *Management Sciences,* vol. 18, no. 4, III (1971), pp. 3-24.

[3]Duncan and Phillips, *Retailing.*

[4]Charles Horgan, "A Systems Approach to Manpower Planning in Department Stores," *Journal of Retailing,* vol. 44, no. 3 (Fall 1968), pp. 13-30; Gary Johnson, "The Thinking Man's Approach to Sales Force Scheduling," *Stores,* (November 1969), pp. 37-38.

[5]Bell and Paul, "Quantitative Determination of Manpower Requirements in Variable Activities."

[6]George C. Engel, "How You Can Get More Volume from Your Sales Staff," *Stores,* January 1965, pp. 62-63.

[7]Bell and Paul, "Quantitative Determination of Manpower Requirements in Variable Activities."

[8]R. L. Day, "Linear Programming in Media Selection," *Journal of Advertising Research,* vol. 2 (June 1962), pp. 40-44; Duncan and Phillips, *Retailing;* Engel, "More Volume"; Horgan, "Systems Approach to Manpower Planning in Department Stores."

[9]D. B. Brown, "A Practical Procedure for Media Selection," *Journal of Marketing Research,* vol. 4 (1967), pp. 262-64.

[10]A. Charnes, W. W. Cooper, and R. Ferguson, "Optimal Estimation of Executive Compensation by Linear Programming," *Management Science,* vol. 1, no. 2 (January 1955), pp. 138-51; Charnes *et al.,* "A Goal Programming Model for Media Planning" and "Note on an Application of a Goal Programming Model for Media Planning," *Management Science,* vol. 14, no. 8 (April 1968), pp. 423-36; Charnes, Cooper, and R. J. Nilhaus, "A Goal Programming Model for Manpower Planning," Management Science Research Report No. 115 (also see No. 188) (Carnegie-Mellon University, August 1968); S. M. Lee and M. Bird, "A Goal Programming Model for Sales Effort Allocation," *Business Perspectives,* vol. 6, no. 4 (Summer 1970).

[11]In addition to those cited elsewhere in this chapter, see the following: E. M. L. Beale, P. A. B. Hughes, and S. R. Broadbent, "A Computer Assessment of Media Schedules," *Operational Research Quarterly,* vol. 17 (1966), pp. 381-412; D. M. Ellis, "Building Up a Sequence of Optimum Media Schedules," *Operational Research Quarterly,* vol. 17 (1966), pp. 413-24; A. M. Lee, "Decision Rules for Media Scheduling: Static Campaigns," *Operational Research Quarterly,* vol. 13 (1962), pp. 229-42, and "Decision Rules for Media Scheduling: Dynamic Campaigns," *Operational Research Quarterly,* vol. 14 (1963), pp. 355-72; W. T. Morgan, "Practical Media Decisions and the Computer," *Journal of Marketing,* vol. 27 (July 1963), pp. 26-30; P. A. Naert, "Optimizing Consumer Advertising and Markup in a Vertical Market Structure," *Management Science,* vol. 18, no. 4, III (1971), pp. 64-72; A. C. Rohloff, "Quantitative Analysis of the Effectiveness of TV Commercials," *Journal of Marketing Research,* vol. 3 (1966), pp. 239-45; M. W. Sasieni, "Optimal Advertising Expenditure," *Management Science,* vol. 18, no. 4, III (1971), pp. 64-72; C. J. Taylor, "Some Developments in the Theory and Application of Media Scheduling Methods," *Operational Research Quarterly,* vol. 14 (1963), pp. 291-305; W. I. Zangwill, "Media Selection by Decision Programming," *Journal of Advertising Research,* vol. 5 (September 1965), pp. 30-36.

[12]Day, "Linear Programming in Media Selection."

[13]J. F. Engel and M. R. Warshaw, "Allocating Advertising Dollars by Linear Programming," *Journal of Advertising Research,* vol. 4 (September 1964), pp. 42-48.

[14]P. Kotler, "Toward an Explicit Model for Media Selection," *Journal of Advertising Research,* vol. 4 (March 1964), pp. 34-41.

[15]D. B. Brown and M. R. Warshaw, "Media Selection by Linear Programming," *Journal of Marketing Research,* vol. 2 (1965), pp. 83-88.

[16]S. F. Stasch, "Linear Programming and Media Selection—a Comment," *Journal of Marketing Research,* vol. 4 (1967), pp. 205-06; "Linear Programming and Space-Time Considerations in Media Selection," *Journal of Advertising Research,* vol. 5 (1965), pp. 40-46.

[17]F. M. Bass and R. T. Lonsdale, "An Exploration of Linear Programming in Media Selection," *Journal of Marketing Research,* vol. 3 (1966), pp. 179-88.

[18]This example is presented to demonstrate the application potential of goal programming for advertising media scheduling. Therefore, only model formulation is presented here,

without the resulting solution. For a more theoretical study, see Charnes, *et al.,* "Goal Programming Model for Media Planning" and "Note on an Application."

[19]The weighted exposure unit can be defined as "the audience of the vehicle among relevant market segments adjusted for the distribution of the audience among market segments relative to the assumed distribution of the product among market segments." For a detailed explanation of the measure see Bass and Lonsdale, "Exploration of Linear Programming in Media Selection," p. 180.

## CHAPTER 11

[1]W. T. Morris, "Diversification," *Management Science,* vol. 4 (July 1958), pp. 382-91.

[2]W. I. Zangwill, "Non-Linear Programming via Penalty Functions," *Management Science,* vol. 13, no. 5 (January 1967), pp. 344-58.

[3]C. H. Jones, "Parametric Production Planning," *Management Science,* vol. 13, no. 10 (July 1967), pp. 843-66; E. S. Buffa and W. H. Taubert, "Evaluation of Direct Computer Search Method for the Aggregate Planning Problem," *Industrial Management Review,* vol. 8 (Fall 1967), pp. 19-36.

[4]G. D. Eppen and F. J. Gould, "A Lagrangian Application to Production Models," *Operations Research,* vol. 16 (July-August 1968), pp. 819-29; H. Everett, "Generalized Lagrange Multiplier Method for Solving Problems of Optimum Allocation of Resources," *Operations Research,* vol. 11 (May-June 1963), pp. 399-417.

[5]J. C. Hetrick, "A Formal Model for Long-Range Planning," *Long Range Planning,* vol. 1 (March 1969), pp. 16-23; S. M. Farag, "A Planning Model for the Divisionalized Enterprise," *Accounting Review,* vol. 43 (April 1968), pp. 312-20.

[6]G. A. Steiner, *Top Management Planning* (London: The Macmillan Co., 1969); H. I. Ansoff, "A Model for Diversification," *Management Science,* vol. 4 (July 1958), pp. 392-414; J. P. Crecine, "A Computer Simulation Model for Municipal Budgeting," *Management Science,* vol. 13, no. 11 (July 1967), pp. 786-815.

[7]R. Petrovic, "Optimization of Resource Allocation in Project Planning," *Operations Research,* vol. 16 (May-June 1968), pp. 559-68; L. E. Briskin, "A Method of Unifying

Multiple Objective Functions," *Management Science,* vol. 12, no. 10 (June 1966), pp. 406-16.

[8]E. H. Bowman, "Consistency and Optimality in Managerial Decision-Making," *Management Science,* vol. 9, no. 5 (January 1963), pp. 1-10; C. H. Jones, "Parametric Production Planning," *Management Science,* vol. 13, no. 10 (July 1967), pp. 843-66; C. C. Holt, F. Modigliani, and H. A. Simon, "A Linear Decision Rule for Production and Employment Scheduling," *Management Science,* vol. 2 (October 1955), pp. 1-30.

[9]Among the books and articles that apply linear programming to management planning, the following are especially worthy of study: R. Bellman, "On the Computational Solution of Linear Programming Problems Involving Almost-Block-Diagonal Matrices," *Management Science,* vol. 3 (July 1957), pp. 403-06; E. H. Bowman, "Consistency and Optimality in Managerial Decision-Making," *Management Science,* vol. 9, no. 5 (January 1963), pp. 1-10; A. Charnes and W. W. Cooper, "Management Models and Industrial Applications of Linear Programming," *Management Science,* vol. 4 (October 1957), pp. 38-91; A. Charnes, W. W. Cooper, and O. Fergusen, "Optimal Estimation of Executive Compensation by Linear Programming," *Management Science,* vol. 1, no. 2 (January 1955), pp. 138-51; W. W. Cooper, *An Introduction to Linear Programming* (New York: John Wiley & Sons, Inc., 1954); G. B. Dantzig, "Linear Programming under Uncertainty," *Management Science,* vol. 1 (April-July 1955), pp. 197-206, "On the Status of Multistage Linear Programming, *Management Science,* vol. 6 (October 1959), pp. 53-72, "Recent Advances in Linear Programming," *Management Science,* vol. 2 (January 1956), "Thoughts on Linear Programming and Automation," *Management Science,* vol. 3 (January 1957), pp. 131-39; B. P. Dzielinski *et al.,* "Simulation Tests of Lot-Size Programming," *Management Science,* vol. 9, no. 3 (January 1963), pp. 229-58; R. B. Fetter, "A Linear Programming Model for Long-Range Capacity Planning," *Management Science,* vol. 7 (July 1961), pp. 507-17; W. W. Garvin *et al.,* "Applications of Linear Programming in the Oil Industry," *Management Science,* vol. 3 (July 1957), pp. 407-30; A. Henderson and R. Schlaifer, "Mathematical Programming," *Harvard Business Review,* vol. 32 (May-June 1954), pp. 75-100; J. P. Magee, *Production Planning and Inventory Control* (New York: McGraw-Hill, 1967); A. S. Manne, "Programming of Economic Lot Sizes," *Management Science,* vol. 4 (January 1958), pp. 115-35, "Linear Programming and Sequential Decisions," *Management Science,* vol. 6 (April 1960), pp. 259-67; P. Masse and R. Gilbrat, "Applications of Linear Programming to Inventories in the Electric Power Industry," *Management Science,* vol. 3 (January 1957), pp. 149-66; R. E. McGarrah, *Production and Logistics Management* (New York: John Wiley & Sons, Inc., 1963); C. McMillan, *Mathematical Programming* (New York: John Wiley & Sons, Inc., 1970); J. Rogers, "A Computational Approach to the Economic Lot Scheduling Problem," *Management Science,* vol. 4 (April 1958), pp. 264-91; S. K. Sengupta *et al.,* "On Some Theorems of Stochastic Linear Programming with Applications," *Management Science,* vol. 10 (October 1963), pp. 131-42; W. F. Sharpe, "A Linear Programming Algorithm for Mutual Fund Portfolio Selection," *Management Science,* vol. 13, no. 7 (March 1967), pp. 499-510; D. J. Smalter, "Analytical Techniques in Planning," *Long Range Planning,* vol. 1 (September 1968), pp. 25-33; G. E. Thompson, "On Varying the Constraints in a Linear Programming Model of the Firm," *American Economic Review,* vol. 58 (June 1968), pp. 485-95; P. A. Wardle, "Forest Management and Operations Research: A Linear Programming Study," *Management Science,* vol. 11, no. 12 (August 1965), pp. 260-70; D. J. Wilde, *Optimum Seeking Methods* (Englewood Cliffs, N. J.: Prentice-Hall, 1964).

## CHAPTER 12

[1] See Harry Williams, *Planning for Effective Resource Allocation in Universities* (Washington, D. C.:  American Council on Education, 1966), p. 11.

[2] U. S. Department of Health, Education, and Welfare, *Earned Degrees Granted* (Washington, D. C.:  U. S. Government Printing Office, 1966), p. 2.

[3] John Walsh, "Research and Graduate Education," in *Contemporary Issues in American Education* (Washington, D. C.:  U. S. Government Printing Office, 1965), pp. 132-38.

[4] *Ibid.*

[5] HEW, *Earned Degrees Granted,* pp. 12-13.

[6] Reprinted from *Management Science,* vol. 18, no. 8 (April 1971), pp. 395-408, by permission of the Institute of Management Science. The authors would like to express their appreciation to Professors William W. Cooper and James E. Bruno for helpful suggestions. Any errors are solely the responsibility of the authors.

[7] Annual current expenditure (1962-63 dollars) by institutions of higher education rose from $3.6 billion in 1954-55 to an estimated $9.7 billion in 1964-65—an increase of 169 percent. It is expected that current expenditures will reach $20.1 billion in 1974-75 —an increase of 107 percent over the 1964-65 figure. See Walsh, "Research and Graduate Education."

[8] D. Abbey and C. R. Jones, "On Modeling Education Institutions," *Bulletin of the Institute of Management Science,* vol. 15 (August 1969), p. 67.

[9] Office of Institutional Research, *OIR Series,* University of Toronto, 1966; R. M. Oliver, "An Equilibrium Model of Faculty Appointments, Promotions and Quota Restrictions," Ford Foundation Research Program in University Administration Report No. 68-3 (Berkeley: University of California, 1968); R. M. Oliver, D. S. Hopkins, and R. Armacost, "An Academic Productivity and Planning Model for a University Campus," University of California Administrative Studies Project in Higher Education Report No. 3 (1970); V. E. Palmour and R. V. Wiederkehr, "A Decision Model for Library Policies on Serial Publications," 17th International Conference of TIMS, London, 1970; Robert L. Smith, "Accommodating Student Demand for Courses by Varying the Classroom Size Mix," University of California Administrative Studies Project in Higher Education Report No. 4 (1970); Robert G. Spiegelman, "A Benefit/Cost Model to Evaluate Educational Programs," *Socio-Economic Planning Sciences,* vol. 1, pp. 443-60.

[10] J. Gani, "Formulae for Projecting Enrollments and Degrees Awarded in Universities," *Journal of the Royal Statistical Society,* A 126 (1963), pp. 400-09; K. T. Marshall, R. M. Oliver, and S. S. Suslow, "Undergraduate Enrollments and Attendance Patterns," University of California Administrative Studies Project in Higher Education Report No. 4

(1970); R. M. Oliver, "Models for Predicting Gross Enrollments at the University of California," Ford Foundation Research Program in University Administration Report No. 68-3 (Berkeley: University of California, 1968); Oliver and K. T. Marshall, "A Constant Work Model for Student Attendance and Enrollment," Ford Foundation Research Program in University Administration Report No. 69-1 (Berkeley: University of California, 1969).

[11]D. J. Bartholomew, "A Mathematical Analysis of Structural Control in a Graded Manpower System," Ford Foundation Research Program in University Administration Paper P-3, December 1969; J. Halpern, "Bounds for New Faculty Positions in a Budget Plan," Ford Foundation Research Program in University Administration Paper P-10 (Berkeley: University of California, 1970); R. M. Oliver, "An Equilibrium Model of Faculty Appointments, Promotions and Quota Restrictions," Ford Foundation Research Program in University Administration Report No. 68-3 (Berkeley: University of California, 1968); S. M. Row and G. B. Weathersby, "A Control Theory Solution to Optimal Faculty Staffing," Ford Foundation Research Project in University Administration Paper P-11, University of California, 1970; A. Young and G. Almond, "Predicting Distributions of Staff," *Computer Journal,* no. 4, pp. 246-50.

[12]R. H. Crandall, "A Constrained Choice Model for Student Housing," *Management Science,* vol. 16, no. 2 (October 1969), pp. 112-20; F. F. Leimkuhler and M. D. Cooper, "Analytical Planning for University Libraries," Ford Foundation Research Program in University Administration Paper P-1 (Berkeley: University of California, 1970); G. Menges and G. Elstermann, "Capacity Models of University Management," 17th International Congress of TIMS, London, 1970; P. M. Morse, *Library Effectiveness: A Systems Approach* (Cambridge, Mass.: M.I.T. Press, 1968); Gordon Williams *et al.,* "Library Cost Models: Owning Versus Borrowing Serial Publications," U. S. Department of Commerce, National Bureau of Standards, 1968).

[13]J. E. Swanson, *Financial Analysis of Current Operations of Colleges and Universities* (Ann Arbor, Mich.: Institute of Public Administration, University of Michigan, 1966); G. B. Weathersby, *The Development and Applications of a University Cost Simulation Model* (Berkeley: Graduate School of Business Administration and Office of Analytical Studies, University of California, 1967).

[14]R. F. Barton, "On Optimization in the American University," Texas Tech University, June 25, 1970; R. M. Durstine, "Modeling the Allocation Process in Education," Center for Studies in Education and Development, Graduate School of Education, Harvard University, 1970; W. Judy and J. B. Levine, *A New Tool for Educational Administrators: A Report to the Commission on the Financing of Higher Education* (Toronto: University of Toronto Press, 1965); W. Judy *et al.,* Campus V Documentation, vols. 1-6, Systems Research Group, 1970; H. E. Koenig *et al.,* "A Systems Model for Management Planning and Resource Allocation in Institutions of Higher Education," Michigan State University, 1968; L. P. Nordell, "A Dynamic Input-Output Model of the California Educational System," Office of Naval Research Technical Report No. 25 (Berkeley: Center for Research in Management Science, University of California, 1967); Robert K. Thompson, "Higher Education Administration: An Operating System Study Utilizing a Dynamic Simulation Model," in *Corporate Simulation Models,* ed. Schrieber (Seattle: University of Washington Press, 1970); Ismail B. Turksen and Albert B. Holzman, "Micro Level Resource Allocation Models for Universities," 37th ORSA Meeting, Washington, 1970; G.

B. Weathersby and M. C. Weinstein, *A Structural Comparison of Analytical Models for University Planning,* Ford Foundation Research Project in University Administration Paper P-12 (Berkeley: University of California, 1970); H. Williams, *Planning for Effective Resource Allocations in Universities* (Washington, D. C.: American Council on Education, 1966).

[15]Weathersby and Weinstein, *Structural Comparison of Analytical Models.*

[16]For a multiyear planning model based on the goal programming approach, see S. M. Lee, A. Lerro, and B. McGinnis, "Optimization of Tax Switching for Commercial Banks," *Journal of Money, Credit, and Banking,* vol. 3, no. 2 (May 1971), pp. 293-303; S. M. Lee and W. Sevebeck, "An Aggregative Model for Municipal Economic Planning," *Policy Sciences,* vol. 2, no. 2 (June 1971), pp. 99-115.

[17]The terminal degree represents Ph.D., D.B.A., J.D., and LL.B.

[18]This example is based on actual operational data at the College of Business, Virginia Polytechnic Institute and State University. We would like to thank Dean H. H. Mitchell for his help in preparing this study. The average salaries of the faculty are, however, arbitrarily determined.

[19]The desired faculty/student ratio is based on the AACSB accredition regulations and the dean's academic policy.

[20]The first draft of this numerical example was presented at the 11th American Meeting of the Institute of Management Science, 1970.

[21]The dean of the college may have administrative options of cutting enrollments or phasing out programs to balance the budget and desired academic goals. This is based on the specific conditions of the particular university or college. The dean has, however, indirect means of controlling enrollment. For example, he can add more difficult courses as part of the requirements or decrease such courses. Such action will affect the loads placed on other departments or colleges in the university.

[22]The authors are happy to report that the dean of their college has used the information generated by this study in his academic planning.

# CHAPTER 13

[1]Reprinted from *Policy Sciences,* vol. 2, no. 2 (1971), with the permission of Elsevier Publishing Co.

[2]For the general economist perhaps the best introduction to goal programming is Y. Ijiri, *Management Goals and Accounting for Control* (Chicago: Rand-McNally, 1965). For the application of goal programming to different areas, see A. Charnes and W. W. Cooper, *Management Models and Industrial Applications of Linear Programming* (New York: John Wiley & Sons, Inc., 1961); A. Charnes *et al.,* "An Extended Goal Programming Model for Manpower Planning," Management Science Research Paper No. 188, Carnegie-Mellon University, June 1969; V. Jaaskelainen, "A Goal Programming Model of Aggregate Production Planning," *Swedish Journal of Economics,* vol. 71, no. 2 (1969), pp. 14-29; S. M. Lee and E. Clayton, "A Goal Programming Model for Academic Resource Allocation," *Management Science,* vol. 18, no. 8 (April 1972), pp. 395-408.

[3]Most of the economic studies dealing with municipal governments have been concentrated in problems associated with urban developments, pollution, transportation, migration patterns, or a certain segment of the total municipal economic system.

[4]J. P. Crecine, "A Computer Simulation Model of Municipal Budgeting," *Management Science,* vol. 13, no. 11 (July 1967), pp. 786-815.

[5]Since the Department of Defense adopted a program of budgeting now known as PPBS and the directive of Johnson's Bureau of Budget (BOB) in August, 1965, it has become mandatory for virtually all agencies and departments of the federal government to use the PPB procedure in submitting their budgets and program memoranda to the BOB. For the development of PPBS, see *Budgeting for National Objectives* (New York: Committee for Economic Development, 1966); H. P. Hatry and J. F. Cotton, *Program Planning for State, County, and City* (Washington, D. C.: George Washington University, 1967); R. N. McKean, *Efficiency in Government Through Systems Analysis* (New York: John Wiley & Sons, Inc., 1958); A. Smithies, "Conceptual Framework for the Program Budget," in *Program Budgeting,* ed. D. Novick (Cambridge, Mass.: Harvard University Press, 1965).

[6]Many terms are used interchangeably for the same meaning, such as cost-benefit analysis, cost-utility analysis, cost-effectiveness analysis, systems analysis, and the like. The Committee for Economic Development, however, distinguishes cost-benefit analysis from cost-effectiveness analysis as: "Cost-benefit analyses provide the means for comparing the resources to be employed on a specific project (cost) with the results (dollar benefits) likely to be obtained from it. Cost-effectiveness analyses; on the other hand, are designed to measure the extent to which resources allocated to a specific objective under each of several alternatives actually contribute to accomplishing that objective, so that different ways of gaining the objective may be compared" (Budgeting for National Objectives, pp. 20-30).

[7]R. N. McGregor, "Capital Budgeting in a Small City," *Municipal Finance,* November 1961, pp. 96-100.

[8]Since 1950, the population within the corporate limits of Blacksburg has grown from 3,358 persons to an estimated 9,000 in 1966, representing an increase of 168%. This trend of rapid population increase has been brought about primarily by the expansion of Virginia Polytechnic Institute and the location of several new industries in the Blacksburg area. From *Business Zoning Study of Blacksburg, Virginia* (Blacksburg Planning Commission, 1967).

[9]Virginia Polytechnic Institute, *Central Business District Study, Blacksburg, Virginia* (Department of Urban and Regional Planning, 1966.

[10]Younger and Associates, Inc., "Administrative Survey for Town of Blacksburg, Virginia– Draft" (Blacksburg: 1968).

[11]The priority structures used in the model are provided by the mayor and town manager of Blacksburg.

## CHAPTER 14

[1]"Special Report: The $60-Billion Crisis over Medical Care," *Business Week,* January 17, 1970, pp. 50-64.

[2]Leroy E. Burney, "Impact Forces on Rising Costs of Health Care and Health Care Facilities," and Joseph W. Degen, "The Conference," in *Conference on Costs of Health Care Facilities* (Washington, D. C.: National Academy of Sciences, 1968); "Special Report," *Business Week.*

[3]Figures here and in the following paragraph are from "Special Report," *Business Week.*

[4]James L. Souder, "The Planning Process in Health Care," in *Conference on Costs of Health Care Facilities* (Washington, D. C.: National Academy of Sciences, 1968).

[5]Phillip R. Lee, "Charge to the Conference," in *Conference on Costs of Health Care Facilities* (Washington, D. C.: National Academy of Sciences, 1968).

[6]Joseph B. Davis, "Increasing Productivity of Physicians," *Group Practice,* vol. 18 (July 1968), pp. 19-23.

[7]Carl W. Anderson, "Administration of Group Practice–Business Manager's Point of View," *Group Practice,* vol. 18 (July 1968), pp. 43-49.

[8]This study is based on the author's paper "An Aggregative Budget Planning Model for Health Care Clinics," delivered at the 12th American Meeting for the Institute of Management Science, October 1971.

[9]Anderson, "Administration of Group Practice"; *Physicians' Visits* (Vital and Health Statistics, Series No. 10, No. 49).

[10]Davis, "Increasing Productivity."

[11]J. M. Strickey, "Administration of Group Practice–A Physician's Point of View," *Group Practice,* vol. 18 (July 1968), pp. 36-42.

[12]To simplify the model design, in this study some of the details of the clinic operation are omitted, such as exact salaries for each employee, vacation, health insurance cost, and other fringe benefits. However, in actual model design these can be easily included in the goal programming model.

[13]This study is based on the author's paper "An Aggregative Resource Allocation Model for Hospital Administration," which was presented at the Third Annual Conference of the American Institute for Decision Sciences, October 1971, St. Louis.

[14]The term "professional" is used in this hospital for laboratory technician, pathologist, electrocardiology technician, cardiologist, radiology technician, radiologist, pharmacist, and anesthesiologist.

## CHAPTER 15

[1]Edwin M. Bartee, "Problem Solving with Ordinal Measurement," *Management Science,* vol. 17, no. 10 (June 1971), pp. 622-33.

[2]*Ibid.,* p. 623.

[3]*Ibid.,* p. 624.

[4]The procedure of a paired comparison method presented in this chapter is derived from William T. Morris, *The Analysis of Management Decisions* (Homewood, Ill.: Richard D. Irwin, Inc., 1964), pp. 164-65.

[5]For a more detailed analysis of this topic see C. Churchman *et al., Introduction to Operations Research* (New York: John Wiley & Sons, Inc., 1957), pp. 597-624; F. S. Hillier and G. J. Leiberman, *Introduction to Operations Research* (San Francisco: Holden-Day, Inc., 1967), pp. 12-22; and H. M. Wagner, *Principles of Management Science* (Englewood Cliffs, N. J.: Prentice-Hall, Inc., 1970), pp. 533-46.

[6]R. L. Ackoff and M. W. Sasieni, *Fundamentals of Operations Research* (New York: John Wiley & Sons, Inc., 1968), pp. 79-84.

# INDEX